HEATING AND COOLING WITH GROUND-SOURCE HEAT PUMPS IN COLD AND MODERATE CLIMATES

HEATING AND COOLING WITH GROUND-SOURCE HEAT PUMPS IN COLD AND MODERATE CLIMATES

Fundamentals and Basic Concepts

Vasile Minea

CRC Press is an imprint of the
Taylor & Francis Group, an **informa** business

First edition published 2022
by CRC Press
6000 Broken Sound Parkway NW, Suite 300, Boca Raton, FL 33487-2742

and by CRC Press
2 Park Square, Milton Park, Abingdon, Oxon, OX14 4RN

Library of Congress Cataloging-in-Publication Data
Names: Minea, Vasile, author.
Title: Heating and cooling with ground-source heat pumps in cold and moderate climates : fundamentals and basic concepts / Vasile Minea.
Description: First edition. | Boca Raton : CRC Press, 2022. | Includes bibliographical references and index. | Summary: "This book focuses on fundamentals and design principles of vertical and horizontal indirect and direct expansion closed-loop, and ground- and surface water ground-source heat pump systems. It also details the thermodynamic aspects of mechanical and thermo-chemical compression cycles of geothermal heat pumps, as well as the energetic, economic and environmental aspects associated with the use of ground-source heat pump systems for heating and cooling residential and commercial/institutional buildings in moderate and cold climates"-- Provided by publisher.
Identifiers: LCCN 2021058608 (print) | LCCN 2021058609 (ebook) |
ISBN 9780367469870 (hardback) | ISBN 9781032231884 (paperback) |
ISBN 9781003032540 (ebook)
Subjects: LCSH: Ground source heat pump systems. | Heating. | Air conditioning. |
Architecture and climate.
Classification: LCC TH7417.5 .M56 2022 (print) | LCC TH7417.5 (ebook) |
DDC 697--dc23/eng/20220118
LC record available at https://lccn.loc.gov/2021058608
LC ebook record available at https://lccn.loc.gov/2021058609

ISBN: 978-0-367-46987-0 (hbk)
ISBN: 978-1-032-23188-4 (pbk)
ISBN: 978-1-003-03254-0 (ebk)

DOI: 10.1201/9781003032540

Typeset in Times
by MPS Limited, Dehradun

Contents

Preface – Volume 1

Ground-source heat pump systems use the constant very low temperature of shallow ground/soil/rocks and groundwater or surface waters as heat/sink sources for efficient and environmentally sustainable space and hot water heating, and space cooling/dehimidifying of residential, commercial/institutional, and even equivalent industrial buildings. Depending on climate, the temperatures of earth (at a few meters below the surface), groundwater or surface water range from 7°C to 21°C, temperatures much higher during the winter and much lower in the summer than those of ambient air, especially in cold and moderate climates.

Therefore, the energy efficiencies of ground-source heat pump systems in both heating and cooling modes of operation are higher than those of air-source heat pumps, and offer substantial potential to reduce greenhouse gas emissions for building heating and cooling applications, which may explain the growing interest in using such systems in many countries, especially in recent decades. Even though the installation costs of ground-source heat pump systems are generally higher than those of air-source heat pump systems for the same heating and cooling capacities, the additional costs are usually recovered via energy and maintenance cost savings in 7–12 years. Additionally, the system technical life is estimated at up to 25 years for the indoor components, and up to 50 (or even more) years for the ground-coupled heat exchangers.

Volume 1 of this book contains 22 chapters, and focuses on thermo-physical properties of ground/soil/rocks, groundwater and surface water, as well as on fundamentals and basic concepts of vertical and horizontal closed-loop (indirect, secondary fluid) and direct (mono-fluid) expansion, and on open- and closed-loop groundwater and surface water ground-source heat pump systems. It emphasizes some thermodynamic aspects of mechanical vapor compression geothermal heat pumps, as well as technical, energetic, and environmental advantages and limitations associated with the use of ground-source heat pump systems for heating and cooling residential, commercial/institutional, and equivalent industrial buildings in cold and moderate climates. This volume provides information related to the main geothermal energy resources and main types of ground/soil/rocks, about outdoor and indoor design temperatures and several types of conventional and non-conventional HVAC systems for both residential and commercial/institutional buildings, heat transfer inside and outside ground-coupled heat exchangers, and methods to evaluate the earth effective thermal conductivity. Volume 1 also describes the actual context and targets of heat pumping technology and summarizes some future R&D needs, aiming to improve the energy and economic efficiency of ground-source heat pump systems in order to respond to actual and future customer requirements.

Volume 1 is intended to serve both the practicing engineers involved in the selection and/or design of sustainable ground-source heat pump systems as well as the scientist researchers and students as a work that covers the wide field of ground-source heat pump principles. The main scope is to contribute to increasing the number of successful implementations of ground-source heat pump systems, especially in cold and moderate climates.

Vasile Minea
Retired - Hydro-Québec Research Institute,
Laboratory of Energy Technology, LTE (Québec) Canada

Biography

Vasile Minea is a PhD graduate of civil, industrial, and agricultural installation engineering faculty from the Bucharest Technical Construction University, Romania. He worked as a professor at that university for more than 15 years, teaching courses such as HVAC systems for civil, agricultural, and industrial buildings, as well as thermodynamics, heat transfer, and refrigeration. During this period, his R&D works focused on new technologies for heat exchangers, residential and industrial heat pumps, development and experimentation of advanced compression-absorption/resorption heat pumps concepts, as well as on the usage of solar energy for comfort cooling processes and industrial cold water and ice production. Since 1987, Dr. Minea has been working as a scientist researcher at the Hydro-Québec Research Institute, Laboratory of Energy Technologies in Québec, Canada. His research activity mainly focused on heat recovery from large-scale commercial/institutional HVAC systems, various industrial processes, and low-temperature geothermal energy resources from the aid of mechanical vapor compression heat pumps. Dr. Minea has extensively studied ground-source heat pump concepts as indirect (secondary fluid) and direct expansion (mon-fluid) using vertical and horizontal ground-coupled heat exchangers, some of them being integrated in low-energy buildings located in cold and moderate climates. Most of his R&D theoretical and experimental results have been published in many conference proceedings (as *ASHRAE Winter and Summer Annual Conferences, International Congress of Refrigeration, IEA Heat Pump Conferences, and CLIMA-REHVA Conferences*) and in prestigious journals such as *ASHRAE Journal, International Journal of Refrigeration, Applied Thermal Engineering,* and *IEA Heat Pump Centre Newsletter.*

1 Introduction

Volume 1 of this book focuses on fundamentals, basic concepts, and design principles of vertical and horizontal indirect (secondary fluid) and direct (mono-fluid) expansion closed-loop, and groundwater and surface water very low-temperature (low-enthalpy) ground-source heat pump systems. It details some thermodynamic aspects of mechanical vapor compression geothermal heat pumps, as well as technical, energetic, economic, commercial, and environmental information associated with the use of ground-source heat pump systems for heating and cooling residential, commercial/institutional, and equivalent industrial buildings in moderate and cold climates.

Chapter 2 first summarily presents some information related to outdoor and indoor design temperatures as well as principles of heating and cooling loads for both residential and commercial/institutional buildings in cold and moderate climates. Second, the most commonly used software tools in industry and academia for whole building simulation including or not means to simulate the major ground-source heat pump systems and their sub-systems are summarized.

Chapter 3 presents several types of HVAC systems for both residential and commercial/institutional buildings. The main components of each selected concept are identified as well as their principal advantages and drawbacks.

Chapter 4 defines the deep (high- and moderate-temperature) and shallow (low- and very-low temperature) geothermal energy resources; provides short information about their origins, depths, and temperatures; as well as on their usages and thermal conversion technologies.

Chapter 5 presents the main types of ground/soils and shortly describes some of their thermos-physical properties of interest for ground-source heat pump systems.

Chapter 6 succinctly presents some laboratory as well as in-field currently used methods to evaluate the ground/soil/rocks effective (average) thermal conductivity. Details are included on typical mobile apparatus, and some testing methods.

Chapter 7 refers to the classification of ground-source heat pump systems currently used in cold and moderate climates according to their application fields and heat/sink energy sources.

Chapter 8 first presents the basic thermodynamic parameters of real subcritical mechanical vapor compression geothermal heat pump, and most-used diagrams and refrigerants. Second, the thermodynamic cycles of typical brine-to-air geothermal heat pumps with vertical closed-loop (indirect, secondary fluid) ground-coupled heat exchangers operating in heating and cooling modes, are described. Finally, some means to express the energy and exergy efficiencies of geothermal heat pumps are detailed.

Chapter 9 presents some aspects related to typical construction of refrigerant-to-air condensers of geothermal heat pumps, and some basic relations (as overall heat transfer coefficients and heat transfer rates), required for their thermal design.

DOI: 10.1201/9781003032540-1

Chapter 10 refers to some construction (as tube and fins materials, refrigerant and air flow distribution, and condense draining) of conventional finned-tube air-to-refrigerant evaporators for geothermal heat pumps, and heat transfer (as thermal resistance, overall heat transfer coefficient, heat transfer rate, and heat transfer surface) aspects.

Chapter 11 first describes the conventional distributed and central geothermal heat pump systems usually applied in large-scale commercial and institutional buildings. It provides information about current materials for piping, as well as some basic notions concerning brine and water pumping.

Chapter 12 presents the basic configurations and operating modes of residential and commercial/institutional vertical closed-loop (indirect, secondary fluid) ground-source heat pump systems, as well as some fundamental elements of heat transfer inside and outside the boreholes.

Chapter 13 describes the heat transfer inside and outside borehole heat exchangers provided with single U-shaped vertical tubes. Also summarized are some aspects of analytical (as infinite line- and cylindrical-source) and numerical (as long- and short-term) analyzing methods.

Chapter 14 summarizes several characteristics of closed-loop (indirect, secondary fluid) horizontal ground-source heat pump systems applied to residential and non-residential buildings, as the complex influences heat and mass transfer with the ambient air, heat transfer around buried ground-coupled heat exchangers, as well as flow regimes and convective heat transfer of thermal carrier fluids inside horizontally buried pipes.

Chapter 15 presents the basic concepts and their operating principles, some performance indicators and rating conditions, refrigerant flow patterns, pressure drops and heat transfer (vaporization and condensation) principles, as well as the main advantages and limitations of both horizontal and vertical closed-loop, direct expansion (mono-fluid) ground-source heat pump systems.

Chapter 16 first presents the basic configurations as well as the operating principles of closed-loop vertical thermos-syphon ground-source heat pump systems, and the characteristics of the most used working fluids. Second, the density, pressure, temperature, and velocity profiles are identified, as well as the heat transfer mechanisms and the thermal resistance. Finally, some advantages and drawbacks of these systems, and future R&D needs are summarized.

Chapter 17 first classifies the open-loop groundwater systems, and then presents a number of the most relevant properties of aquifers and the potential problems that may impact the safe operation of geothermal heat pumps. Some advantages and limitations of open-loop groundwater heat pump systems are finally summarized.

Chapter 18 presents the basic configurations of open-loop, dual, and multiple-well groundwater heat pump systems for residential and commercial/institutional buildings. The main components of production and return wells are described, and the characteristics of groundwater submersible pumps.

Chapter 19 refers to basic concepts and configurations of open-loop single-well (standing column) groundwater heat pump systems that can be applied to residential and commercial/institutional buildings. Some characteristics of groundwater flow and heat transfer outside and inside standing column wells are presented.

Chapter 20 first presents the basic concepts of open- and closed-loop surface water ground-source heat pump systems for residential and commercial/institutional buildings. The main physical properties of stationary surface water and stratification/mixing/turnover phenomenon are then shortly described. Finally, the heat transfer and energy balance at the surface bodies are detailed.

Chapter 21 summarizes some of technical, energetic, operational, economic, and environmental advantages and limitations of ground-source heat pump systems.

Chapter 22 first describes the actual context and targets of heat pumping technology, and then summarizes some of future R&D needs, aiming to improve the energy and economic efficiency of the technology in order to respond to global requirements of the humanity.

2 Outlook for Building Heating and Cooling Loads, and Simulation Tools

2.1 INTRODUCTION

To accurately design and size the ground-source heat pump systems in any geographical location, it should first determine the buildings' heating and cooling loads (i.e., average amounts of energies that must be added to or removed from spaces during each month of the year) to maintain the indoor temperatures and relative humidity at the design values.

Building heating and cooling loads significantly affect the size of ground-coupled heat exchangers, piping, air handlers, and distribution ductwork (Bose et al. 1985; DOE 1995; ASHRAE 2011).

For both residential and commercial/institutional buildings, the calculation of heating and cooling loads requires the knowledge of elements as the followings: (i) building type, size, and space usage; (ii) outdoor and indoor thermal conditions; (iii) occupancy (number of people) and activity of space occupants; and (iii) characteristics of the equipment emitting heat and/or moisture in the conditioned spaces.

The main contributors to building heating and cooling energy requirements are: (i) heat transmission through building structures; (ii) un-controlled air infiltration; (iii) controlled fresh air ventilation; (iv) solar radiation; (v) lights; (vi) electrical appliances and motors; and (vii) materials entering/leaving the buildings.

2.2 OUTDOOR AND INDOOR DESIGN CONDITIONS

Outdoor design conditions for building annual heating and cooling load calculations in the coldest and hottest months of the year should be selected according to local climate data and code requirements.

For most North American locations, for example, the winter (December, January, and February, i.e., a total of 2,160 hours) and summer (June through September, i.e., a total of 2,928 hours) design temperatures usually selected are the outdoor dry-bulb air temperatures that exceeded 97.5% and 2.5% of the total hours during the winter and summer months, respectively. The mean coincident wet-bulb and dew-point temperatures are tabulated along the dry-bulb design temperature values, and, generally, the most unfavorable combination of outdoor air temperatures and relative humidity in winter, assumed at 70–80% for most locations), and

DOI: 10.1201/9781003032540-2

wind mean velocities occurring at the specific dry-bulb temperatures is commonly used (ASHRAE 2011). Common indoor conditions for heating and cooling are 20°C dry-bulb temperature and 30% relative humidity, and 24°C dry-bulb temperature and 50–65% relative humidity, respectively.

The concept of ''indoor thermal comfort'' involves the interaction of many factors, generally pertaining to sedentary physical activity levels of healthy adults at atmospheric pressure equivalent to altitudes up to 3,000 m and for periods not less than 15 minutes. According to ANSI/ASHRAE Comfort Standard 55 (2010), in spaces where the occupants have activity levels that result in metabolic rates between 1.0 and 1.3 met. Met the energy produced per unit surface area (about 1.8 m^2) inside the body due to metabolic activity of an average person seated at rest (e.g., 58.2 W/m^2) and where clothing provides between 0.5 and 1.0 clo (where clo is an unit used to express the thermal insulation provided by clothing ensembles; 1 clo = 0.155 $°C{\cdot}m^2/W$ of thermal insulation).

In Figure 2.1, two comfort zones are shown: one for 0.5 clo (when the outdoor environment is warm) and one for 1.0 clo (when the outdoor environment is cool) of clothing insulation, respectively. It can be seen that: (i) the comfort zone moves left with higher clothing, higher metabolic rate, and higher radiant temperature; and (ii) the comfort zone moves right with lower clothing, lower metabolic rate, and lower radiant temperature. The graph shown in Figure 2.1 applies to operative temperatures only, defined as the uniform temperatures of imaginary black enclosures in which the occupants would exchange the same amount of heat by radiation plus convection as in actual non-uniform environments. It represents the range of operative temperatures for 80% occupant acceptability. In other words, this is based on a 10% dissatisfaction criterion for general (whole body) thermal comfort, plus an additional 10% dissatisfaction that may occur on average from local (partial body) thermal discomfort.

The operational temperature, a simplified measure of human thermal comfort derived from air dry-bulb temperature, mean radiant temperature, and speed, is defined as follows:

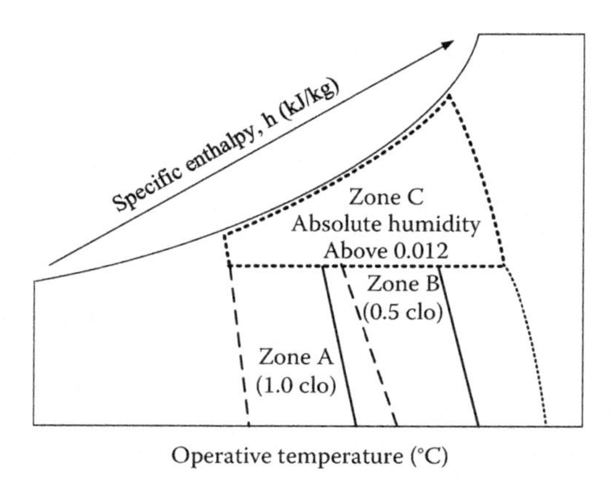

FIGURE 2.1 Ranges of operative temperature and absolute humidity for comfort zones.

$$OT = \frac{\bar{T}_{rad} + \bar{T}_{air}\sqrt{10 * w_{air}}}{1 + \sqrt{10 * w_{air}}} \quad (2.1)$$

where

OT is the operative temperature (°C)

$\bar{T}_{rad}$ is the mean radiant temperature (°C)

$\bar{T}_{air}$ is the mean dry-bulb air temperature (°C)

w_{air} is the air velocity (m/s)

The operative temperature can be also expressed as follows:

$$OT = \frac{h_{rad} \cdot \bar{T}_{rad} + h_{conv} \cdot \bar{T}_{air}}{h_{rad} + h_{conv}} \quad (2.2)$$

where

h_{rad} is the radiative heat transfer coefficient (W/m^2K)

h_{conv} is the convective heat transfer coefficient (W/m^2K)

$\bar{T}_{rad}$ is the mean radiant temperature (°C)

$\bar{T}_{air}$ is the mean dry-bulb air temperature (°C)

2.3 RESIDENTIAL BUILDINGS

Residential buildings can be categorized as follows (ASHRAE 2011): (i) detached one- or two-story single-family with exposed walls in four directions, a roof, rooms reasonably open, unitary air conditioning systems (for entire house or for each floor), generally provided with distribution of air to each room and centralized mixed air return; and (ii) multifamily with maximum of three exposed walls and a roof, each living unit has a single unitary air conditioning system.

Most of heating, ventilating, and air conditioning systems of residential buildings present particular features as (ASHRAE 2011): (i) heating and cooling thermal loads are primarily imposed by heat losses and heat gains through structural components, and by air leakage/infiltration, and ventilation; (ii) are designed to meet the maximum thermal load by assuming normal occupancy; (iii) use heating and cooling units (as air-source or geothermal heat pumps) of relatively small (nominal cooling) capacities (between 18 and 32 kW for heating, and between 5 and 18 kW for cooling); (iv) present greater duct leakages and/or heat losses because of frequent duct installation in attics or other unconditioned buffer spaces; (v) the heat losses or heat gains between conditioned and unconditioned spaces are based on estimated temperatures in the unconditioned spaces; and (vii) the HVAC equipment operates at partial thermal loads during most of the season.

2.3.1 HEATING LOADS

Heating load calculations for residential buildings involve factors as: (i) maximum heat loss of each room or space to be heated, and the simultaneous maximum heat loss for the entire building; (ii) heat losses through the building structures are considered instantaneous since heat transfer is essentially conductive depending on the local climate, building type and usage; a combined allowance of 20–25% for piping warm-up and heat losses to unheated spaces is commonly applied; (iii) ignore solar and internal gains, and building's heat storage; (iv) wind is a major determinant of outdoor air infiltration that causes both sensible and latent heat losses; (v) temperatures of adjacent unconditioned buffer spaces (such as garages, attics, basements, or enclosed porches) are required; (vi) for uninsulated adjacent spaces, the temperatures can be considered as outdoor temperatures; (vii) in cold climates, humidification is required to maintain comfortable indoor relative humidity under heating conditions; and (viii) air leakage rates, usually specified as air exchanges per hour, depend on the building envelope's effective leakage area and driving pressure caused by buoyancy (stack effect), and wind; the only accurate procedure for determining air leakage rate is by measurement using a pressurization test (commonly called a blower door test).

2.3.2 Cooling Loads

Calculation of sensible and latent (usually estimated at 20–30% of the sensible part) cooling loads for residential dwellings, mainly due to heat gains through structural components (as walls and ceilings), windows, as well as due to lighting, infiltration and ventilation, and miscellaneous sources, such as cooking, laundry, bathing, and human occupancy (usually assumed to be 0.066 kW of sensible heat gain per person), depends on factors such as: (i) outdoor design dry-bulb and mean coincident wet-bulb temperatures; (ii) extremely hot events of short duration; and (iii) amount of conditioned air required to ensure the proper design of the air handling units and other air distribution systems.

2.4 COMMERCIAL AND INSTITUTIONAL BUILDINGS

Annual heating and cooling loads of typical commercial/institutional buildings can be determined as functions of (ASHRAE 2011; DOE 1995): (i) meteorological typical data for a given location; (ii) building total and zone surfaces; (iii) people number and occupancy; and (iv) equipment and lighting loads in the different thermal zones.

Prior sizing of the ground-source heat pump systems and the distributed and/or central equipment for commercial/institutional buildings, actions as the following should be achieved: (i) divide the building into core and perimeter thermal zones with similar temperature settings and loads due to people, equipment, and lighting as well as to solar gains and infiltration loads; (ii) evaluate the peak heating and cooling loads in each thermal zone; (iii) select individual geothermal heat pumps in the thermal zones; (iv) calculate building's monthly heating and cooling loads preferably using hour-by-hour building energy simulation software over 8,760 hours (Figure 2.2), as well as the total building heating and cooling loads (block loads); and (v) calculate the length of vertical or horizontal ground-coupled heat exchanger lengths (or groundwater or surface water thermal and flow

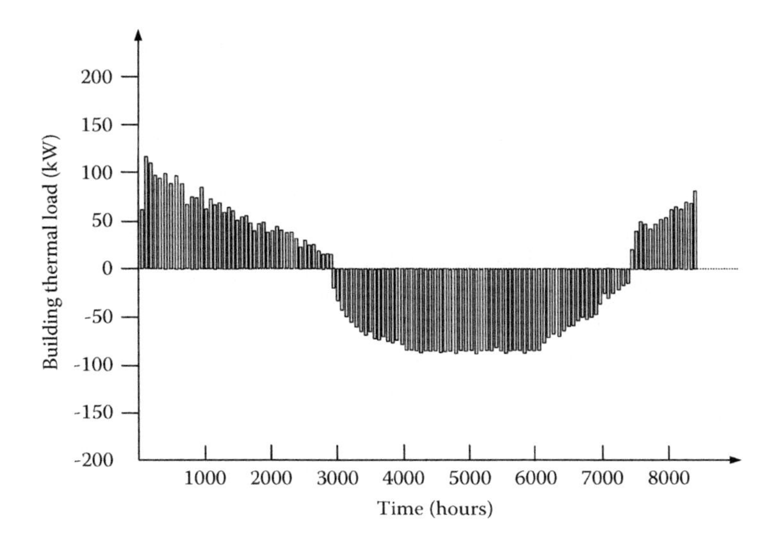

FIGURE 2.2 Example of typical monthly heating and cooling loads for commercial/institutional buildings over one year (8,760 hours) (Notes: positive and negative loads are heating and cooling loads, respectively; graph not for a real building and not drawn to scale.)

requirements) to maintain the brine (or water) temperature within minimum and maximum temperature limits for optimum operation of geothermal heat pumps.

Some general guidelines for zoning the building into thermal zones are: (i) perimeter (exterior) zones (directly affected by outdoor conditions during summer and winter) should have the same orientation or cardinal direction; the perimeter zone's peak cooling loads generally occur during peak solar load, usually in July or August; and (ii) interior (core) zones should not have significant exterior surface slightly affected by outdoor climate conditions; typically, the interior (core) zones have only cooling loads as a result of people, lights, equipment, and ventilation (in summer).

It can be noted that at certain times of the day during the heating-dominated season, some perimeter zones may require space heating, while others may require space cooling. The block heating and cooling loads, needed to determine the maximum heat extracted or rejected during the heating and cooling from/to ground/soil/rocks (or groundwater or surface water), consist of the sum of the individual zone thermal loads at the time of the peak building thermal loads.

2.4.1 Heating Loads

In the case of commercial/institutional buildings (such as stores, offices, theaters, and assembly halls), the heat supplied by occupants, lights, motors, and machinery should always be determined. In any evaluation, however, night, weekend, and any other unoccupied periods must be considered in order to verify if the ground-source heat pump systems have sufficient heating capacity to bring the building to the required indoor temperature before the occupants arrive.

The general procedure for calculation of design heating load for commercial/institutional buildings that, usually, are estimated for the winter design temperature occurring at night and, therefore, no credit is taken for the heat given off by internal sources (people, lights, etc.), can be summarized as follows (ASHRAE 2011; DOE 1995): (i) select outdoor weather design conditions, as dry- and bulb-, and dew-point temperatures, relative humidity, and wind direction and speed; (ii) select indoor design conditions (as dry-bulb temperatures and relative humidity) to be maintained in each thermal zone at outdoor design conditions during the coldest month; (iii) estimate temperatures in unheated spaces adjacent to the thermal zones that should be at intermediate temperatures between indoor and outdoor temperatures; (iv) compute the global heat transfer coefficients (U-values) of outside walls, ceilings, and windows, as well indoor doors, and basement or grade-level slab elements; (v) determine net areas of outside walls, windows, and roofs next to the heated and unheated spaces; (vi) compute heat transmission losses for each kind of wall, glass, floor, ceiling, and roof in the building by multiplying the heat-transfer coefficient in each case by the area of the surface and the temperature difference between indoor and outdoor air or adjacent unheated space; (vii) compute sensible and latent heat loads of outdoor air (ventilation) entering the thermal zones; (viii) sum the losses caused by conductive heat transfer (through walls, floors, ceilings, glasses, or other surfaces) and infiltration (through cracks and crevices, around doors and windows, or through open doors and windows), ventilation, and other thermal loads as supply and return duct losses to unheated spaces; and (ix) ignore internal gains in perimeter zones; core zones, however, can require cooling during the heating peak and should be accounted for.

The peak heating loads are needed to size the geothermal heat pumps in the thermal zones, to size the piping in the distribution system, and to size the circulating pumps, plate heat exchangers (in closed-loop groundwater or surface water geothermal systems), and other pumping components; this generally occurs during the early morning.

2.4.2 Cooling Loads

Space cooling loads, which must be removed to maintain the indoor air temperatures at the designed values, result from conduction, convection, and radiation heat transfer processes through the commercial/institutional buildings' envelopes and from internal heat sources (also called heat gains).

According to the mode in which heat is transferred to the building space, the following heat gains are generally due to (ASHRAE 2011; DOE 1995): (i) thermal conduction through exterior walls and roofs, as well as through interior partitions and doors; (ii) solar radiation and thermal heat conduction through transparent surfaces (windows) of which energy performances are determined by parameters as heat transfer coefficients (U-values), visible transmission factors that represent the percentages of the available visible light that is allowed to pass through the windows, and the amount of air leakage through the window; (iii) interior lighting, people (occupants), and various (electrical, gas, or steam) appliances and equipment (e.g., electrical motors); for energy savings, it is recommended to use multiple-pane, low-E, and gas-filled, high-energy efficient label windows, and avoid aluminum frame windows and use wood, vinyl, or fiberglass as frame insulating materials; it can be noted that in modern low-energy buildings, envelopes have improved in response to more restrictive energy codes, and internal loads have increased because of factors such as increased use of computers and the advent of dense-occupancy spaces (e.g., call centers); and (iv) ventilation and infiltration of outdoor air, and diffusion of moisture through building materials (walls, roofs, etc.), a natural phenomenon, is often neglected.

Most components of the cooling loads vary in magnitude over a wide range during a 24-hour period, often out of phase with each other. This means that, at any instant, there is an appreciable difference between the net instantaneous heat gain and net actual cooling load caused by the heat storage and subsequent release of heat by the building structure and internal materials (e.g., furniture). Instantaneous and delayed heat gains are: (i) sensible, added directly to the conditioned space by conduction, convection, and/or radiation; and (ii) latent, added to the building by water vapor emitted by occupants, shower areas, swimming pools, or other equipment; to maintain a constant absolute humidity, water vapor must condense on the cooling coils and be removed at the same rate as it is added to the space; unlike sensible loads, which correlate to supply air quantities required in a space, latent loads usually only affect the size of geothermal heat pumps' evaporators (i.e., cooling coils).

The space heat gain by radiation is partially absorbed by the surfaces and contents of the space and does not affect the room air until sometime later. The radiant energy must first be absorbed by the surfaces that enclose the space (i.e., walls, floor, and ceiling) and the material in the space. As soon as these surfaces and objects become warmer than the space air, some of their heat will be transferred to the air in the room by radiation and convection. Since their heat storage capacity determines the rate at which their surface temperatures increase for a given radiant input, it governs the relationship between the radiant portion of heat gain and the corresponding part of the space cooling load. The sum of space instantaneous heat gains at any given time does not necessarily equals the cooling load for the space at that same time because some absorbed and stored energy contributes to space cooling load by re-radiation after a time lag, i.e., after the heat sources have been switched off or removed. For example, because of heat storage, the rate of cooling load from lighting at any given moment can be quite different from the heat equivalent of power supplied instantaneously to those appliances (Figure 2.3). There is thus always significant delay between the time a heat source is activated, and the point when re-radiated energy equals that being instantaneously stored. This time lag must be considered when calculating cooling loads, because the cooling loads required can be much lower than the instantaneous heat gains being generated, and the space's peak loads may be significantly affected. Accounting for the time delay effect is the major challenge in cooling load calculations. Several methods have been developed to take the time delay effect into consideration.

Peak cooling loads of building core (interior) zones occur during the last hour of occupancy, while those of perimeter (exterior) zones generally occur at the time of the peak solar load; if internal loads are not significant, generally in July or August,

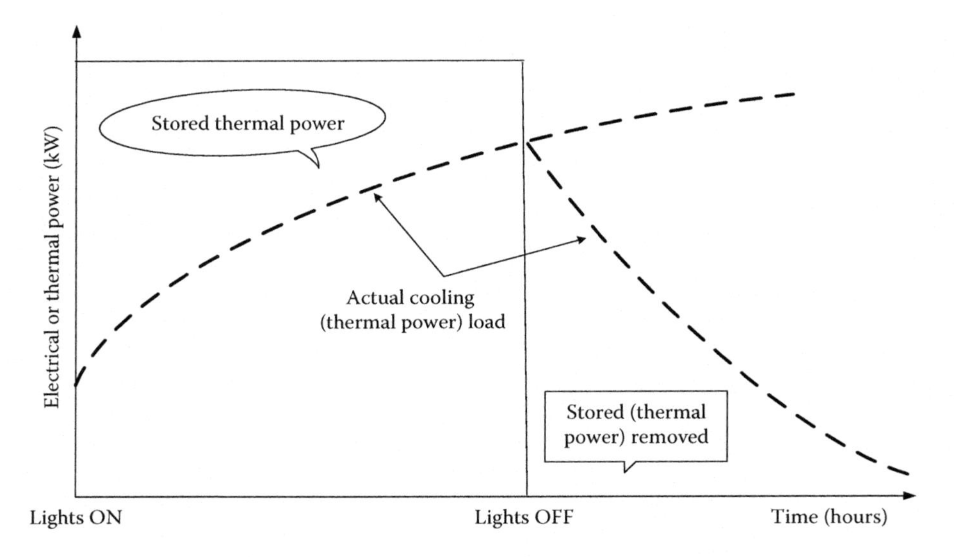

FIGURE 2.3 Effect of thermal storage from lights on cooling load.

when the outdoor temperature reaches a maximum. The design cooling load calculation should be done at three different times on the design day, usually July 21 (or, depending on the region, other hottest day of the year). If building summertime occupancy is low, as in some schools, the hottest day of the year with full occupancy should also be evaluated. For commercial/institutional buildings, the cooling load calculation can be performed by using heat balance methods as cooling load temperature differences (CLTD), solar cooling load (SCL) factors, and internal cooling load factors (CLF) methods (Spitler 2014).

2.5 BUILDING SIMULATION SOFTWARE TOOLS

Energy simulation software tools, using a mathematical model created on the basis of fundamental physical principles and sound engineering practice, are allowed to determinate with accuracy some variables that can support designers to reduce the cost of buildings' energy consumptions and quantify several aspects related to construction, operation, and control of buildings. They are the result of combined development efforts between academia, governmental institutions, industry, and professional organizations (de Wilde 2014).

Building thermal simulation software tools for buildings are used to (Augenbroe and Hensen 2004; Wang and Zhai 2016): (i) determine the thermal load profiles and energy demand for heating, cooling, ventilation, lighting, and main and auxiliary equipment (e.g., pumps, fans, elevators); (ii) provide energy balances for thermal zones and the whole building; (iii) determine the temperature trends in thermal zones and comfort indicators (as the radiant temperatures, CO_2 concentration, and relative humidity); (iv) determine the appropriate size of HVAC system components (as air handling units, heat exchangers, boilers, chillers, water storage tanks, heat pumps, and energy recovery devices); (v) analyze the energy consumption and calculate the cost of the energy used; and (vi) optimize control strategies (e.g., controller setup for shading, window opening, heating, cooling, and ventilation for increased operation performance, etc.).

Among the main uncertainties of building performance simulation tools concerning the results' reliability are mentioned (de Wilde 2014; Dodoo et al. 2017): (i) the quality of input data, (ii) the competence of the simulation engineers, and (iii) the applied methods in the simulation engine.

In the future, among the most important tasks that should be addressed can be mentioned (Clarke 2015; Heo et al. 2012; Reddy 2006): (i) better concept promotion; (ii) standardization of input data and accessibility of model libraries; (iii) standard performance assessment procedures; (iv) better embedding of building performance simulation in practice; (v) operational support and fault diagnosis with building performance simulation; (vi) education, training, and user accreditation; (vii) adjusting the assumed simulation inputs to match the observed data from utilities; (vii) increasing the accuracy and, thus, reducing the errors (i.e., the discrepancy between simulation results and the actual measured performance of the building) that normally occur due to approximations in model inputs, such as occupancy behavior.

At the beginning of the 1990s, the engineering equation solver (EES) was introduced (Klein 1993).

Necessary input data for whole-building energy simulations are as follows: (i) information about local climate (weather): ambient air temperature, relative humidity, direct and diffuse solar radiation, and wind speed and direction; (ii) information about the site: location and orientation of the building, shading by topography and surrounding buildings, and ground properties; (iii) information about the building shape and geometry; (iv) information about the building envelope: materials and constructions, windows and shading, thermal bridges, infiltration, and openings; (v) information about internal gains: lights, equipment, and occupants, including occupancy patterns and thermostat settings; (vi) information about ventilation system: transport and conditioning (heating, cooling, humidification) of air; (vii) information about room units for heating, cooling, and ventilation; (viii) information about the central units for transformation, storage, and delivery of energy to the building; and (ix) information about the control of window opening, shading devices, ventilation systems, room HVAC units, and mechanical room main components.

Among the most complete simulation software tools available, there is Energy Plus, ESP-r (Energy Simulation Software tool), IDA ICE (Indoor Climate Energy), IES-VE (Integrated Environmental Solutions – Virtual Environment), and TRNSYS. EnergyPlus software (http://energyplus.net/, accessed September 2, 2019) is a free, open-source, cross-platform, console-based program for whole building running on the Windows, Mac OS X, and Linux operating systems.

Energy Plus is one of the most-known energy simulation software tools. Its development began in 1996, sponsored by the Department of Energy (DOE) from the United States of America (USA). Initially, the U.S. government was developing two different software tools, BLAST and DOE-2, which were abandoned after many discussions and represented a first step and the working basis of the Energy Plus.

The Energy Plus has the features and capabilities of BLAST and DOE-2, however is an entirely new software tool that combines the heat balance of BLAST with a generic HVAC system. The Energy Plus aims to develop and organize software tools in modules that can easily work together or separately. It is important to outline that in Energy Plus does not exist a visual interface that allow users to see and concept the building. In this case third-party software tools, i.e., Design Builder needs to be used. Energy Plus is a thermal simulation software tool that allows the analysis of energy throughout the building and the thermal load and it is used by engineers, architects, and researchers to model the energy use and water use in buildings. The software tool simulates models for heating, cooling, lighting, ventilation, and other flows of energy and water use.

It includes some modular structures to model energy consumption for heating, cooling, ventilation, and lighting, and not found in other programs (e.g., a slab and basement program to model heat transfer from the ground/soil, solar hot water systems, water-to-water geothermal heat pumps, heat recovery devices, and demand controlled ventilation), allowing the user to create new simulation structures.

Some of the notable features and capabilities of Energy Plus software include (https://energyplus.net/): (i) integrated, simultaneous solution of thermal zone conditions and HVAC system response that does not assume that the HVAC system can meet zone loads and can simulate unconditioned and under-conditioned spaces; (ii) heat balance-based solution of radiant and convective effects that produce surface temperatures thermal comfort and condensation calculations; (iii) sub-hourly, user-definable time steps for interaction between thermal zones and the environment; with automatically varied time steps for interactions between thermal zones and HVAC systems; these allow Energy Plus to model systems with fast dynamics while also trading off simulation speed for precision; (iv) combined heat and mass transfer model that accounts for air movement between zones; (v) advanced fenestration models including controllable window blinds, electrochromic glazing, and layer-by-layer heat balances that calculate solar energy absorbed by window panes; (vi) illuminance and glare calculations for reporting visual comfort and driving lighting controls; (vii) component-based HVAC that supports both standard and novel system configurations; (viii) a large number of built-in HVAC and lighting control strategies and an extensible runtime scripting system for user-defined control; (ix) functional mockup interface import and export for co-simulation with other engines; and (x) standard summary and detailed output reports as well as user definable reports with selectable time-resolution from annual to sub-hourly, all with energy source multipliers.

TRNSYS is a transient system simulation software tool with a modular structure that has been specially designed to develop complex systems related to energy, outlining the problem in a number of smaller components. The components ("Types") may range from a simple heat pump to a multi-zone of a building complex. The components are configured through the graphical user interface known as TRNSYS Simulation Studio. In the simulation software tool energy TRNSYS, the construction of the building can be achieved by the introduction of data on a dedicated visual interface, known for TRNBuild. The software tool sets the time intervals, which may vary from 15 minutes to an hour, but may be able to perform simulations in the time interval of 0.1 seconds.

The library software tool, in addition to a multi-zone, allows the use of many commonly used components, including solar panels, photovoltaic systems, HVAC systems, cogeneration systems, and hydrogen, among others. It also allows the creation of routines to manipulate weather data and other data by changing the simulation results. The modular nature of this software tool facilitates the addition of mathematical models to the software tool. The components can be shared among multiple users without having to recompile the software tool due to the use of DLL technology. In addition, TRNSYS allows the user to incorporate other components developed in software tools such as Matlab, and enable non-users to view and do parametric studies in simplified representations of web pages.

eQUEST® (eQuest, Quick energy simulation tool; http://doe2.com/equest/) is a free user interface, and sophisticated tool built to be used with the DOE-2 building energy modeling program. It allows a detailed comparative analysis of building design and technologies as ground-source heat pumps and provides results as the annual building energy consumption and the hourly building heating and cooling demands with an affordable level of effort, one of the most commonly used building energy modeling, user-friendly program in North America (Hanam 2010).

The software tool ESP-r (Energy Simulation Software Tool) is a mathematical software intended to accurately determine the building energy and environmental performances by using several complex equations to deal with aspects such as building geometry, construction type, operation schedules, heat dissipation and heat gains, occupants' influences, and climate conditions. Shading, insulation, HVAC systems, electricity and renewable energy embedded systems, lighting, natural ventilation, combined heat and power generation, and photovoltaic systems are included in pre-determined models. The models created and the time of building simulations can be viewed or exported to other software, Energy Plus.

The thermal simulation software tool, IDA Indoor Climate Energy, is a general modular simulation platform that uses equations in the neutral model format language. The user defines the tolerances that control the accuracy of the solution, thus allowing the isolation of numerical modeling approaches. Extensions can be added to the initial model, and the variables, parameters, and equations can be easily inspected.

The dynamic simulation software tool, IES VE (Integrated Environmental Solutions – Virtual Environment), provides a variety of building geometric representations for simulation. The software allows interaction with other energy simulation software tools, and can achieve optimizations, taking into account criteria such as human comfort and energy.

REFERENCES

ASHRAE. 2011. *Handbook Fundamentals*. SI Edition. Supported by ASHRAE (American Society of Heating, Refrigeration and Air-Conditioning Engineers, Inc.), Atlanta, GA.

Augenbroe, G., J. Hensen. 2004. Simulation for better building design. *Building and Environment. Building Simulation for Better Building Design* 39(8):875–877.

Bose, J.E., J.D. Parker, F.C. McQuiston. 1985. *Design/Data Manual for Closed-Loop Ground-Coupled Heat Pump Systems*. Oklahoma State University. American Society of Heating, Refrigerating and Air-Conditioning Engineers, Inc., Tullie Circle, NE, Atlanta, GA.

Clarke, J. 2015. A vision for building performance simulation: A position paper prepared on behalf of the IBPSA Board. *Journal of Building Performance Simulation* 8(2):39–43.

de Wilde, P. 2014. The gap between predicted and measured energy performance of buildings: A framework for investigation. *Automation in Construction* 41:40–49.

Dodoo, A., U. Tettey, A. Yao, L. Gustavsson. 2017. Influence of simulation assumptions and input parameters on energy balance calculations of residential buildings. *Energy* 120:718–730.

DOE. 1995. *Commercial/Institutional Ground-Source Heat Pump Engineering Manual.* Prepared by CANETA Research Inc. for U.S. Department of Defense U.S. Department of Energy, Oak Ridge National Laboratory, Oak Ridge, TN.

EnergyPlus (http://apps1.eere.energy.gov/buildings/energyplus/).

Hanam, B. 2010. *Development of an Open Source Hourly Building Energy Modeling Software Tool.* University of Waterloo, Waterloo.

Heo, Y., R. Choudhary, G.A. Augenbroe. 2012. Calibration of building energy models for retrofit analysis under uncertainty. *Energy and Buildings* 47:550–560.

Klein, S.A. 1993. Development and integration of an equation-solving program for engineering thermodynamics courses. *Computer Applications in Engineering Education* 1(3):265–275.

Reddy, T.A. 2006.Literature review on calibration of building energy simulation programs. *ASHRAE Transactions* 112(1):226–240.

Spitler, J.D. 2014. *Load Calculation Application Manual.* 2nd Edition. ASHRAE, Atlanta, GA.

Wang, H., Z. Zhai. 2016. Advances in building simulation and computational techniques: A review between 1987 and 2014. *Energy and Buildings* 128:319–335.

3 Conventional Building HVAC Systems

3.1 INTRODUCTION

Conventional heating, ventilating, and air-conditioning (HVAC) systems generally consist of: (i) heating equipment (such as boilers, furnaces, or air/water-source heat pumps); (ii) cooling units (such as air conditioners or air/water-source heat pumps); and (iii) piping/duct network for air, water, or combined air/water distribution.

3.2 RESIDENTIAL AND SMALL COMMERCIAL/INSTITUTIONAL BUILDINGS

3.2.1 Air-Source Heat Pump and Furnace Split Systems

An air-source heat pump and furnace split HVAC system (also called forced-air systems) typically consists of (Figure 3.1): (i) an indoor unit, such as an air handler (a large metal box comprising an air blower; a natural gas-, propane-, or oil-burned furnace; or an electric air heater), a refrigerant indoor coil with expansion valve (EX-1) and check valve (CV-1), as well as filters, sound attenuators, and dampers; (ii) an outdoor unit (either an air conditioner or an air-source heat pump) comprising a refrigerant compressor (C), outdoor coil, a fan and refrigerant tubing with reversing valve (RV), expansion valve (EX-2), check valve (CV-2), and controls; (iii) programmable or non-programmable thermostats; and (iv) a ductwork ventilation system that distributes the conditioned air to room diffusers and grilles, and returns it to the air handler.

Among the advantages of such residential HVAC split systems are the following: (i) relatively small equipment, and usually low annual energy and operating costs and (ii) up to 97% energy efficiency. The main disadvantages are: (i) risks of furnace explosion and carbon monoxide leaks; (ii) risks for dispersing dust, odors, and excessively drying the air inside the house; (iii) require ductwork for forced-air distribution; and (iv) low heating energy efficiency in cold climates at subfreezing outdoor temperatures, thus, requiring electric resistances (or fossil-burned furnaces) as back-up heating devices.

3.2.2 Dual (Hybrid)-Energy Source Heat Pump Systems

In cold and very cold climates, the compact dual (hybrid)-energy source heat pump systems could become more effective. When outdoor temperatures are above, for example, –12°C, the building is heated by the air-source heat pump. At outdoor temperatures below –12°C, the heat pump uses a heat source such as a fossil fuel

DOI: 10.1201/9781003032540-3

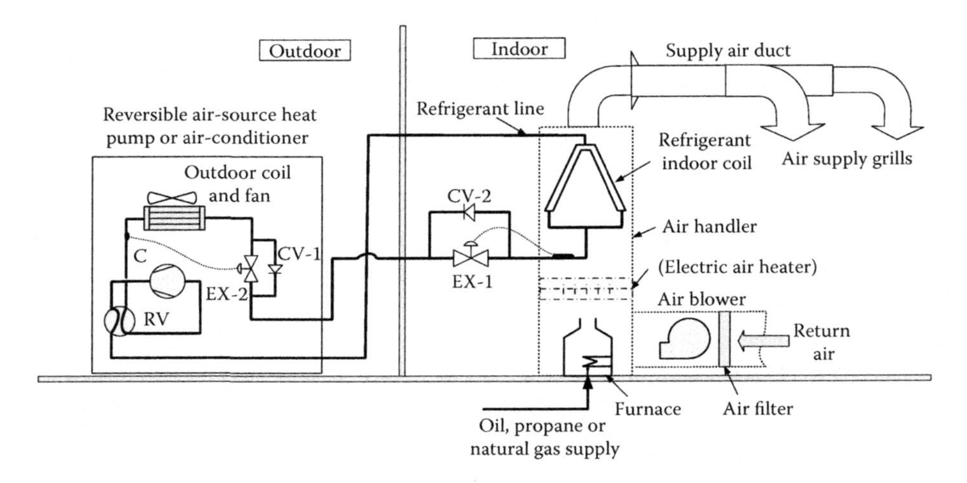

FIGURE 3.1 Schematic of a residential or small commercial/institutional fossil fuel furnace and air-conditioning split HVAC system; C, compressor; CV, check valve; compressor; EX, expansion valve; RV, reversing valve (Note: Schematic showing the geothermal heat pump operating in heating mode).

(natural gas and propane). Compact dual (hybrid)-energy source heat pump systems could include special heat transfer components such as the following (Vander Vaart 1982; Parent et al. 1991; Minea et al. 1995; Minea 2011): (i) heat-augmented A-coil heat exchangers (see Figure 3.2), (ii) brine A-coil heat exchangers (see Figure 3.3), and (iii) add-on plate heat exchangers (see Figure 3.4).

3.2.3 Heat-Augmented Heat Exchanger

The dual-energy heat pump with an A-coil-type "heat-augmented" heat exchanger (Figure 3.2) provides additional energy to the building at very low outdoor

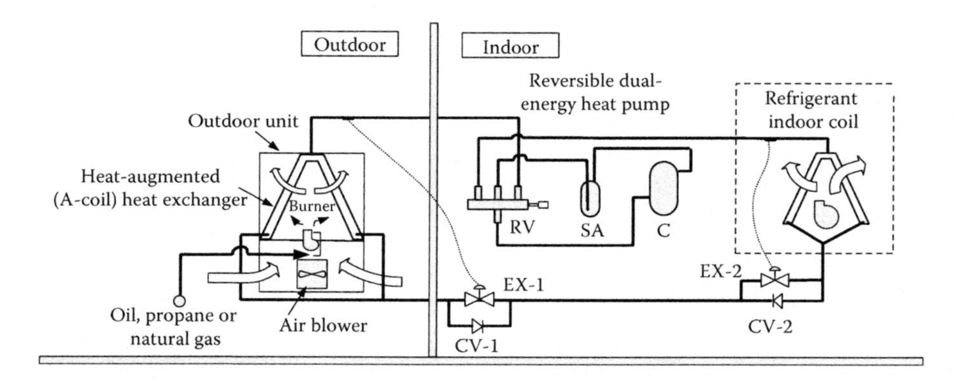

FIGURE 3.2 Dual-energy source heat pump with fossil fuel "heat-augmented heat exchanger"; C, compressor; EX, expansion valve; RV, reversible valve; SA, suction accumulator.

temperatures (e.g., below –12°C in cold-climate countries such as Canada) to reduce the electrical peak power demand for space heating. The fossil fuel (oil, propane, or natural gas) burner is located inside the outdoor unit under the A-coil (Vander Vaart 1982). The heat pump does not operate in the back-up heating mode and, as in the previous system, in the defrost mode, the fuel burner supplies heat to the outdoor evaporator without reversing the thermodynamic cycle.

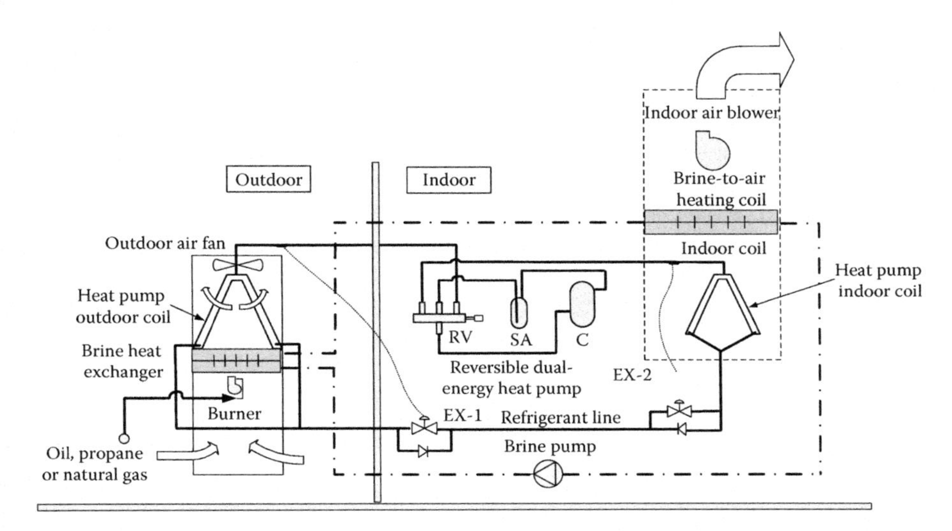

FIGURE 3.3 Dual-energy source heat pump with fossil fuel "heat-augmented coil" and brine (glycol/water mixture)-to-air heating coil; C, compressor; EX, expansion valve; RV, reversible valve; SA, suction accumulator.

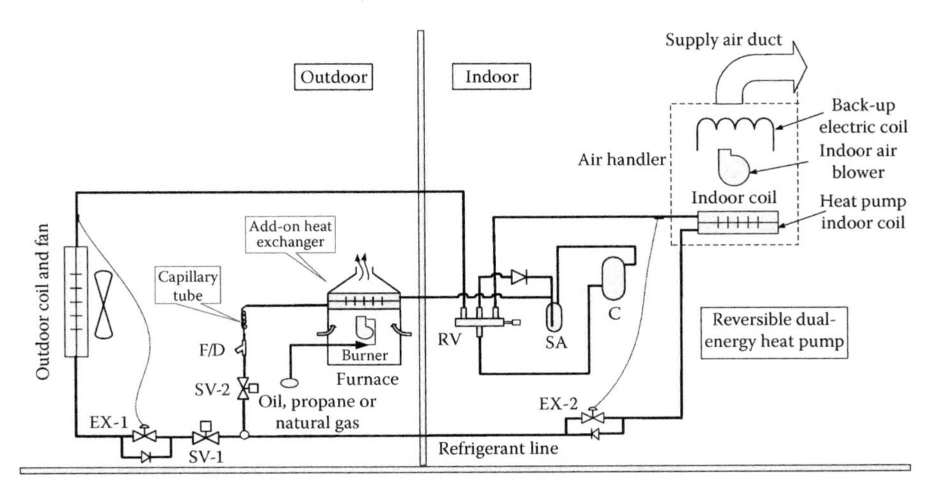

FIGURE 3.4 Schematic representation of a dual-energy source heat pump; C, variable speed compressor; EX, expansion valve; F/D: refrigerant filter/drier; RV, reversible valve; SA, suction accumulator; SV, solenoid valve.

A third operating mode allows the heat pump to operate at temperatures below those that would render conventional heat pumps totally inefficient. The outdoor unit contains a heat-augmented coil and a fuel-fired burner that generates heat to vaporize the liquid refrigerant. The heat-augmented coil operates as an evaporator in the air-to-air (conventional) using a bidirectional capillary tube as an expansion device, and as a condenser in the cooling mode. The defrost mode is provided without reversing the thermodynamic cycle because the fuel-fired burner supplies heat to defrost the outdoor evaporator. Thus, the heat-augmented (or back-up) heating mode, the third operating mode, renders the heat pump fully operative at temperatures well below those that would render conventional heat pumps virtually inoperative and/or inefficient in the heating mode of operation (Vander Vaart 1982). In this mode, the outdoor air blower shuts down. The heat-generating source can be a fossil fuel source placed at the bottom of the "A-coil" located outside the building. The refrigerant flow is similar to that of the conventional heating mode. When the outdoor blower is de-energized, the gas burner is energized by igniting the gas. The flames are spread out across the bottom of the A-coil and the heat is absorbed almost totally, while the burning process approaches 100% efficiency. The compressor can use the relatively high-pressure vapor phase of the refrigerant, which would be impossible to achieve in the absence of the additional heat provided by the fossil heat source.

3.2.4 Brine Heat Exchanger

Another concept consists of a conventional air-to-air heat pump equipped with a parallel glycol closed-loop and heat exchanger (Figure 3.3) (Vander Vaart 1982; Parent et al. 1991).

3.2.5 Add-On Heat Exchanger

The refrigeration circuit of a conventional air-to-air heat pump has been modified by including an add-on heating unit (Figure 3.4) (Minea et al. 1995; Minea 2011). The dual-energy source heat pump includes an indoor unit with a variable-speed compressor, a suction accumulator, a four-way reversing valve, and controls (Figure 3.4). The second cabinet, located outdoors, contains a horizontal finned plate heat exchanger with an air fan, as well as an expansion valve EX-1 with a bypass and check valve CV-1.

The third cabinet of this add-on unit, also located outdoors, operates at low (usually, below –12°C) ambient temperatures. It includes a small furnace with a gas-fired burner and a compact add-on combustion gas-to-refrigerant heat exchanger. It preheats, vaporizes, and superheats the refrigerant in the back-up heating mode. Being located outdoors, the burned gases do not contaminate the indoor atmosphere. Finally, the fourth cabinet located inside the house includes a finned indoor coil with an indoor air blower, expansion valve EX-2, and bypass with check valve CV-2. Two additional solenoid valves, SV-1 and SV-2, make it possible to bypass the outdoor coil and to supply low-pressure refrigerant liquid to the add-on heat exchanger via the capillary tube installed upstream of the solenoid valve SV-2. Finally, the check valve CV-3 allows the refrigerant vapor leaving the add-on coil to bypass the four-way reversible valve (RV) and to flow, via suction accumulator (SA), to the compressor suction line in the back-up heating mode.

3.2.6 Mini-Split Air-Source Heat Pump Systems

Mini-split systems (also called ductless because they don't require air ducts; thus, ideal for heating and cooling new and removed/retrofit constructions that don't have existing ductwork, as well as for room additions where extending or installing distribution ductwork is not feasible), contain two main components: (i) an outdoor unit and (ii) an indoor air-to-air heat exchanger(s) usually suspended from the ceiling or mounted high on indoor walls to send conditioned air directly into the living spaces directly connected to the outdoor unit via refrigerant lines. Single-zone mini-split heat pump systems have one outdoor unit and one indoor unit (Figure 3.5).

Multi-zone mini-split heat pump systems have an outdoor unit serving up to four indoor units by using the same outdoor heat pump unit (Figure 3.6) to independently

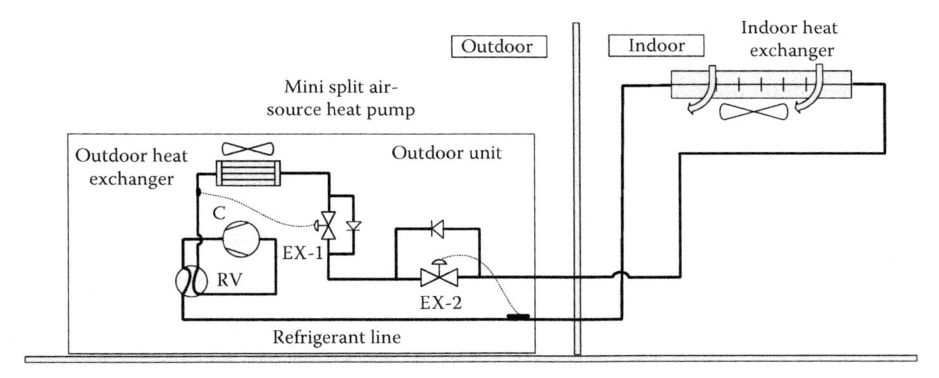

FIGURE 3.5 Schematic representation of a residential and small commercial/institutional mini-split single-zone heat pump; C, variable-speed compressor; EX, expansion valve; RV, reversible valve.

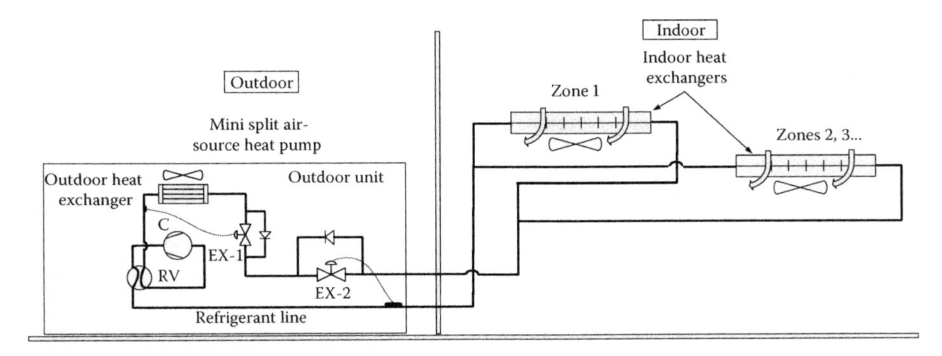

FIGURE 3.6 Schematic representation of a multi-zone mini-split heat pump system for residential and small commercial/institutional buildings; C, variable speed with inverter compressor; EX, expansion valve; RV, reversing valve.

control the temperatures in each individual room or conditioned space. Such air-source heat pumps operate similarly to standard heat pumps, though most use inverter technology, thus being able to run most of the time when heating or cooling are required, even at very low capacities, ramping up or down as needed to maintain even indoor temperatures. Additional advantages of mini-split ductless air-source heat pump systems are as follows: (i) indoor units can be hidden, preserving the room aesthetic (ii) are ideal for larger homes that only require certain occupied spaces of the building to be conditioned at a given time via separate thermostats and remote controls.

3.3 LARGE-SCALE COMMERCIAL/INSTITUTIONAL BUILDINGS

Conventional large-scale commercial/institutional HVAC systems include, among many other components (Seyam 2018): (i) heating and cooling heat exchangers; (ii) fossil-fueled or electric boilers or furnaces that produce hot water distributed to air heating coils, (iii) electric or absorption chillers that generate cool water distributed to air cooling coils, (iv) wet or dry cooling tower(s), (v) humidification and de-humidification devices, (vi) controls, (vii) roof-top units that supply conditioned air into the buildings' spaces, (viii) air- or water-source heat pumps, (ix) mixing-air plenums, (x) outdoor air intake ducts, (xi) air filter(s), (xii) supply and exhaust fan (s), (xiii) air supply and return ducts, and (xiv) terminal air distribution devices.

Depending on the location, the heating and cooling equipment of HVAC systems for large-scale commercial/institutional buildings can be: (i) central, serving the entire building; and (ii) decentralized, each serving single or adjacent building thermal zones. Central and decentralized HVAC systems can be selected based on criteria as: (i) climate conditions (dry- and wet-bulb temperatures, relative humidity, wind, etc.); (ii) building type, age, architectural configuration, and the profile of monthly heating and cooling loads; (iii) availability of energy conventional (fossil, electrical) and renewable (geothermal, solar, wind) sources; (iv) predicted life cycle operating and maintenance costs; and (v) system-estimated reliability and flexibility.

While central HVAC systems may imply higher first (capital) and operating costs, decentralized concepts, of which the main equipment may be located on the buildings' roofs, adjacent ground surface, or in the buildings' basements or the living spaces for more flexibility, could require higher operating costs. HVAC systems for large-scale commercial/institutional buildings can be also sub-classified into (Figure 3.7): (i) all-air system,; (ii) all-water systems, (iii) air-water systems, (iv) water-source heat pump systems, and (v) heating and cooling panel systems.

3.3.1 All-Air HVAC Systems

All-air conventional central HVAC systems usually combine several devices, such as supply and return air fans, pre-heat, cooling, and reheat heat exchangers, air mixing plenum, filter(s), and outdoor air, in one air handling unit (Figure 3.8). All-air HVAC systems can be: (i) single or multi-zone, (ii) with terminal reheat units, (iii) with dual duct(s), and (iv) variable air volume.

Single-zone all-air conventional HVAC systems consist of air handling units (that can be wholly integrated where heat and cooling sources are available or separate

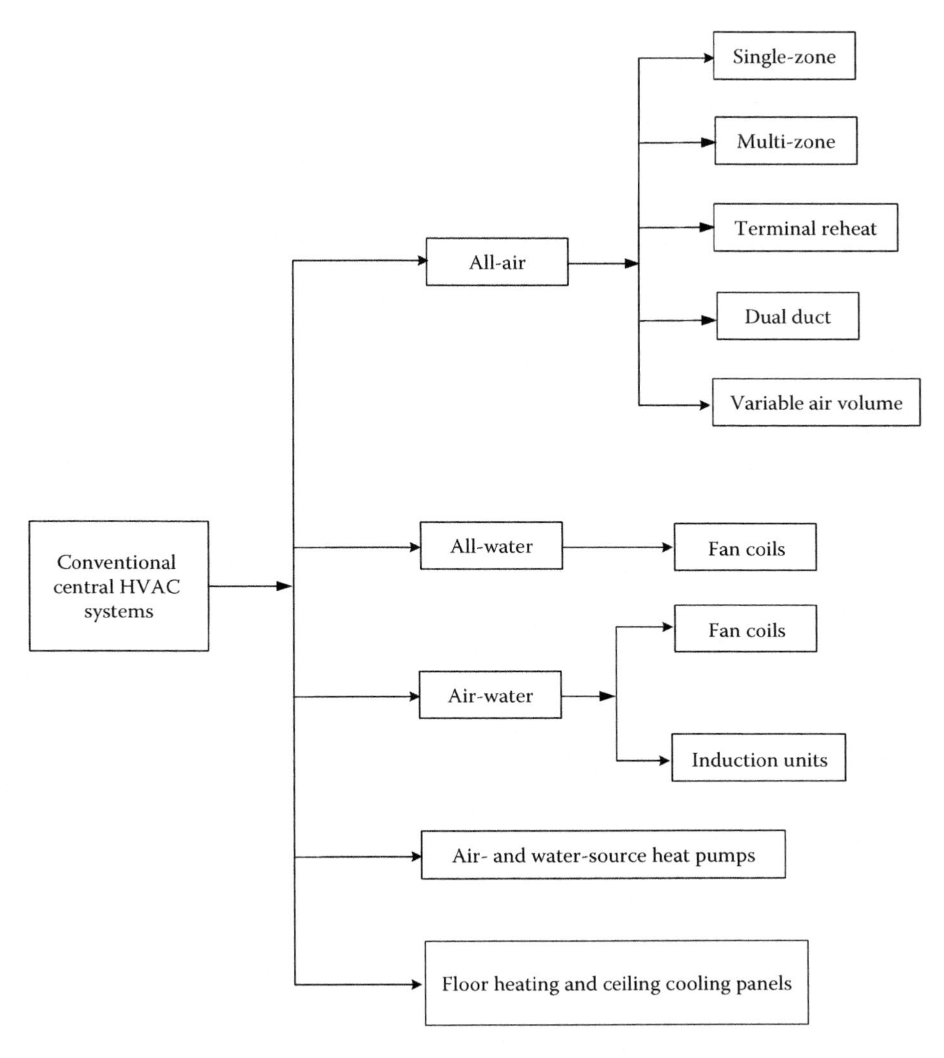

FIGURE 3.7 Classification of conventional central HVAC systems.

where heat and cooling sources are detached), heat and cooling sources, distribution ductwork, and air delivery devices (Figure 3.9). The integrated package is most commonly rooftop units connected to ductwork to deliver the conditioned air to several spaces with the same thermal loads. Such systems have relatively low cost compared to other conventional HVAC systems and are simple to design, operate, control (with one on-off thermostat to meet the required thermal load of the single zone), and maintain.

In multi-zone all-air conventional HVAC systems, individual, separate air ducts are provided for each building's zone. Cold or hot (or return) air flowing through cooling and heating coils are mixed at the air handling units to achieve the thermal requirement for each zone of which conditioned air cannot be mixed with that of other zones (Figure 3.10).

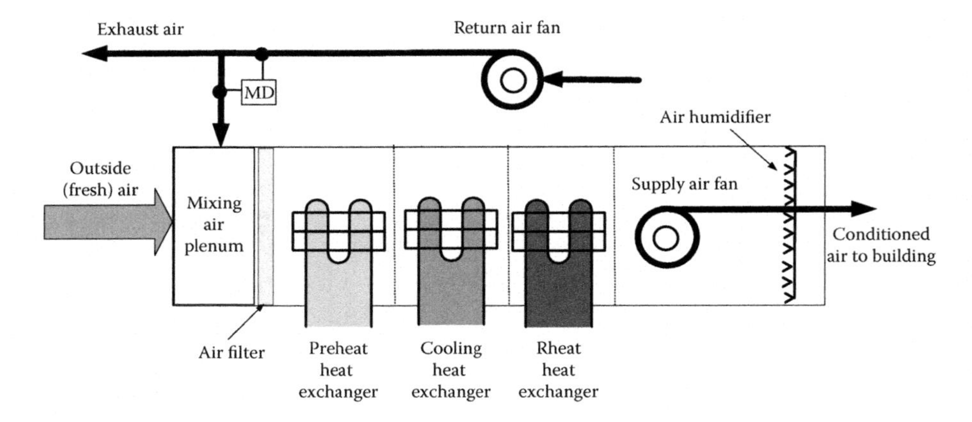

FIGURE 3.8 Equipment arrangement for all-air central HVAC system; MD, motorized air damper.

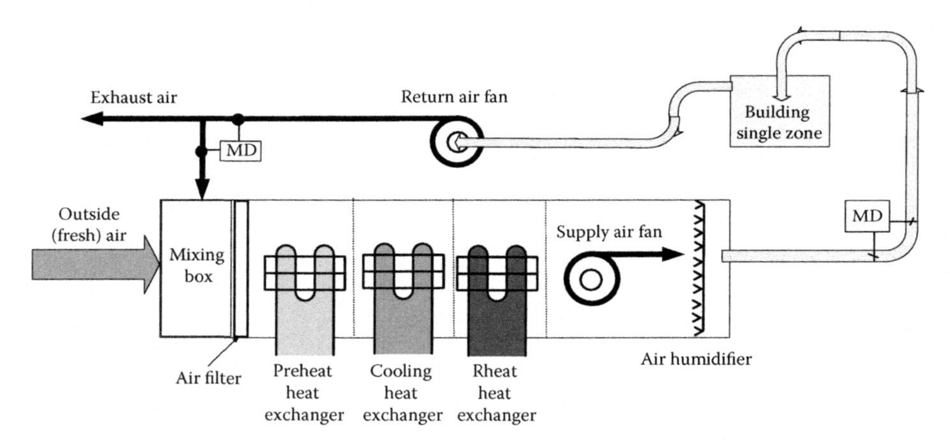

FIGURE 3.9 Single-zone all-air conventional HVAC system; MD, motorized air damper.

The advantage of such multi-zone HVAC concepts is to adequately condition several thermal zones without energy waste associated with the terminal reheat systems. The terminal reheat all-air multi-zone systems (Figure 3.11) consist of adding to the single zone (see Figure 3.9) heating equipment such as a hot water coil to the downstream of the supply air near each building thermal zone, equipment controlled by a thermostat to adjust the heat output. In other words, the the supply air from air handling units is cooled to the lowest cooling point, and the terminal reheat adds the required heating load. The advantage of terminal reheat consists of the fact that it is flexible and can be installed and removed to accommodate thermal charges in multiple building zones.

The dual-duct all-air systems are modified multi-zone concepts where central air handling units provide two, hot and cold, air streams (Figure 3.12) through separate, parallel ducts throughout the served areas. Each thermal zone has a terminal mixing

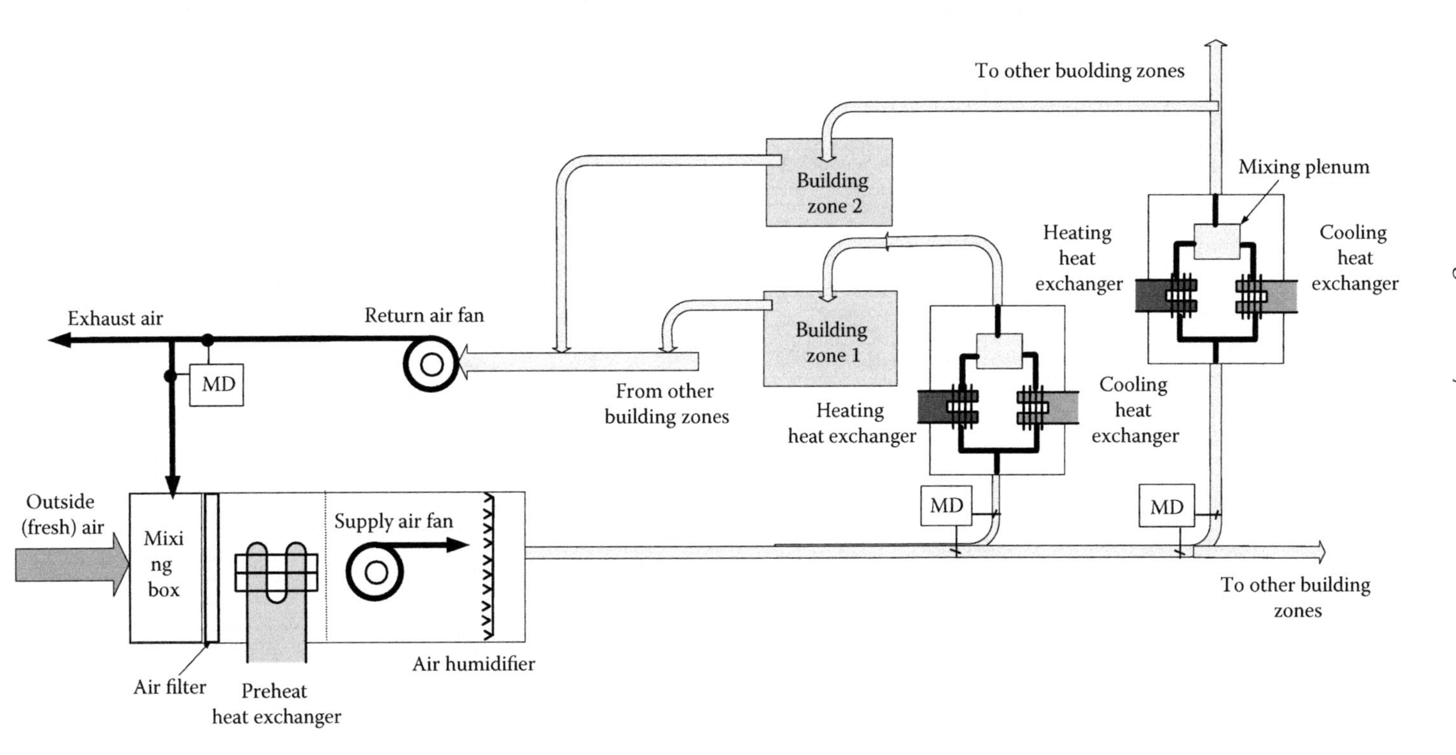

FIGURE 3.10 Multi-zone all-air conventional HVAC system; MD, motorized air damper.

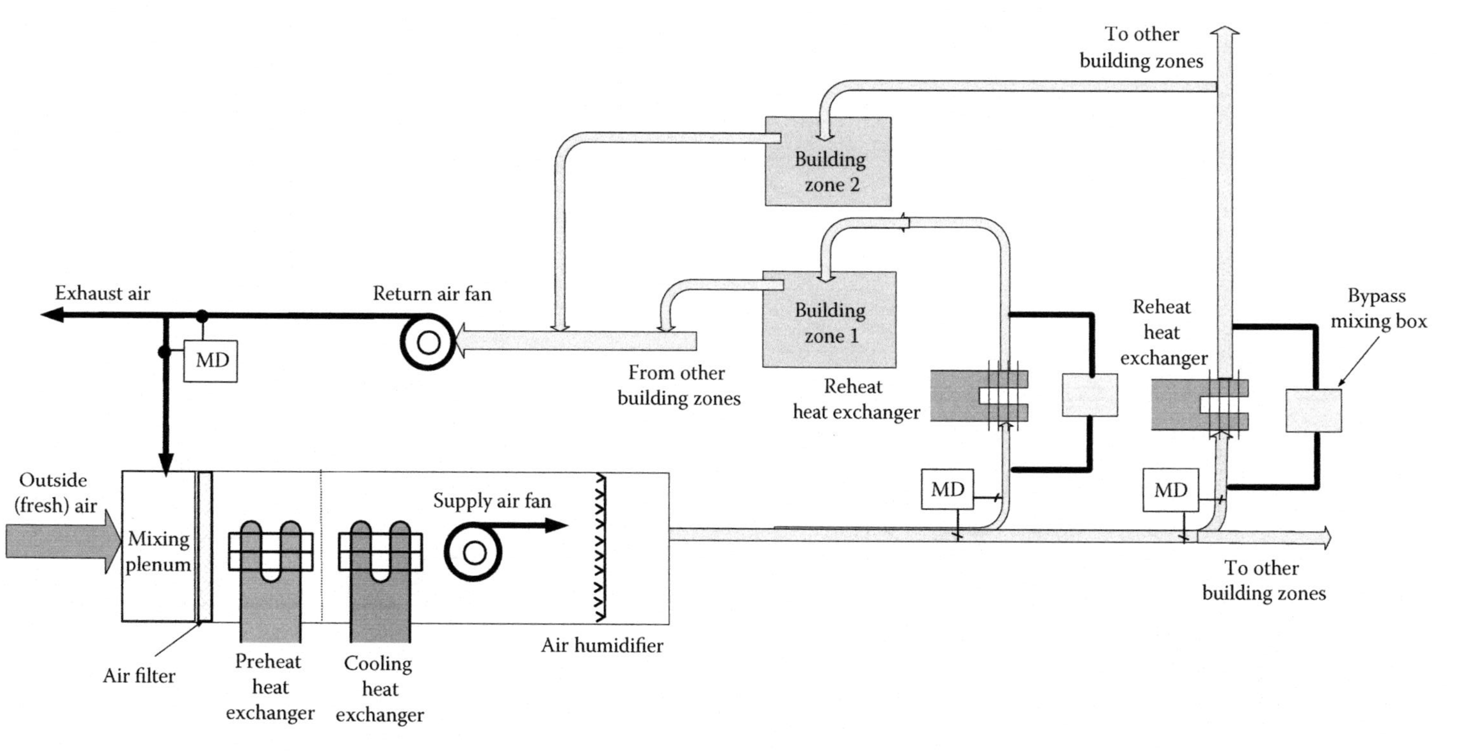

FIGURE 3.11 Single-duct all-air conventional HVAC system with reheat terminal devices and bypass units; MD, motorized air damper.

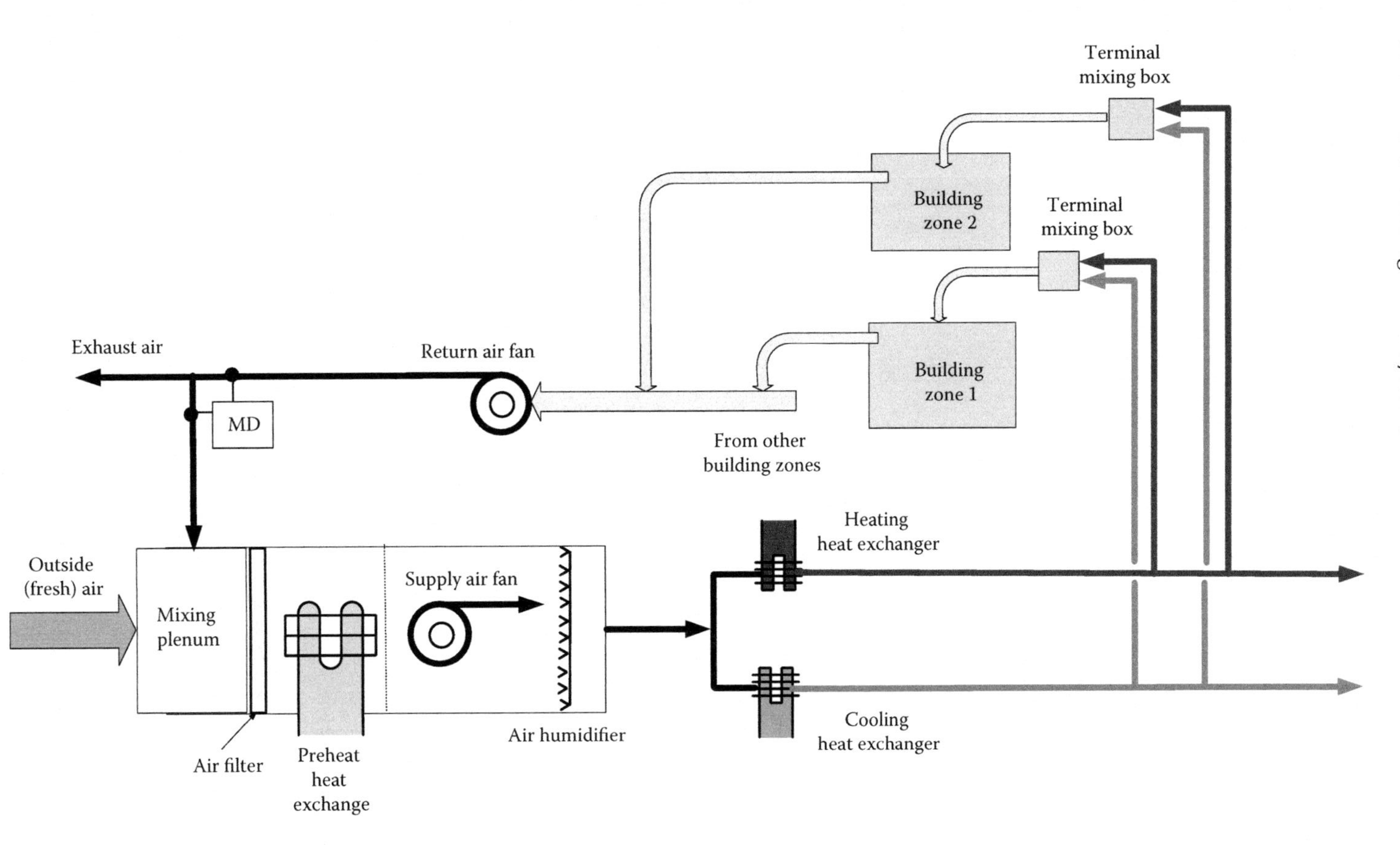

FIGURE 3.12 Dual-duct all-air conventional HVAC system.

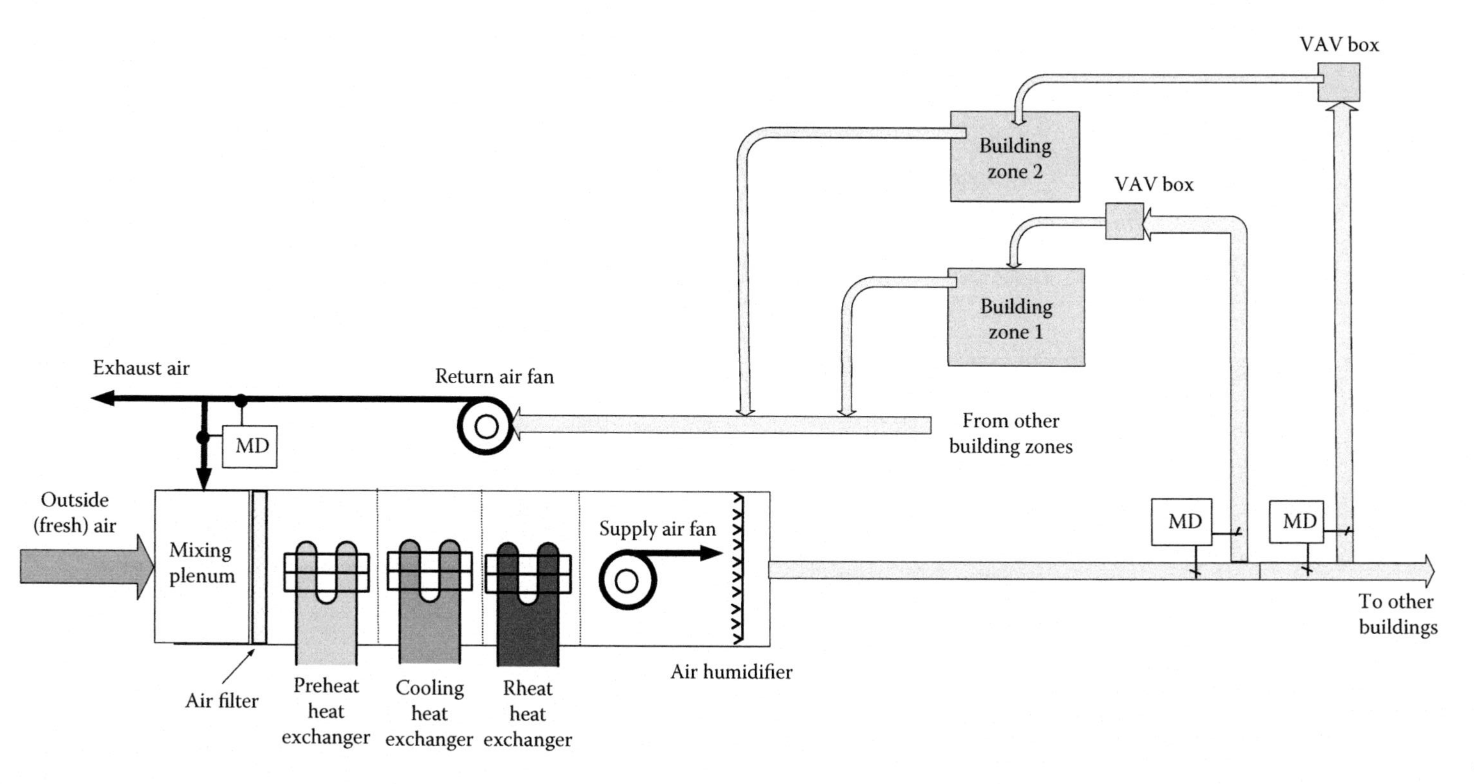

FIGURE 3.13 All-air conventional HVAC system with variable air volume terminal units; MD, motorized air damper; VAV, variable air volume.

box controlled by a zone thermostat to adjust the supply air temperature by mixing the supply hot and hot air.

Variable air volume (VAV) systems use single air supply and return ducts with varying airflow to keep temperatures at set points in spaces that require different airflow rates due to changes in zone thermal loads. Such systems consist of central air handling units that provide supply air to terminal control boxes located in each zone to adjust the supply air volume (Figure 3.13). The temperature of the supply air of each zone is controlled by manipulating the supply airflow rate.

3.3.2 All-Water HVAC Systems

In all-water HVAC systems, heated and cooled water is distributed from a central mechanical room to conditioned spaces via vertically or horizontally installed fan-coil units (Figure 3.14) that include water-to-air heat exchangers, air blowers (fans), and standard controls. Such water heating-only systems may also include heat delivery devices such as baseboard radiators and convectors. In decentralized HVAC systems, the fan-coil units can be placed inside rooms or other heated spaces. In central HVAC systems, the fan coils can be connected to boilers to produce heat or to water chillers to produce cooled water.

3.3.3 Air-Water HVAC Systems

Air-water HVAC systems combine advantages of both all-air and all-water concepts. The water carries out up to 80–90% of the building heating load. In such systems, the supply air and water are provided from central air handling units (e.g., boilers and chillers) (Figure 3.15).

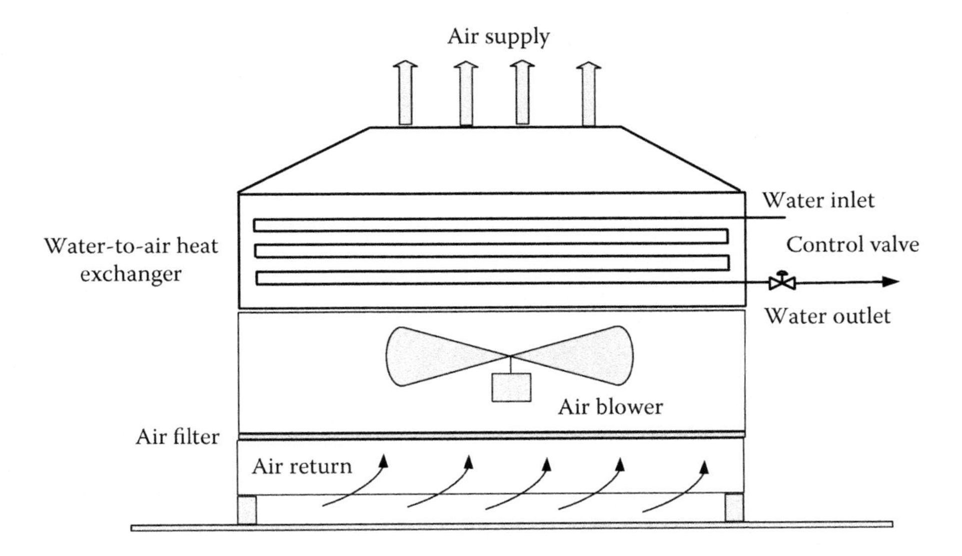

FIGURE 3.14 All-water system with fan coil.

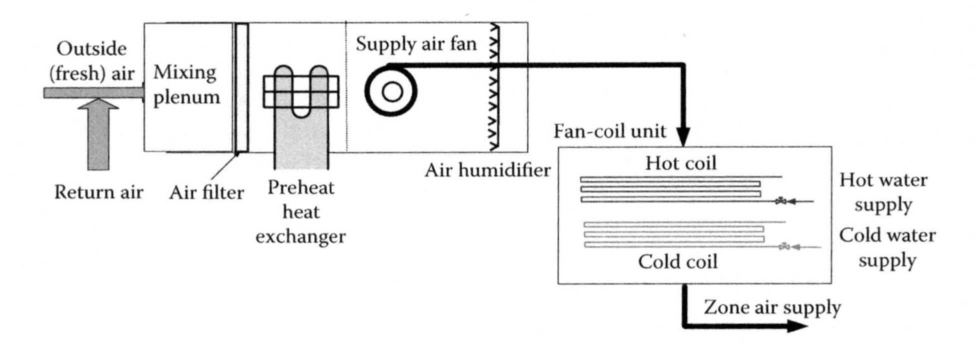

FIGURE 3.15 Air-water HVAC system using a four-pipe fan-coil unit.

REFERENCES

Minea, V. 2011. Dual energy heat pumps. In Proceedings of the 10th IEA (International Energy Agency) Heat Pump Conference, May 16–19, Tokyo, Japan.

Minea, V., A. Laperrière, D. Parent. 1995. Developmemt of a residential dual energy heat pump for cold climates. In Proceedings of the *19th International Congress of Refrigeration*, Den Hague, The Netherlands, August 20–25.

Parent, D., V. Minea, J. Demontigny. 1991. *Heat Pump Laboratory Pilot with Non-Electric Backup*. Final report. Hydro-Québec Research Institute, Shawinigan, Québec, Canada.

Seyam, S. 2018. Types of HVAC systems (https://www.intechopen.com/books/hvac-system/types-of-hvac-systems. Accessed March 13, 2020. DOI: 10.5772/intechopen.7842).

Vander Vaart, G. 1982. *To Kool-Fire Limited. Heat-Augmented Heat Exchanger*. USA Patent 4,311,192 (I-19-82, Cl. 165-29.000).

4 Geothermal Energy Resources

4.1 INTRODUCTION

Geothermal energy is a natural, renewable energy resource stored at any ground/soil depth, as well as in groundwater and surface water (lakes, ponds, rivers, seas). At depths accessible today for practical usage of geothermal energy (let's say, between 2 m and 2–3 km), the average temperatures in the Earth's crust varies from below 5°C (in northern regions) to more than 360°C (e.g., in volcanic zones). According to Earth's average temperatures, the practically usable geothermal resources can be subdivided into the following categories (adapted from ASHRAE 2011) (Table 4.1): (i) high temperatures (or high-enthalpy) (150°C ≤ T ≤ 360°C), mainly used for electric power generation; (ii) medium (intermediate) temperatures (90°C < T < 150°C), also used for electric energy generation by using advanced power generation by using advanced (e.g., ammonia-water absorption or Organic Rankin) cycles; (iii) low temperatures (30°C < T ≤ 90°C), mainly used for direct use applications such as urban and industrial district heating; (iv) very low temperatures (5°C < T < 30°C), generally used for heating and cooling residential, commercial/institutional, and industrial buildings by using geothermal heat pumps.

A distinction must be made between deep geothermal energy, which describes the heat derived from heat generation within the Earth and stored in hot rocks or trapped in steam or superheated liquid water, and shallow geothermal energy, which refers to heat stored within the first about 200 m of the surface that does not come primarily from the center of the deep strata of the Earth, but largely influenced by solar incident radiation and ambient weather conditions at the ground/soil surface (DOE 1995).

High- and medium- (intermediate-) temperature geothermal resources are geographically limited to geological active areas where the thermal gradient is exceptionally high and, thus, steam or superheated water can be extracted for use in combined electricity and heat production, and/or for urban and industrial district heating, respectively. On the other hand, low- and very-low-temperature (shallow) geothermal energy resources are abundant over vast geographic regions in moderate and cold-climate regions (e.g., United States, Canada, Sweden, and Japan) where the Earth's average density of geothermal heat flux near the surface is about 0.14 W/m^2, resulting in a temperature increase (geothermal thermal gradients) between 1 and 3°C per 100-m depth. Under appropriate conditions, accessible high-, intermediate-, low-, and very-low-temperature geothermal resources can be utilized for both electrical power generation via combined heat and power in cogeneration plants, and the direct or indirect (via geothermal heat pumps) use (Tester et al. 2005). Utilization of these geothermal energy resources has the potential to provide

DOI: 10.1201/9781003032540-4

TABLE 4.1
Types and Characteristics of Geothermal Energy Resources

Type	Temperature (°C)	Main heat source	Depth (m)	Usual usage	Technology
High temperature	150–360	Very deep aquifers/steam reservoirs with very high temperature gradients	3,000–5,000	Electricity and heat generation	Enhanced geothermal systems using advanced (e.g., Organic Rankin or ammonia-water) electricity generation cycles
Medium temperature	90–150	Deep aquifers	2,000–4,000	Electricity and/or heat generation	Enhanced geothermal systems using advanced (e.g., Organic Rankin or ammonia-water) electricity generation cycles
Low temperature	30–90	Deep aquifers	1,000–3,000	Heating	Urban and/or industrial district heating networks
Very low temperature	5–30	Ground/soil/rocks	2–200	Heating and cooling	Reversible geothermal heat pumps
		Groundwater and surface water bodies			

long-term, secure base-load energy (as an alternative for capacity addition and/or replacement of existing base load fossil fuels) being able to play a significant role in mitigating climate change by reducing greenhouse gas emissions.

4.2 HIGH-TEMPERATURE (DEEP) GEOTHERMAL ENERGY

High-temperature (deep) geothermal energy, associated with the Earth volcanic activity near plate tectonic boundaries or at crustal and mantle hot spot anomalies, is generally stored within the deep Earth crust's rocks or in fluids as saturated or superheated steam, or superheated water that fill the ground pores and fractures. This kind of geothermal energy is naturally transferred from the deeper parts of the Earth crust to the Earth's surface by conduction and convection through open spaces in ground/rock formations where superheated or saturated steam and/or superheated hot water can circulate.

The main sources of high-temperature geothermal energy are (ASHRAE 2011): (i) igneous resources, associated with magma resulting from volcanic activity; (ii) hydrothermal convection sources, i.e., hot fluids near the surface of the Earth, resulting from deep circulation of water in areas of high heat rates; (iii) geo-pressured resources, consisting of hot water in deep sedimentary strata, where pressures are greater than 75 MPa; (iv) radiogenic heat generated by constant or variable radioactive decay of uranium and thorium existing in granitic rocks; and (v) deep aquifers that can occur in deep sedimentary basins.

The engineered (or enhanced) geothermal systems (EGS) are used today for electrical power generation (Tester et al. 2005). In such systems, the fluid extracted is injected back at some distance from the production well(s) to pass again through the deep aquifer and recover some of the energy stored in granitic and/or sedimentary rock formations in order to produce electricity at the Earth's surface.

4.3 MEDIUM-TEMPERATURE GEOTHERMAL RESOURCES

Medium- (intermediate-) temperature (90–150°C) geothermal energy is found in continental settings, where above-normal heat production through radioactive isotope decay increases terrestrial heat flow or where aquifers are charged by water heated through circulation along deeply penetrating fault zones. Enhanced (engineered) geothermal systems using advanced generation cycles (e.g., Organic Rankin Cycle or ammonia-water absorption) are able to directly convert the deep high-temperature geothermal energy to electricity.

4.4 LOW-TEMPERATURE GEOTHERMAL RESOURCES

Low-temperature geothermal energy resources may provide abundant heat for direct use in individual buildings and/or district heating networks (including swimming pools), as well as for industrial (e.g., drying, desalination, fish farms) and agricultural (i.e., greenhouse) heating processes.

4.5 VERY-LOW (SHALLOW) GEOTHERMAL ENERGY

Depending on the hydrogeological characteristics of each geographic location, the most common very-low-temperature geothermal energy resources are as follows (Sachs et al. 1998; Goetzler et al. 2012): (i) ground/soil (widely available worldwide), (ii) groundwater (aquifers), and (iii) surface water (e.g., lakes/ponds, rivers, seas, municipal wastewater, sewage effluents, and underground mine water). Very-low-temperature geothermal resources (5–30°C) are predominately supplied by solar radiation because about 46% of total solar energy is stored in shallow underground depths extending from 0 to 200 m below Earth's surface where the effects of natural Earth temperature gradients can be considered negligible; even the ground temperature increases by around 1°C for every 33 m of depth (Figure 4.1).

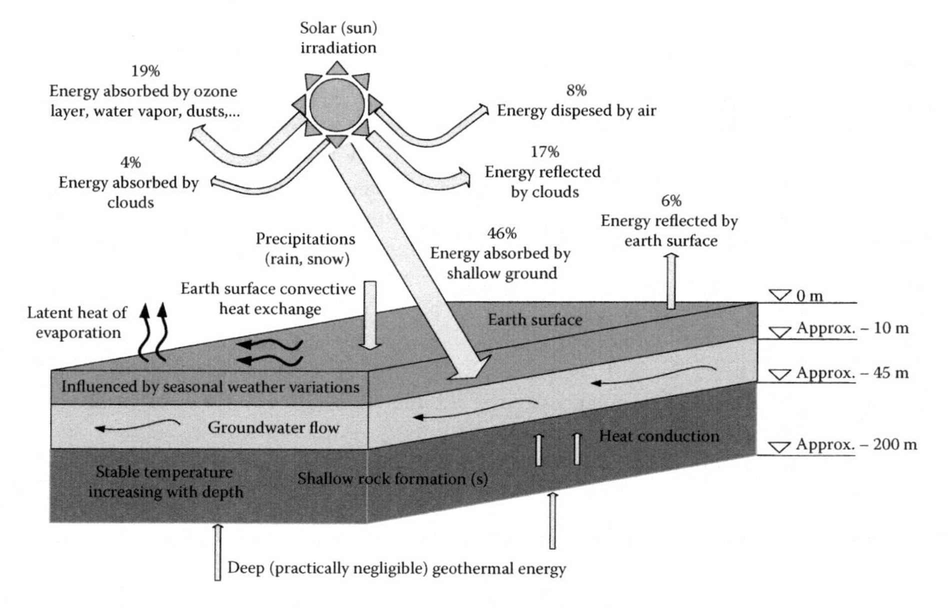

FIGURE 4.1 Average distribution of solar incident energy at Earth's surface.

The average solar radiation that is adsorbed by the ground is in the order of 1,500 kWh/m^2 annually, while the geothermal heat flow is restricted to around 0.6 kWh/m^2. In practice, this means that the major portion of extracted heat from the shallow underground is derived from solar energy, rather than geothermal heat from below. This basic knowledge, on how heat transfer in the underground works, suggests that shallow geothermal applications can be regarded as a huge and almost unlimited solar energy. However, putting single closed-loop vertical systems too close to each other will lead to continuous chill down of the underground. Depending on geological and climate conditions and how much energy is extracted, the "safe" distance varies between 20 and 30 m. Under normal conditions, the temperature at a depth of approximately 10 m reflects the average temperature in the air (+14.3°C on average). However, at places with snow in the winter, the ground temperature will be a few degrees higher since the snow will insulate the surface. At greater depths, the ground temperature will increase due to thermal heat flow. This flow creates a geothermal gradient that on average is around 3°C/100 m. In countries with old crystalline rocks, the gradient is often much less, while countries with clayey rocks have a higher gradient. The heat flow represents around 0.07 W/m^2. However, the variation is rather large and depends greatly on geographical position and local geological conditions. On an annual basis, the average solar energy is of approximately 1,100 kW per meter of Earth's surface. In layers closer (e.g., up to 10–20 m) to the ground/soil surface, the temperature depends on weather conditions. Increasing with the depth, the influence of ambient conditions are damped and delayed. In moderate and cold climate regions, the use of geothermal heat pumps (see Chapter 8) allows extracting heat at temperatures as low as 5–30°C in order to heat air or water at temperatures up to 40–45°C, providing useful heating to residential, commercial/institutional, and similar industrial buildings/facilities.

REFERENCES

ASHRAE. 2011. *ASHRAE Handbook, HVAC (Heating, Ventilating and Air Conditioning) Applications*, SI Edition, Supported by ASHRAE (American Society of Heating, Refrigerating and Air-Conditioning Engineers) Research, Atlanta, GA.

DOE. 1995. *Commercial/Institutional Ground-Source Heat Pump Engineering Manual.* Prepared by Caneta Research Inc. for U.S. Department of Defense U.S. Department of Energy, Oak Ridge National Laboratory.

Goetzler, W., M. Guernsey, R. Kar. 2012. *Research and Development Roadmap: Geothermal (Ground-Source) Heat Pumps*. Report DOE/EE-0810. Department of Energy Office of Energy Efficiency & Renewable Energy, Washington, DC.

Sachs, H.M., A.I. Lowenstein, H.I. Henderson, S.W. Carlson, J.E. Risser. 1998. Innovative commercial ground source heat pump system sources and sinks: Engineering and economics. In Proceedings of the *ACEEE (American Council for an Energy Efficiency in Buildings) Summer Study on Energy Efficiency in Buildings*, Pacific Grove, CA (https://www.bookdepository.com/1998-Aceee-Summer-Study-on-Energy-Efficiency-Buildings).

Tester, J.W., E.M. Drake, M.W. Golay, M.J. Driscoll, W.A. Peters. 2005. *Sustainable Energy – Choosing Among Options*, MIT Press, Cambridge, MA, 850 pp.

5 Ground/Soil Types and Thermo-Physical Properties

5.1 INTRODUCTION

To achieve high energetic and economic performances for any closed-loop (indirect, secondary fluid, or direct expansion) ground-source heat pump system, reliable hydro-geological data, including the complex, non-homogeneous, multi-phase structure, and thermal properties of ground/soils must be determined via laboratory and/or in-situ geotechnical investigations. Depending on geographic location, the ground/soil/rocks thermal property conditions strongly influence the drilling costs, the depth and number of boreholes and/or trenches, and the rate of heat transfer between the surrounding earth and ground-coupled (vertical or horizontal) heat exchangers.

5.2 GROUND/SOIL TYPES

The composition of ground/soil in the vicinity of ground-coupled heat exchangers is one of the fundamental factors affecting the design of closed-loop ground-source heat pump systems, more specifically the amount of heat extraction and rejection, as well as the seasonal energy storage (Bose et al. 1985). In the United States, for example, ground/soils are divided into 12 texture classes (U.S. Department of Agriculture–USDA) (Figure 5.1): sand, loamy sand, sandy loam, loam, silt loam, silt, sandy clay loam, clay loam, silty clay loam, sandy clay, silty clay, and clay (ASTM International D2487/1975). On the other hand, rocks, naturally occurring as grains of silicate and aggregate minerals held together by chemical bonds, constitute the basic units of which the solid earth is composed. The bedrocks are an attractive very-low-temperature geothermal energy source that keeps a near-constant temperature over the year regardless of ambient temperature variations.

According to their formation, the rocks are divided into three major classes (https://www.britannica.com/science/rock-geology, accessed June 6, 2020): (i) igneous (e.g., basalt and obsidian), formed when magma within the earth cooled and hardened; (ii) sedimentary (as conglomerate and limestone), formed from particles of sand, shells, pebbles, and other fragments of material; and (iii) metamorphic (gneiss and marble), formed under the surface of the Earth from the metamorphosis that occurs due to intense heat and pressure. These three classes are subdivided into numerous groups and types (as granite, gneiss, schist, limestone, and silt-stone) on

DOI: 10.1201/9781003032540-5

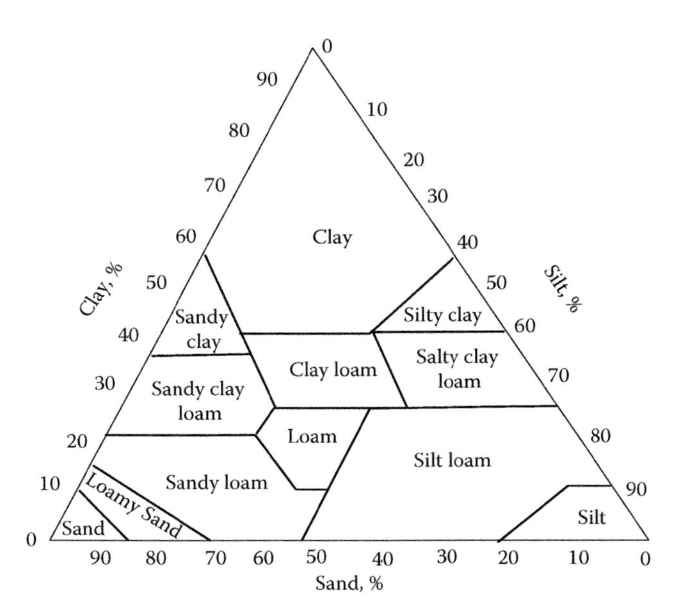

FIGURE 5.1 Ground/soil texture triangle (https://www.researchgate.net/figure/USDA-Soil-Texture-Triangle_fig2_279631053, accessed June 6, 2020).

the basis of various factors, the most important being the chemical, mineralogical, and textural properties.

The heat transfer effectiveness between a ground-coupled heat exchanger and the surrounding ground/soil/rocks is strongly dependent on the thermodynamic characteristics of the surrounding earth structure. The main parameters defining thermal properties of the ground/soil/rocks and their ability to transfer and store geothermal energy are: (i) far-field, undisturbed temperature; (ii) density; (iii) mass specific heat; (iv) thermal resistivity and stability; and (v) moisture content, thermal conductivity, and thermal diffusivity. These thermal properties are generally difficult to measure, in particular for vertical ground-coupled heat exchangers, as they usually pass through several earth layers that need to be identified appropriately.

5.3 DENSITY

Because the moisture content of ground/soil/rocks (mixtures of solid particles, water, and air) varies from 0% (dry) to 100% (saturated), two types of bulk density, also called apparent or volumetric density, that consider both solids and pore spaces, should be defined (Bose et al. 1985): (i) dry and (ii) wet density.

The bulk density of ground/soils depends greatly on the mineral makeup of soil and the degree of compaction. The density of quartz is around 2.65 kg/m^3, but the bulk density of a mineral ground/soil is normally about half that density, between 1.0 and 1.6 kg/m^3. In contrast, ground/soils rich in soil organic carbon and some friable clays tend to have lower bulk densities (<1.0 kg/m^3) due to a combination of the low density of the organic materials themselves and increased porosity. For instance, peat ground/soils have bulk densities from 0.02 kg/m³ to 0.98 kg/m³.

To determine the dry bulk density, a ground/soil sample is oven dried and weighed, giving the mass of solids (M_{solid}) and of water (M_{water}) lost during oven drying, with a total mass ($M_{total} = M_{solid} + M_{water}$).

Dry bulk density is defined as:

$$\rho_{dry\ bulk} = M_{solid}/V_{total} \tag{5.1}$$

where

M_{solid} is the mass of solid particles of ground/soil (kg).

V_{total} is the sample total volume that includes the volume of particles, inter-particle voids, and internal pores (m^3).

The ground/soil dry bulk density is inversely related to the material porosity: the more pore space the lower the value for bulk density.

Wet bulk density is defined as:

$$\rho_{wet\ bulk} = M_{total}/V_{total} = (M_{solid} + M_{water})/V_{total} \tag{5.2}$$

where

M_{total} is the total mass of solids plus water (kg).

V_{total} is the sample total volume (m^3).

Table 5.1 shows some typical values for the dry and/or density of some common ground/soils and rocks (ASHRAE 2011).

5.4 MASS SPECIFIC HEAT

The ground/soil specific heat defines the amount of energy stored in a material sample per unit mass or volume, and per unit change in temperature. In practice, values for the ground/soil specific heat are expected to be as high as possible to minimize the storage volume and enhance the heat exchange. The mass specific heat defines the amount of energy stored in a material per unit mass per unit change in temperature. It does not depend on the ground/soil microstructure; thus, in most cases, it is satisfactory to calculate the specific heat of ground/soils from the specific heat of the different constituents according to their volume ratios.

The average mass specific heat of a ground/soil is the sum of the specific heats of the various components that make up the ground/soil. Its value can be determined by adding the individual component values based on specific volume ratios of each (Bose et al. 1985):

$$\bar{c}_{g,mass} = c_{solid} * v_{solid} + c_{water} * v_{water} + c_{air} * v_{air} \tag{5.3}$$

where

$\bar{c}_{g,mass}$ is the mass specific heat (J/kg·K).

c_{solid} is the specific heat of solid phase (J/kg·K).

TABLE 5.1
Density of Some Dry and Wet Ground/Soils and Rocks

Type (-)	Density (kg/m^3)
Dense rock	3,200
Average rock	2,000
Heavy soil, damp	2,100
Heavy soil, dry	2,000
Light soil, damp	1,600
Light soil, dry	1,440
Clay very dense	2,000
Clay loose	1,800
Clay silty dry	1,600
Clay wet	1,800
Sand dry	1,400–1,700
Sand wet	1,900–2,000
Sand dense	2,000
Loam dry	1,700
Loam wet	2,000
Granite	2,600–2,700

$xv_{solid} = 1 - \varkappa$ is the specific volume for the solid phase (-) (usually, mineral and organic components).

$\varkappa$ is the soil porosity (i.e., the ratio of the volume of voids to the total volume of the soil) (-).

c_{water} is the specific heat of water (J/kg·K).

$v_{water} = \varkappa * S$ is the specific volume for the pore water (-).

c_{air} is specific heat of the pore air (J/kg·K).

S is the degree of saturation (-).

$v_{air} = \varkappa * (1 - S)$ is the specific volume for the pore air (-) (practically negligible).

The heat capacity (J/K) of the ground/soil is the sum of the product of the specific heat and mass of each main component:

$$C_{g,i} = \sum \mathrm{m_{g,i} \cdot c_{g,i}} \tag{5.4}$$

As the mineral and organic solid components in the ground/soil have somewhat similar mass specific heats, only the water content remains as a relevant variable, at least in the short term. In the long term, consolidation or shrinkage under external thermal loads, or due to excessive heat extraction, may play a role because the volume ratios change. The ground/soil overall thermal capacity increases with the groundwater content and decreases in the case of freezing. That is because the

TABLE 5.2
Mass Specific Heat for Some Common Ground/Soil/Rocks

Material (-)	Average temperature (°C)	Mass specific heat (kJ/kg·K)
Clay (kaolin)	20	0.937
Quartz sand	20	0.799
Mica	20	0.862
Granite	20	0.803
Sandstone	20	0.921
Limestone	20	0.904
Quartz	20	0.787
Dense rock	n/a	0.84
Average rock	n/a	0.84
Heavy soil, damp	n/a	0.96
Heavy soil, dry	n/a	0.84
Light soil, damp	n/a	1.05
Light soil, dry	n/a	0.84

specific heat of ice is lower (c_{ice} = 1, 884 J/kg·K) compared with that of c_{water} = 4, 186 J/kg·K for water. Table 5.2 shows some typical values for the mass specific heat of rocks and ground/soils (ASHRAE 2011), and for some common substances (as solids, water, and air) found in ground/soils (Bose et al. 1985). The specific heat can be determined in the laboratory by mixing water and ground/soil of different temperatures (commonly, 0°C for ground/soil and 20°C for water) and measuring and the steady-state temperature of the mixture. If the total thermal energy of both components remains constant, and the specific heat of one component is known (e.g., c_{water}), then the mass specific heat of the ground/soil ($\bar{c}_{g,mass}$) can be determined. During the ground/soil freezing, the ground-coupled heat exchanger absorbs not only geothermal heat through the ice core but also the ice solidification enthalpy (latent heat), which increases its mass specific heat and heat capacity.

5.5 THERMAL RESISTIVITY AND THERMAL STABILITY

Ground/soil thermal resistivity and thermal stability are functions of the material composition and moisture content. Thermal resistivity, a reciprocal parameter of thermal conductivity of the ground/soil (expressed in K·m/W and defined as the temperature difference – expressed in °C or K – between opposite faces of a cubic meter sample of ground/soil caused by a thermal power transfer of one watt) varies with moisture content and is also affected by the operation of underground (vertical or horizontal) heat exchangers in both heating and cooling modes. Dry ground/soils are thermally more resistive and mechanically more fragile. The ability of the ground/soil to maintain its thermal resistivity in the presence of an underground

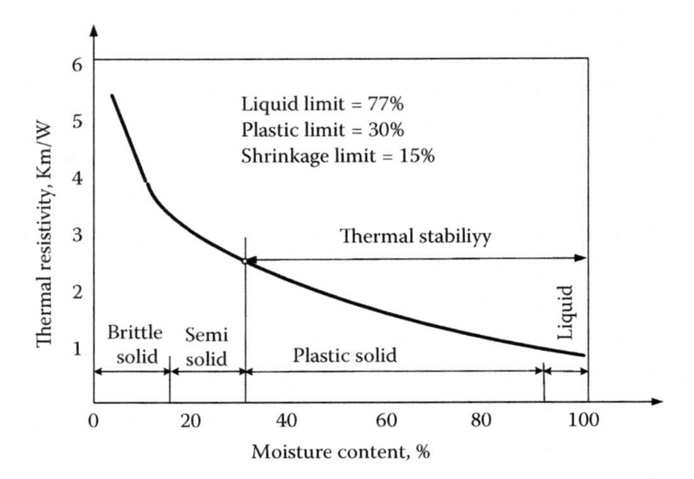

FIGURE 5.2 Influence of the ground/soil type and moisture content on thermal resistivity.

heat source is known as thermal stability, a property defined in terms of critical moisture content representing a zone where small changes in moisture content result in small changes in ground/soil thermal resistivity (Figure 5.2). For example, the ground/soil thermal resistivity increases due to its drying caused by heat released by ground-coupled heat exchangers operating in the cooling mode. The ground/soils are thermally unstable when a small decrease in moisture content drives an important increase in thermal resistivity, as in the case of highly organic ground/soils.

5.6 MOISTURE CONTENT

The combined heat and mass transfer in the immediate vicinity of vertical (and horizontal) ground-coupled heat exchangers depends on factors such as: (i) ground/soil type and mineralogical structure and texture, and the quality of the particles' contact; (ii) performance of ground-source heat pump systems; (iii) temperature gradient primarily accomplished by conduction and, to a certain degree, by moisture migration; and (iv) moisture content gradient since moisture tends to migrate towards the heat sink (during heat extraction in heating mode) and away from the heat source (during heat injection in cooling mode); during summer, excessively dry soil resulting from heat rejection must be avoided. Thermally, ground/soil moisture content may exist in both unfrozen and/or frozen states around the vertical or horizontal ground-coupled heat exchangers. This is because the ground/soil temperature in the immediate vicinity of the ground-coupled heat exchangers can drop below the freezing point (0°C) during winter operation, and reach temperatures above 20–30°C during summer operation. Mass (or gravimetric) and volumetric moisture content are variable properties of ground/soil materials that have significant effects on ground/soil thermal properties because, when groundwater replaces the air between particles, it reduces the contact resistance. The larger moisture contents are the larger ground/soil specific heat, thermal conductivity, and latent heat.

The mass moisture content can be expressed with respect to the sample's dry mass as follows:

$$MC_{dry} = (m_{water}/m_{dry})\cdot 100(\%) \tag{5.5}$$

where

MC_{dry} is the ground/soil mass moisture content (%).
m_{water} is the mass of water (kg).
m_{dry} is the mass of solid dry material (kg).

For ground/soils that change in volume with moisture content, the mass moisture content can be expressed in terms of the mass of water per unit mass of the moist specimen:

$$MC_{wet} = m_{water}/m_{wet}\cdot 100(\%) \tag{5.6}$$

where

m_{water} is the mass of water (kg).
m_{wet} is the mass of wet material before drying (kg).

Volumetric moisture content is defined as:

$$MC_{vol}\theta = V_{water}/V_{total,\ wet} = V_{water}/(V_{solid} + V_{water} + V_{air}) \tag{5.7}$$

where

MC_{vol} is the volumetric moisture content (%).
V_{water} is the volume of water (m^3).
$V_{total,\ wet}$ is the total volume of wet ground/soil material (m^3).
V_{solid} is the volume of solid particles, vegetation tissue, etc. (m^3).
V_{air} is the volume of air (m^3).

When the ground/soil freezes because of geothermal heat extraction by the ground-coupled heat exchangers in the heating mode, the thermo-physical properties of the ground/soil are significantly altered. It the case of ground/soil with grainy texture, for example, the ground/soil thermal diffusivity depends on its moisture content (Figure 5.3) (Bose et al. 1985).

Table 5.3 shows some approximate critical moisture content (defined as the average ground/soil moisture content at which the drying rate begins to decline) values for granular, silt, and clay ground/soils (Bose et al. 1985).

Additional aspects that concern the importance of ground/soil moisture content are (Bose et al. 1985; Hailey et al. 1990; Leong et al. 1998): (i) it is beneficial to keep the ground/soil moisture value as high as possible above dry soil conditions; (ii) for almost all ground/soils, there is a sharp increase of the heat extraction rate between 0% and 12.5% of saturation, followed by a moderate increase between 12.5% and 25% of saturation; (iii) any decrease in moisture content from 12.5% of saturation and complete dryness (0%) has a negative effect on the geothermal heat

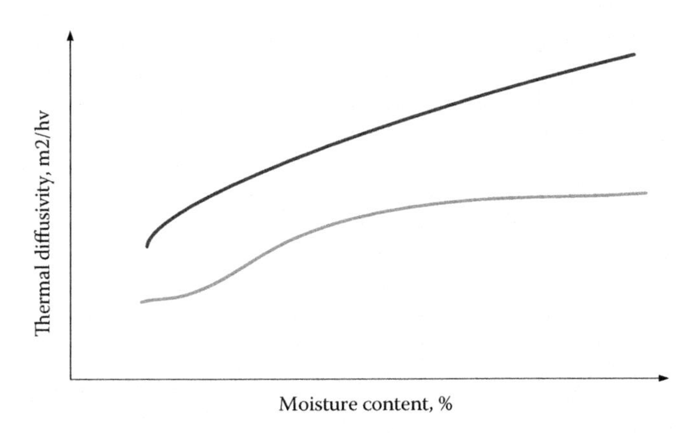

FIGURE 5.3 Typical tendency variation of ground/soil thermal diffusivity with its moisture content.

TABLE 5.3
Critical Moisture Content for Some Common Ground/Soils

Soil type (-)	Dry mass density (kg/m^3)	Critical moisture content (%)
Granular	2,000	≤12
Silt	1,800	12–16
Clay	1,700	16–22

pumps' coefficients of performance; (iv) moisture content above 25% of saturation leads to generally better performances of geothermal heat pumps; (v) heat extraction rate increases with the ground/soil moisture content that, ideally, should be between 25% and full (100%) saturation; (vi) the effect of moisture content variation above 50% of saturation on ground-source heat pump performance is relatively insignificant; however, the reduction of the ground/soil moisture content from 50% saturation to 0% (dryness) leads to an increasing demand for supplemental (additional, emergency) heating; that means that it is beneficial to keep the ground/soil moisture content as high as possible above dry ground/soil conditions; (vii) when the ground/soil temperature near a ground-coupled heat exchanger is above 35–40°C, it may lead to a belt around the ground-coupled heat exchangers behaving like an annular zone of insulation, which may provide a very sharp decline (by up to 35%) of geothermal heat pumps' coefficients of performance; (viii) according to ground/soils' texture classes (see Figure 5.1), the largest value of heat extraction rates are for sand at all degrees of saturation, followed by silty loam and silty clay; (ix) the freezing process of ground/soils around the ground-coupled heat exchangers produces almost constant values of the brine temperature, heat extraction, and the coefficient of performance.

The moisture content may undergo a seasonal change depending on the climate and it may also depend upon the history of heating and cooling that the ground/soil coupling device has undergone. When a buried pipe is transferring heat to the ground/soil (i.e., rejecting heat in the cooling mode), moisture tends to be driven away from the vicinity of the pipe. When the pipe is cooling the earth, the soil moisture is drawn toward the pipe. Thus, one might expect ground/soil dry-out and increased resistance to heat transfer to be a definite possibility near the end of a dry cooling season, but to be of no concern to heating only heat pump systems.

5.7 THERMAL AND HYDRAULIC CONDUCTIVITY

The rate of heat transfer between the surrounding ground/soil and the vertical (and horizontal) ground-coupled heat exchangers depends strongly on the ground/soil thermal conductivity (Farouki 1986) and actual temperature gradients. Therefore, after the determination of the type (or types) of ground/soil and/or rocks surrounding the borehole(s), the determination of the thermal conductivity (defined as the amount of heat transferred per unit time through a unit cross-sectional area under a unit temperature gradient applied in the direction of this heat flow) of the earth formation surrounding the ground-coupled heat exchangers, is one of the most important helping to optimize the total length and the number of boreholes or trenches. The concept of "equivalent" thermal conductivity should be used to account for the complex heat transfer between buried pipes and ground/soils, due to freezing and thaw of ground/soil, and flow of groundwater. It mainly depends on the ground/soil moisture content, density and temperature, ground/soil texture, dry/wet bulk density, mineralogical components, chemical properties and porosity, in addition to the size and distribution of constitutive particles.

Fourier's law ($\dot{q} = -k_g dT/dx$) shows that the proportionality factor that relates the rate at which heat is transferred by conduction to the temperature gradient is the ground/soil thermal conductivity:

$$\bar{k}_g = |\dot{q}| \cdot dx/dT \tag{5.8}$$

where

$\bar{k}_g$ is the average (or effective) ground/soil thermal conductivity (W/m·K).

$\dot{q}|$ is the heat flux density (W/m^2).

x is the distance (m).

T is the temperature (K).

Ground/soil thermal conductivity mainly depend on: (i) mineral content; the thermal conductivity of quartz is higher than other minerals that constitute the ground/soil; cohesive ground/soils have less mineral content; therefore, coarse-grained ground/soils tend to have higher thermal conductivity than fine-grained ground/soils; (ii) porosity; the thermal conductivity of minerals is higher than that of air and water; therefore, an increase in porosity increases the volumetric fraction of air and water, which results in a decrease of soil thermal conductivity; (iii) degree of saturation; an increase in degree of saturation increases the volumetric fraction of water and decreases that of air; the thermal conductivity of air (k_{air} = 0.025 W/mK) is much lower than that of water (k_{water} = 0.6 W/mK); therefore, an increase of degree of saturation increases ground/soil "effective" thermal conductivity; (iv) temperature; thermal conductivity of ice (k_{ice} = 2.22 W/mK at 0°C) is approximately four times higher than water thermal conductivity (k_{water} = 0.6 W/mK); therefore, when ground/soil temperature drops below freezing, a part of the water will become ice and the thermal conductivity increases.

Ground/soils with low thermal conductivity achieves lower heat transfer rates, while high thermal conductivity provides higher recovering (or dissipating) heat rates, higher (in heating mode) or lower (in cooling mode) temperatures of the brine within the ground-coupled heat exchangers, as well as higher performances for geothermal heat pumps, shorter borehole depths, and, consequently, lower first costs. In addition, high heat rejection rates to the ground/soil (in the cooling mode) had a detrimental impact on the ground/soil thermal conductivity, leading to reduction of heat transfer (Hailey et al. 1990). In practice, values for the ground/soil thermal conductivity are expected to be as high as possible to minimize the storage volume and enhance the heat transfer. For higher performances, it is desirable to have a densely packed soil with a high thermal conductivity and moisture content in the immediate vicinity of ground-coupled heat exchangers. Saturated (100%) and half- saturated (50%) soils display a noticeable change in the thermal conductivity, particularly below the freezing point (0°C). Freezing increases the thermal conductivity significantly because k_{water} = 0.57 W/mK changes to k_{ice} = 2.18 W/mK. Consequently, the thermal conductivity of the ground/soil can only be expressed approximately.

Determined by laboratory or in-field experiments, the thermal conductivity of ordinary ground/soil with temperatures of 10~40°C, is 2.3 W/m·K under the condition of ordinary moisture, 0.55–0.6 W/m·K under the desiccation state, and 2.7 W/m·K under the waterlogged condition. With the increase of the moisture content, the ground/soil thermal conductivity increases, but this trend of increasing is diminished gradually. And when moisture content increases to a certain specific value, the thermal conductivity is relatively invariable. The bigger ground/soil density is, the greater through the experiment to calculate the ground/soil thermal conductivity under different moisture content and different density. In normal conditions, the increase of temperature results in an increase of soil thermal conductivity, and when temperature increases within the range of 20°C, ground/soil thermal conductivity increases about 0.1–0.2 W/m·K. However, once ground/soil is frozen, the property and performance of melt ground/soil and frozen soil is absolutely different. Measured with experiments, the thermal conductivity of inferior clay is 1.616 W/m·K under its melt state and 2.454 W/m·K under its frozen state. So, generally speaking, soil thermal conductivity of the frozen state should be greater compared with that of melting. This is mainly because the heat conduction coefficient of the liquid water is 0.58 W/m·K, while that of ice (0°C) is 2.25 W/m·K. In the ground/soil, freezing increases the thermal conductivity significantly because k_{water} = 0.57 W/mK increases to k_{ice} = 2.18 W/mK. For preliminary design of complex geothermal systems it can be taken with sufficient accuracy from diagrams considering water content, saturation density, and texture of the ground/soil. The thermal conductivity can vary from 0.25 W/mK for dry soil to 2.5 W/mK for wet

TABLE 5.4
Thermal Conductivity of Some Common Ground/Soils and Rocks

Type (-)	Thermal conductivity (W/m·K)
Heavy soil, damp	1.3
Light soil, damp	0.86
Light soil, dry	0.35
Dense rock	3.46
Average rock	2.42
Sand and clay	0.52–3.30

soil (Rawlings and Sykulski 1999). Generally, sandy soils have higher thermal conductivity because of a high content of quartz. Rocks that are rich in quartz, like sandstone, have high thermal conductivity, indicating that heat readily passes through them. Rocks that are rich in clay or organic material, like shale and coal, have low thermal conductivity. The rocks have significantly higher values for thermal conductivity. The thermal conductivity of the ground/soil is relatively low. The ground thermal conductivity significantly varies for the same type of rock. For example, the thermal conductivity of quartzose sandstone wet rock varies from about 3.1 to about 7.8 W/h. Table 5.4 shows some typical values for the thermal conductivity of soils and rocks (ASHRAE 2011).

In countries such as Sweden and Canada, the bedrock consists mostly of crystalline rocks (formed from the slow crystallization of magma below Earth's surface), such as granite (a light-colored igneous rock with grains large enough to be visible with the unaided eye; composed mainly of quartz and feldspar with minor amounts of mica, amphiboles, and other minerals) and gneiss (a metamorphic rock with a banded or foliated structure, typically coarse-grained and consisting mainly of feldspar, quartz, and mica; a metamorphic rock, generally made up of bands that differ in color and composition, some bands being rich in feldspar and quartz, others rich in hornblende or mica). Such crystalline rocks have a mean thermal conductivity of 3.47 W/m·K with standard deviations of 0.380 and 0.465, respectively. The thermal conductivity of crystalline rocks varies with the percentage of quartz (Figure 5.4) (Sundberg 1988).

Generally, the ground/soil thermal conductivity increases with the moisture content because the contact resistance within the material is reduced when water replaces the air in between particles. Additionally, as heat is extracted, moisture tends to migrate towards the ground-coupled heat exchangers which increases the thermal conductivity, being a dominant factor responsible for seasonal thermal conductivity variations. Experiments show that an augmentation of the ground/soil thermal conductivity by 0.17 W/m·K may lead to a reduction of geothermal heat pump operating time by 1.3%, due to the effect of thermal conductivity on ground heat storage (Drown and Den Braven 1992). In the presence of groundwater flow,

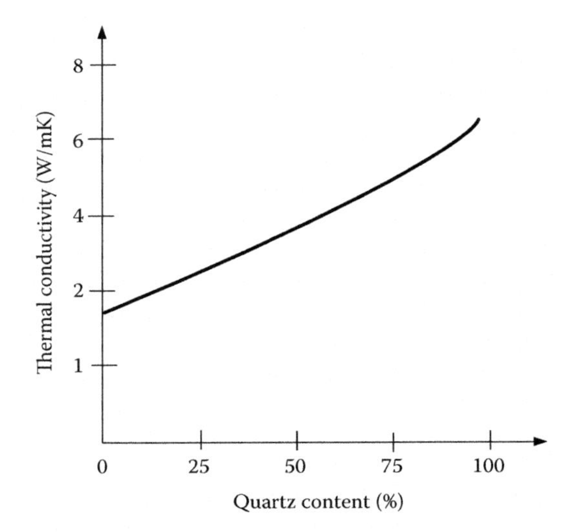

FIGURE 5.4 Relation between thermal conductivity and quartz content.

the heat exchange is more complex because the groundwater flow determines the changes of the temperature field in the direction of the movement, enhancing the heat exchange. The hydraulic conductivity of the ground/soil, which describes the ease with which the groundwater can move through the ground/soil/rocks spaces or fractures, is, thus, another parameter to consider.

Hydraulic conductivity (that is a quantitative measure of ground/soil's ability to transmit water when subjected to a hydraulic gradient) requires determining the groundwater velocity and the horizontal pressure gradient:

$$w_D = K \cdot \Delta p \tag{5.9}$$

where

w_D is Darcy velocity (m/s).

K is the hydraulic conductivity (a proportionality constant) that defines the linear relationship between the quantity of groundwater moving through a cross-sectional area per unit of time and the hydraulic gradient.

Δp is the horizontal pressure gradient (Pa).

5.8 THERMAL DIFFUSIVITY

The ground/soil average thermal diffusivity combine into a single expression of the ground average thermal conductivity ($\bar{k}_{ground}$), density ($\bar{\rho}_{ground}$) and the specific heat ($\bar{c}_{ground}$) as follows:

$$\bar{\alpha}_{ground} = \bar{k}_{ground} / (\bar{\rho}_{ground} \cdot \bar{c}_{ground}) \tag{5.10}$$

TABLE 5.5
Thermal Diffusivity of Some Common Ground/Soils and Rocks

Type (-)	Thermal diffusivity (m^2/h)
Dense rock	0.00465
Average rock	0.00372
Heavy soil, damp	0.00232
Heavy soil, dry	Not available
Light soil, damp	0.00186
Light soil, dry	0.00102

where

$\bar{\alpha}_{ground}$ is the ground average thermal diffusivity (m^2/s)
k_{ground} is the ground thermal conductivity (W/mK).
$\bar{\rho}_{ground}$ is the ground average density (kg/m^3).
$\bar{c}_{ground}$ is the ground average specific heat (J/kgK).

As can be seen in Equation 5.10, temperature-dependent properties of ground/soils as thermal conductivity (k_{ground}), specific heat (c_{ground}) and density (ρ_{ground}), define the thermal diffusivity (α_{ground}, m^2/s) of a parameter that significantly impacts the design and operation of ground-source heat pump systems. This parameter describes the depth and velocity of penetration of a temperature wave into the ground/soil. In other words, it is a measure of how quickly a material can carry heat away from a heat source. If the ground/soil temperature is measured at least two depths z_1 and z_2, the thermal diffusivity can be calculated either from the amplitude decrease or from the time lag. Typical values for the thermal diffusivity of some common ground/soils and rocks are given in Table 5.5 (ASHRAE 2011).

REFERENCES

ASHRAE. 2011. *Handbook Fundamentals*. SI Edition. Supported by ASHRAE (American Society of Heating, Refrigeration and Air-Conditioning Engineers, Inc.), Atlanta, GA.

ASTM International D2487/1975. Standard Test Method for Classification of Soils for Engineering Purposes (https://www.astm.org/DATABASE.CART/HISTORICAL/D2487-69R75.htm. Accessed May 4, 2020).

Bose, J.E., J.D. Parker, F.C. McQuiston. 1985. *Design/Data Manual for Closed-Loop Ground-Coupled Heat Pump Systems*. Oklahoma State University. American Society of Heating, Refrigerating and Air-Conditioning Engineers, Inc., Tullie Circle, NE, Atlanta.

Drown, D.C., K.R. Den Braven. 1992. Effect of soil conditions and thermal conductivity on heat transfer in ground source heat pumps. In Proceedings of the ASME (American Society of Mechanical Engineers) JSES (Japan Solar Energy Society) International Solar Energy Conference, Maui, Hawaii, 5–9 April.

Farouki, O.T. 1986. *Thermal Properties of Soils*. Trans. Tech. Publications, Series on Rock and Soil Mechanics. Vol. II., 136 pp. Publisher: Clausthal Technical University, Clausthal-Zellerfeld, Germany.

Hailey, S.M., T.P., Kast, D.C. Drown. 1990. Thermal conductivity and soil conditions heat transfer effects on ground source heat pumps. In Proceedings of the 16th Annual Conference of Solar Energy Society of Canada, Halifax, Nova Scotia, June, pp. 317–322.

Leong, H., V.R. Tarnawski, A. Aittomaki. 1998. Effect of soil type and moisture content on ground heat pump performance. *International Journal of Refrigeration* 21(8):595–606.

Rawlings, R.H.D., J.R. Sykulski. 1999. Ground source heat pumps: A technology review. *Building Services Engineering Research and Technology* 20(3):119–129.

Sundberg, J. 1988. Thermal properties of soils and rocks (https://www.researchgate.net/publication/35391832_Thermal_properties_of_soils_and_rocks. Accessed November 26, 2018).

6 Determination of Ground/Soil Effective Thermal Conductivity

6.1 INTRODUCTION

In the vicinity of ground-coupled heat exchangers of ground-source heat pump systems operating in heating and cooling modes, significant variations in hydraulic and thermal parameters of the surrounding ground/soil/rocks occur, which, in addition to local climatic conditions, influence the energy performances of the entire geothermal system (Gehlin 2002).

In the case of vertical ground-coupled heat exchangers, the heat transfer occurs through discontinuous, multi-layered different materials, affected by factors such as ground/soil/rocks undisturbed temperature, density, specific heat, moisture content, and hysteresis effects-memory after past drying-wetting and freezing/thawing cycles. Therefore, it is suggested that average (effective) thermal conductivity of earth formations be considered in the ground-coupled heat exchangers' design (De Vries 1974). Because of a lack of rapid and reliable methods for determining the ground/soil/rock thermal characteristics at a given location, the determination of "effective" (average) thermal conductivity is a significant challenge facing designers of ground-source heat pump systems, mainly because it significantly impacts the required ground-coupled heat exchangers' depths (lengths), geometry, spacing and number, drilling techniques, and initial installation costs, particularly for applications in large-scale commercial/institutional buildings. Underestimating the ground/soil/rock "effective" thermal conductivity may result in overestimation of the ground-coupled heat exchangers size which, in turn, may provide higher initial capital costs. On the other hand, overestimating the ground/soil "effective" thermal conductivity may result in undersizing the ground-coupled heat exchangers, which could achieve too low (in the heating mode) or too high (in the cooling mode) entering/leaving brine temperatures to the geothermal heat pumps, finally leading to less efficient performance and might, eventually, cause compressors' failures.

Without laboratory or in-field measurements, the traditional approach to determine the thermal conductivity of ground/soil/rock formations consists of using tabulated data generally provided within rather large bands of values. For example, for wet quartzose sandstone, the thermal conductivity given in such tables varies from 3.1 W/m·K to 7.8 W/m·K, that make it difficult to find the exact value of the ground/soil/rock thermal conductivity for a particular application. It is also possible to estimate the ground/soil/rocks thermal conductivity based on local information as local drilling contractors, or geological reports and, then, compare with tabulated

DOI: 10.1201/9781003032540-6

data, if available, and "best guess" of the ground/soil composition and far-field temperature (EPRI 1989). Therefore, for large and/or complex institutional/commercial ground-source heat pump systems, ground/soil/rock "effective" thermal conductivity should be determined from more detailed laboratory analyses and/or in-field tests. For large, more complex projects, the ground/soil "effective" thermal conductivity should be determined from laboratory (whereby a soil body has to be exposed to a temperature gradient which may cause a significant moisture transfer in unsaturated soils, and, thus, should be taken into account when interpreting the measured results) or in-field (performed directly with at least one vertical U-shaped heat exchanger, a costly procedure, especially for low capacity, e.g., residential, ground-source heat pump systems) tests.

6.2 LABORATORY METHODS

Laboratory steady-state (when the sample's temperature do not vary during the measurement) and transient (when the temperature of the sample varies with time) experimental methods use ground/soil samples collected from the ground-source heat pump sites to measure their thermal conductivity. In this case, guarded hot plate tests consist of two identical ground/soil specimens placed above and below a flat-plate main heater surrounded by an outer guard heater that eliminates the horizontal heat losses and causes heat from the main heater to be transferred vertically up or down the test specimens. Liquid-cooled heat sinks are placed adjacent to the outer surfaces of the specimens of known thickness, and certain temperature drops are measured across them (American Society for Testing and Materials 1963) (https://www.astm.org/standard/standards-and-publications.html, accessed August 19, 2019). Since the amount of heat transferred per unit time and the test area of the specimen is known, the thermal conductivity can be calculated using Fourier's law for one-dimensional heat conduction (see Equation 5.8). The most rapid and convenient transient method of measuring the "effective" thermal conductivity of ground/soil/rock formations in the laboratory is the "thermal probe" (or "thermal needle") method (De Vries and Peck 1958), based on the theory of the line-source (Kelvin 1882). It consists of a heater producing thermal energy at a constant rate and a temperature-sensing element (e.g., thermocouple or thermistor). The "needle" is inserted into a ground/soil sample to determine the thermal conductivity as a function of temperature gradient.

For ground/soil/rock materials with moisture contents varying from dryness (0%) to saturation (100%) and temperatures from –30 to 95°C, the "effective" thermal conductivity can be calculated with the classical de Vries model based on potential theory analogous to electrical field theory as a weighted sum of the thermal conductivity of all constituents (de Vries, 1963):

$$\bar{k}_g = \frac{\sigma_{air} \cdot \upsilon_{air} \cdot k_{air} + \sum_{i=1}^{n} \sigma_{solid} \cdot \upsilon_{solid} \cdot k_{solid} + \sigma_w \cdot \upsilon_w \cdot k_w}{\sigma_{air} \cdot \upsilon_{air} + \sum_{i=1}^{n} \sigma_{solid} \cdot \upsilon_{solid} + \sigma_{gw} \cdot \upsilon_{gw}} \quad (6.1)$$

where

$\bar{k}_g$ is the ground/soil average thermal conductivity (W/m·K).

σ_{air}, σ_{solid}, and σ_{gw} are mass factors of air, solids, and water, respectively (-).

υ_{air},υ_{solid}, and υ_{water} are volume fractions of air, solids, and water, respectively (-).

k_{air}, k_{solid}, and k_{water} are thermal conductivity of air, solids, and water, respectively (W/m·K).

n is the number of ground/soil solid constituents (-).

In Equation 6.1, the thermal conductivity of solid phase (e.g., sand, silt, clay) and gas phase (air) could be set, for example, as 8.53, 2.93, 2.93, and 0.025 W/m·K, respectively. One of the major advantages of this method is that, for the first preliminary evaluation, only physically based data input is required. The enhancements to the original de Vries model can be mentioned (Tarnawski and Leong 1993): (i) ground/soil/rocks formations can be made of up to 26 mineralogical components whose individual characteristics, such as mass fraction, thermal conductivity, density, shape, and specific heat, must be known; (ii) a general relationship for mineralogical composition of any natural ground/soil/rocks has been introduced; (iii) groundwater is assumed to be a continuous medium over a full moisture content range (dryness to saturation). The main disadvantage of laboratory experiments is that only a small number of samples can be tested and, consequently, the "effective" thermal conductivity values might not be representative of the whole ground-coupled heat exchanger arrays. In other words, the heterogeneity of the ground/soil cannot be taken into account. In addition, the results of laboratory measurements are usually not fully correct, mainly because they neglect the site-specific conditions such as the influence of groundwater movement and the properties of grouting materials.

6.3 IN-FIELD EXPERIMENTAL METHOD

Because of relatively high cost, thermal efficiency, and market impacts, in-field (in-situ) direct measurement methods aiming to determine the ground/soil/rock "effective" ground/soil thermal conductivity at the prefeasibility stage of large-scale (usually over 175 kW of nominal cooling capacity) ground-source heat pump system at a given location is the best and most accurate method, and generally recommended instead of using tabulated/standard data or laboratory tests, even if they are expensive (because of heavy equipment mobilization and of power supply), laborious, and time consuming (Austin et al. 2000).

Compared with data obtained from existing tables, standards, and laboratory tests, the in-field experiments (also known as a thermal response tests) are much more representative because larger volumes of ground/soil/rock material are evaluated under more realistic thermal conditions allowing to reproduce the complex heat transfer taking place in the real geothermal systems. The in-field thermal response tests take into account the total heat transfer between the borehole and the surrounding ground/soil/rock, including groundwater effects and other local disturbances. By using in-situ thermal response tests, parameters such as the following are currently determined (Shonder and Beck 2000): (i) undisturbed ground/soil temperature that is a function of time, season, and borehole depth below the ground/

soil surface; (ii) “effective” thermal conductivity (k_{eff}) for the borehole heat exchange that is the sum of the thermal conductivity of the ground/soil/bedrock and an eventual contribution from groundwater movement, and is represented by heat conduction (in solids) and convection (fluid flows transporting energy); and (iii) thermal borehole resistance ($R_{borehole}$) that describes the heat transfer in the borehole and how effective the closed-loop collector works as a borehole heat exchanger; low thermal borehole resistance means good heat transfer, and low-temperature difference between the borehole wall and the thermal carrier fluid.

For vertical ground-coupled heat exchangers serving buildings of less than 3,000 m^2, at least one in-field test borehole should be drilled. For larger buildings, at least two test holes should be drilled. Each borehole should be drilled to a depth of 15 m below the deepest planned heat exchanger borehole (IGSHPA 2007).

6.3.1 Mobile Apparatus

The in-field ground/soil/rock “effective” thermal conductivity can be measured by using heavy mobile (relative expensive) experimental apparatus. By injecting heat pulses to one or two U-shaped vertical heat exchangers, and measuring associated temperature responses, it becomes possible to create temperature-depth profiles indicating the relative contribution of the different ground/soil/rocks strata, optimize the system from the point of view of thermal carrier fluid operating temperatures, equipment capacity efficiency, and cost (Klof and Gehlin 1996; Kavanaugh 2000).

Thermal response tests with mobile apparatus were first introduced in Sweden and United States in mid-1990 (Austin 1998; Gehlin 2002). A typical experimental testing set includes components such as (Yavuzturk 1999) (Figure 6.1): (i) a trailer that can be towed to the site; (ii) a fossil-fired power generator (that can represent approximately 30% of the total thermal response tests' costs including mobilization and fuel) when electricity is not available on the site; (iii) a purge tank containing water; (iv) a vertical U-shaped ground-coupled (borehole) heat exchanger of approximately the same size and depth as the heat exchanger planned for the site; the type, depth, diameter, and depth of the casing of the test borehole should be the same as the rest of the borehole field is planned to be; it can later be part of the final system; between completion of borehole and start of the thermal response tests, the ground/soil should have time (~5 days) to recover to (quasi-) undisturbed conditions and to allow for (almost) setting of the grouting; (v) an in-line electric heater of variable capacity (e.g., 8–130 W and 0.07–0.95 A) able to operate at various electrical power levels (e.g., between 2.5 and 15 kW); heating load should be as close as possible to the borehole heat exchanger thermal capacity; the heater elements should be rated at 1, 1.5, and 2 kW; the 2 kW water heater element is connected to an electronic power controller, so that by switching individual elements on or off, and by adjusting the controller, the power can be adjusted continuously between 0 and 4.5 kW; during the test, the water inlet and outlet temperatures, and the constant (fixed) heat flux (generally, between 1 kW and 4.5 kW) per borehole are measured at regular time intervals, while keeping constant the water volumetric flow rate and the temperature difference between the borehole inlet and outlet; thus, the outputs of the thermal response tests are the inlet and outlet temperature of the heat-carrier fluid as a function of time; (vi) a constant or variable speed (e.g., 1,400–3,900 rpm) water circulation pump with adjustable pumping rate to circulate heated water through the U-tube in the borehole; the flow in borehole heat exchanger should be turbulent and the pumping rate should be between 1.0 and 2.0 m³/h in most cases; (vii) control devices for water temperature limitation and flow rate; the accuracy of temperature measurement should be at least 0.1 K or better 0.01 K; (viii) a pump to remove all air from the piping system before the heat pulse phase of the experiment begins; (ix) three-way valves to close the connection to the purge system; (x) temperature and flow sensors to monitor and control the flow rate of circulating fluid inside the test borehole; (xi) a needle valve is used to adjust the flow rate; typically, a water flow rate of approximately 0.16 L/s is used; (xii) control devices and data acquisition system; in order to reduce the external influences, the maximum recording interval should be 10 min.

6.3.2 Testing Procedure

In-situ experimental tests aiming to reproduce the heat transfer taking place in ground-coupled heat exchangers impose continuous or intermittent pulses of known and constant heat flux injection (usually, between 1 and 4.5 W) into a ready-to-use borehole heat exchanger and measure the resulting temperature responses (changes) of the heat carrier fluid (usually, water) during a certain period of time while keeping constant the fluid mass flow rate and the temperature difference between the borehole inlet and outlet (Austin 1998).

The typical procedure for in-situ tests is to measure the temperature response of a fluid flowing through a ground-coupled heat exchanger in a single borehole (Figure 6.1).

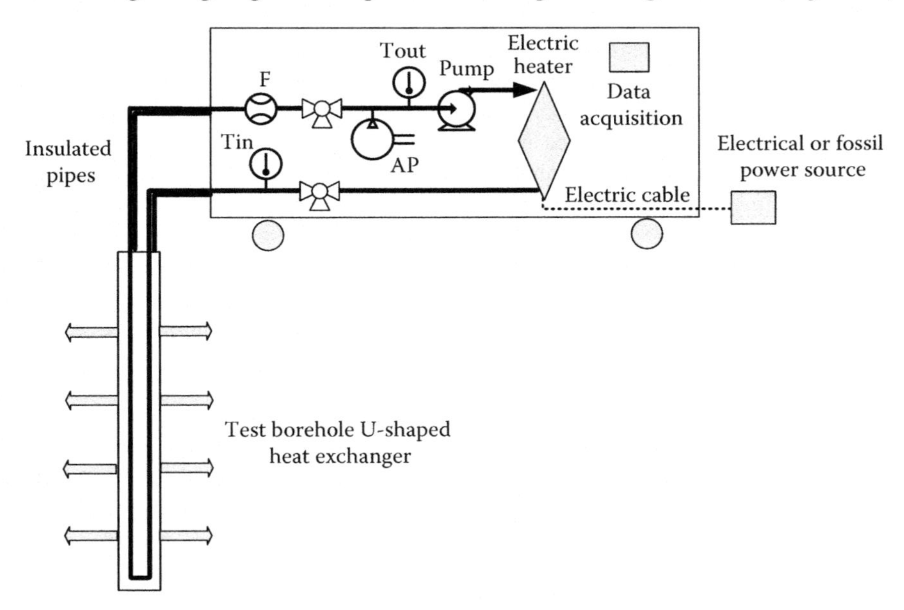

FIGURE 6.1 Schematic of an in-situ ground/soil/rock thermal response mobile test apparatus. AP, air pump; F, flowmeter; T, temperature meter.

Water circulating in the ground-coupled heat exchanger is heated at the surface to inject heat underground and disturb the thermal equilibrium to infer the subsurface thermal conductivity. In other words, they allow creating temperature-depth profiles indicating the relative contribution of the different ground strata. In this way, the system thermal performance and cost can be optimized.

The main steps of in-situ thermal response tests can be summarized as follows: (i) insert a U-tube in the drilled borehole of depth of the actual heat exchangers planned for the project and connect it to a water circulation pump and a water heater; (ii) water circulating in a ground-coupled heat exchanger is heated at the surface to inject heat underground and disturb the thermal equilibrium to infer the subsurface thermal conductivity; (iii) depending on the expected thermal conductivity, a good value for a standard thermal response tests can be around 50–80 W/m. Thus, for a borehole heat exchanger of 100 m, an electric power supply of 5–8 kW is required; a minimum electric power of ~400 V at 16 A (preferably electric resistance heater with adjustable heating load up to approximately 10 kW) is recommended, and the power supply needs to be as constant as possible (less than 5% fluctuation over the test duration). In particular, at a large-scale application site with a temporary power supply, fluctuations can be substantial; power supply should be without interruption – this sounds trivial, but can be a challenge for the duration of several days; (ii) borehole specific thermal power input should be 50 to 80 W/m, which are the expected peak loads on the U-tubes for an actual ground-source heat pump systems; (iv) heat injection should result in a temperature change that is in the order of the planned operating range; with a too-small temperature change, the signal will be low and thus the error in evaluation larger, while with a too-high temperature change, effects not present in the normal operation range might occur (e.g., drying); (v) install the experimental apparatus as shown in Figure 6.1; and (vi) at least once every 10 min, measure the initial average temperature of the thermally undisturbed ground/soil along borehole heat exchanger depth either by "dipping" the borehole with a temperature gauge (e.g., thermocouple) on a graduated tape, and taking readings at every, let's say, 2 m, or, alternatively, by using water circulated throughout the U-tube without any heat input; continuously measure and record the temperature of the thermal (water) carrier fluid entering and leaving the borehole with accuracy of temperature measurement and recording devices of ±0.3 K at least once every 10 min, as well as the energy consumption.

A controlled heat source (usually an electric heater) continuously injects a constant amount of heat into a test borehole heat exchanger through the fluid circulating around the U-shaped pipe for a period of time (at least for 48 hours). The borehole and the surrounding bedrock are heated and cooled at constant rates to absorb heat or cold is measured. The thermal power input to the brine circuit should be more or less the same amount as the expected heating/cooling load for the borehole during normal operation. Thermal response tests should be performed for 48–72 hours, and even more. With in-situ tests shorter than 48 hours, the error in the ground thermal conductivity prediction can be significant. The difference between the ground/soil thermal conductivity estimations of the 20-hour test and the estimations of the 48-hour test was of about 4.6%. A waiting period of five days is suggested for low-conductivity ground/soils (i.e., $k_g \geq 1.7$ W/m·K) after the ground-coupled heat exchanger has been installed and grouted (or water filled) before the thermal response test is initiated. A delay of three days is recommended for higher conductivity formations (i.e., $k_g \geq 1.7$ W/m·K). This period of time is needed to dissipate the heat released during the installation phase (i.e., drilling friction and grouting consolidation). A heat injection rate of 50–80 W/m is recommended by the North American industry guidelines to create a temperature difference of about 3–7°C between the inlet and outlet of the ground heat exchanger (Kavanaugh 2001). The total heating power needed to conduct a thermal response test in a ground heat exchanger that is 150 m long is therefore between 7.5 and 12 kW, which requires an electric current intensity of 31–50 A for a potential difference of 240 V. The power is supplied by a fuel-fired generator or by connecting the thermal response test unit to the electric grid. With heat injection rates of 50–80 W/m, 36–48-hour tests are suggested. However, in practice, it is common to conduct considerably longer tests with heat injection rates exceeding this range. These tests impose pulses of known and fixed heat flux (generally, between 50 and 2,000 W) on the borehole, and measured the resulting temperature responses while keeping constant the brine volumetric flow rate and the temperature difference between the borehole inlet and outlet.

The working principle of a thermal response test is the following: (i) the thermal response test rig is connected to the borehole heat exchanger and water will be circulated in the system (Figure 6.1); (ii) a constant heat injection is provided by a heater (electric resistance, gas, or heat pump); (iii) the thermal energy is transported with the water and introduced into the ground/soil; and (iv) from the development of the temperatures thermal conductivity can be calculated.

To determine the undisturbed ground/soil/rocks temperature, the water is circulated at the beginning of the temperature response test and to measure the temperature development at inlet and outlet without applying any heat load. This parameter gives the background upon which all temperature changes due to heat injection or heat extraction need to be calculated in order to be in equilibrium with the surrounding earth. The main outputs are the inlet and outlet temperatures of the heat carrier fluid as a function of time (see Figure 6.2). From these experimental data, and with an appropriate model describing the heat transfer between the ground/soil and the fluid, the effective thermal conductivity of the surrounding ground/soil is determined. The average fluid temperature ($\bar{T}_{water}$) plotted against time (t) provides a curve similar to that shown in Figure 6.2.

Recommended test specifications could be summarized as follows (Kavanaugh 2000, 2001): (i) thermal property tests should be 50–80 W per meter of borehole, which are the expected peak loads on the U-tube for an actual ground-source heat pump system; (ii) standard deviation of input electrical power should be less than +/– 1.5% of the average value and thermal peak loads less than +/– 10% of average, or resulting temperature variations should be less than +/– 0.3 K from a straight trend line of a log (time) versus average loop temperature; (iii) accuracy of the temperature measurement and recording devices should be +/– 0.3 K; (iv) combined accuracy of the electrical power transducer and recording device should be +/– 2% of the reading; (v) flow rates should be sufficient to provide a differential loop

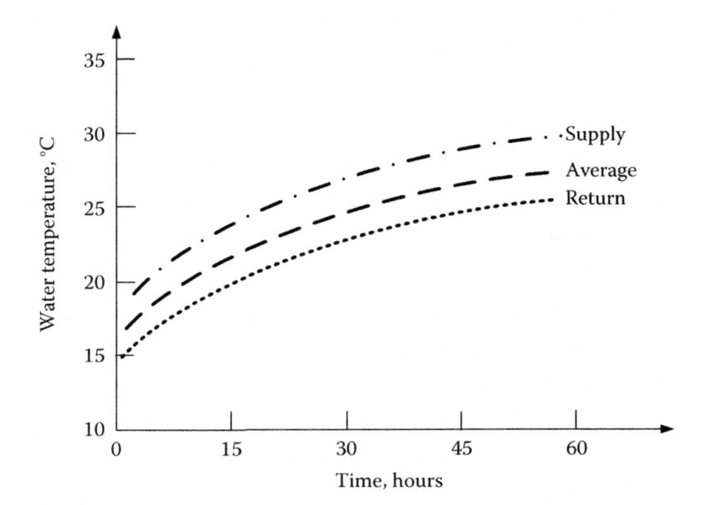

FIGURE 6.2 Typical evolution of fluid temperatures in a thermal response test.

temperature of 3.7–7 K; this is the temperature differential for an actual ground-source heat pump system; (vi) a waiting period of five days is suggested for low-thermal conductivity ground/soils ($k \leq 1.7$ W/mK) after the ground loop has been installed and grouted (or filled) before the thermal conductivity test is initiated; a delay of three days is recommended for higher thermal conductivity ground/soil formations ($k \gtrsim 1.7$ W/mK); (vi) the initial ground/soil temperature measurement should be made at the end of the waiting period by direct insertion of a probe inside a liquid-filled ground-coupled heat exchanger at three locations, representing the average, or by temperature measurement as the liquid leaves the loop during the period immediately after start-up; (vii) data collection should be at least once every 10 minutes; (viii) all aboveground/soil piping should be insulated with a minimum of 13 mm closed-cell insulation or equivalent; test rigs should be enclosed in a sealed cabinet that is insulated with a minimum of 25 mm fiberglass insulation or equivalent; (ix) if a borehole is re-tested, loop temperature should be allowed to return to within 0.3 K of the pre-test initial ground/soil temperature; this typically requires 10–12 days of delay in mid- to high-thermal conductivity formations and 14 days in low-thermal conductivity formations if a complete 48-hour test has been conducted; waiting periods can be proportionally reduced if tests are shorter.

The results of in-situ tests can be adversely influenced by (Austin 1998): (i) oscillations of atmospheric temperature; this adverse influence can be mitigated by increased insulation of the aboveground piping during in-situ measurements; (ii) daily temperature variations have a significant influence on the assessment of the ground/soil "effective" thermal conductivity; because the effect of heat penetration through the ground/soil surface and the correlation between heat carrier fluid temperature and ambient air temperature that depends on depth during the thermal response tests, it is recommended to mitigate their adverse influence by increasing the thermal insulation of the aboveground piping and by installing additional temperature sensors in the borehole itself during in-situ measurements; (iii) the accuracy of the measurements can further be increased by using a higher temperature difference and temperature sensors with lower errors; (iv) heat penetration through the ground surface needs to be taken into account because of its strong influence on estimates of thermal conductivity and thermal resistance of the borehole; it is reasonable to account for this effect by averaging over the length of the borehole since the heat flow between the ground and the heat carrier fluid depends on depth; (v) changes in viscosity of the circulation fluid with temperature may lead to a change in flow rate over the duration of experiments; (vi) thermal response tests for the borehole heat exchangers may be affected by the effect of groundwater flow; (vii) alteration of ground/soil moisture content from complete dryness to 12.5% of saturation strongly influences the geothermal heat pump performances, and any decrease of ground/soil moisture content in this range has a devastating effect on the coefficient of performance; (viii) changes in viscosity of the circulation medium with temperature may lead to a change in flow rate over the duration of an experiment; to adjust for these changes in flow rate that could introduce a bias in the results, the control system can be adapted so that, not the temperature difference itself, but a function of temperature difference and measured flow rate (i.e., the actual energy flux) is kept constant; (ix) the method consisting in averaging the borehole temperature, rather than using its value at the midpoint of the borehole overcomes the known problem of the midpoint temperature method overestimating the value of the borehole temperature, by properly accounting for the influence of the conditions at the ground surface and at the bottom of the borehole heat exchanger; some uncertainty, in addition to sensor errors, is introduced because of the nonhomogeneous nature of the ground/soil (e.g., the time-varying nature of the undisturbed ground/soil temperature, which is affected by seasonal changes near the surface, and downhole variations in the U-tube location and borehole diameter); (x) thermal response tests are sometimes affected by problems that can create detrimental effects on the estimations of ground/soil thermal conductivity and borehole thermal resistance; these issues include problems such as power outage, equipment failure, and fluid leakage; if the problem cannot be resolved quickly, or if it is caused by equipment malfunction or a data logging failure, conducting a retest might be indispensable.

Total uncertainty could be of 9–11% according to parameters that are all independent or nearly independent from each other. Sometimes in-field retesting of boreholes is needed in order to investigate the repeatability and reproducibility of the obtained borehole thermal response results. Before a retest is conducted, the ground/soil temperature should be within 0.3 K of the initial undisturbed ground/soil temperature because if the ground/soil temperature is, let's say, 1.1 K higher than the initial undisturbed ground/soil temperature, the errors may range between 5% and 12% (Kavanaugh, 2001). After a 48-hour in-field test, a minimum recovery time of 10–14 days is suggested to allow the ground temperature to return to within 0.3 K of the undisturbed value. For shorter test durations, the recovery time can be reduced proportionally up to a minimum of 24 hours. The experimentally measured recovery time for the loop temperature to return to 0.3 K of its initial undisturbed value was approximately 10 days. Simulated results suggest a recovery time of approximately 11 days.

Table 6.1 shows that the recovery times for various combinations of ground/soil formation, heat injection rate, and test duration after a thermal response test are

TABLE 6.1
Recovery Times for a 50 W/m of Heat Injection Flux in Different Ground/Soil Formations

Formation	Duration of thermal response test		
	10 hours	25 hours	50 hours
Ground/soil, dry (k_g = 1 W/mK)	6	14	27
Clay, moist (k_g = 1.6 W/mK)	4	9	17
Average rocks and sand, saturated (k_g = 2.4 W/mK)	3	6	11
Dens rock (k_g = 3.4 W/mK)	2	4	8

strongly related to the test duration, the heat injection rate, and the ground formation. For a specific ground formation and a fixed injection rate, increasing the test duration twofold doubles the recovery times. Similarly, for a particular test duration, the recovery times increase proportionally with an increase in the injection rates. The recovery times for medium and low conductivity formations are two to four times longer than for high conductivity formations.

6.3.3 Evaluation of Thermal Conductivity

The temperature difference between the ground/soil/rock wall and the carrier fluid is proportional to the heat flux and the borehole thermal resistance (that, in turn, depends on the ground-coupled heat exchanger material and geometry, and on the thermal properties of the filling medium). Applying a constant thermal load and measuring the fluid temperature changes, it is possible to estimate the ground/soil/rocks average thermal conductivity and borehole thermal resistance by assuming that the heat transfer in the ground/soil/rock is conductive without interaction from groundwater flow. If there is a groundwater movement, it enhances convective heat transfer and, thus, increases the estimated thermal conduction. The difference between the thermal conductivity of the ground/soil/rocks and the uppermost soil layer generally has little impact on the system "overall" thermal performance.

In-field measurements are analyzed using various mathematical heat transfer models to estimate ground/soil "effective" thermal conductivity (Signorelli et al. 2007).

$$k_{g,eff} = \dot{Q}/L\cdot(\bar{T}_{supply} - \bar{T}_{return}) \tag{6.2}$$

where

$k_{g,eff}$ is the "effective" (average) ground/soil/rock thermal conductivity (W/m·K).
$\dot{Q}$ is the total heat flux transfer (W).
L is the borehole length of (m).
$\bar{T}_{supply}$ is the water supply temperature (°C).
$\bar{T}_{return}$ is the water return temperature (°C).

REFERENCES

Austin, W.A. 1998. *Development of an In Situ System for Measuring Ground Thermal Properties*. MSc. Thesis, Oklahoma State University, Oklahoma, 164 pp.

Austin, W.A., C. Yavuzturk, J.D. Spitler. 2000. Development of an in-situ system for measuring ground thermal properties. *ASHRAE Transactions* 106(1): 365–379.

EPRI. 1989. *Soil and Rock Classification according to Thermal Conductivity: Design of Ground-Coupled Heat Pump Systems: Final Report*. Bose, J.E. (Editor), Electric Power Research Institute Special Report, EPRI CU-6600, Palo Alto, CA.

Gehlin, S. 2002. Thermal Response Test-Method Development and Evaluation. Doctoral Thesis. Lulea University of Technology, Lulea, Sweden.

IGSHPA. 2007. *Closed-Loop/Geothermal Heat Pump Systems: Design and Installation Standards*. IGSHPA (International Ground Source Heat Pump Association), Stillwater, OK.

Kavanaugh, S.P. 2000. Field tests for ground thermal properties – Methods and impact on ground-source heat pump design. In Proceedings of ASHRAE Winter Meeting, February 5–9, Dallas, TX.

Kavanaugh, S.P. 2001. *Investigation of Methods for Determining Soil Formation Thermal Characteristics from Short Term Field Tests*. ASHRAE Report-1118, Atlanta.

Kelvin, T.W. 1882. *Mathematical and Physical Papers*. Cambridge University Press, London, UK.

Klof, C., S. Gehlin. 1996. *TED-A Mobile Equipment for Thermal Response Tests*. M.S. Thesis. 198E. Luleå University of Technology, Luleå, Sweden.

Shonder, J.A., J. Beck. 2000. Field test of a new method for determining soil formation thermal conductivity and borehole resistance. *ASHRAE Transactions* 106(1):843–850.

Signorelli, S., S. Bassetti, D. Pahud, D.T. Kohl. 2007. Numerical evaluation of thermal response tests. *Geothermics* 36(2):141–166.

Tarnawski, V.R., W.H. Leong. 1993. Computer analysis, design and simulation of horizontal ground heat exchangers. *International Journal of Energy Research* 17:467–477.

de Vries, D.A. 1963. *Thermal properties of soils*. In *Physics of the Plant Environment*, van Wijk, W.R. (Editor), North-Holland, Amsterdam, The Netherlands.

de Vries, D.A. 1974. *Heat Transfer in Soils. Heat and Mass Transfer in Biosphere. 1. Transfer Processes in Plant Environment*. De Vries, D.A. and Afgan, N.H. (Editors). John Wiley & Sons Inc., New York.

de Vries, D.A., A.I. Peck. 1958. On the cylindrical probe method of measuring thermal conductivity with special reference to soils. *Australian Journal of Physics* 11:225–271; 409–423.

Yavuzturk, C. 1999. *Modeling of Vertical Ground Loop Heat Exchangers for Ground Source Heat Pump Systems*. Submitted to the Faculty of the Graduate College of the Oklahoma State University in partial fulfillment of the requirements for the Degree of Doctor of Philosophy, December, Oklahoma, OK.

7 Classifications of Ground-Source Heat Pump Systems

7.1 INTRODUCTION

The first heat pump was developed and built by von Rittinger (1867), but the first ground-source heat pump was developed and built by Robert C. Webber (https://earthrivergeothermal.com/tag/robert-c-webber/, accessed August 8, 2021) in 1940.

Ground-source heat pump (GSHP) systems (also called geothermal, Earth-coupled, geo-exchange, and Earth energy systems), are unconventional heating, ventilating, and air-conditioning systems that use the ground/soil/rocks, groundwater, or surface water as heat sources (during heating seasons) or heat sinks (during cooling periods).

7.2 CLASSIFICATION ACCORDING TO APPLICATION FIELD

In cold and moderate climates, the ground-source heat pump systems use one or several liquid (water, brine)-source (usually) reversible geothermal heat pumps (see Chapter 8) to extract heat at very low temperatures (generally, between 0°C and 10°C) from ground/soil/rocks (at depths up to 200 m), groundwater or surface water to efficiently space and domestic hot water heating in cold seasons (winters) (e.g., via air handling units or hydronic radiant floors located in residential and commercial/institutional buildings). In hot seasons (summers), the same reversible geothermal heat pump(s) provide space cooling/dehumidifying by rejecting heat to the ground/soil/rocks, groundwater, or surface waters. In areas with suitable ground/soil/rocks conditions (e.g., temperature, moisture content, groundwater flow rate, and velocity), it is possible to daily or seasonally store heat in the earth and/or aquifers during the cooling-dominated seasons and use it during the next heating-dominated periods.

According to their specific application fields and size, the ground-source heat pump systems can be classified as (Figure 7.1): (i) residential and small-scale commercial/institutional systems; (ii) large-scale commercial/institutional and industrial systems; and (iii) district heating and cooling networks. Upon the arrangement of geothermal heat pumps inside the buildings, the large-scale commercial/institutional ground-source heat pump systems are classified as: (i) distributed and (ii) central systems.

7.3 CLASSIFICATION ACCORDING TO HEAT/SINK SOURCES AND COMMON CONFIGURATIONS

Depending on heat/sink sources, as earth (ground/soil/rock), water (groundwater, surface water), and basic configurations (layouts), the ground-source heat pump

DOI: 10.1201/9781003032540-7

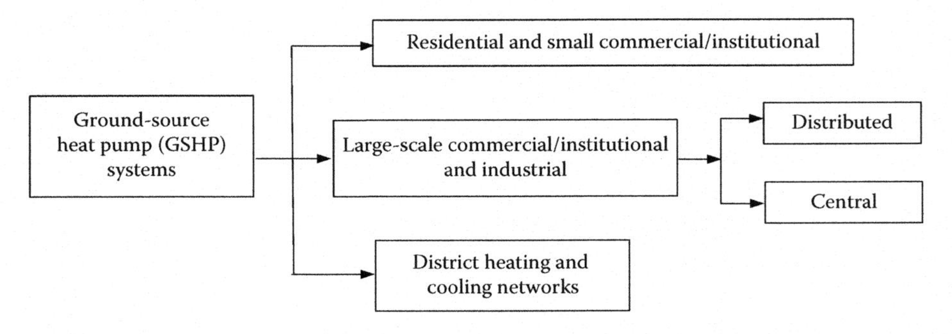

FIGURE 7.1 Classification of ground-source heat pump systems according to application fields.

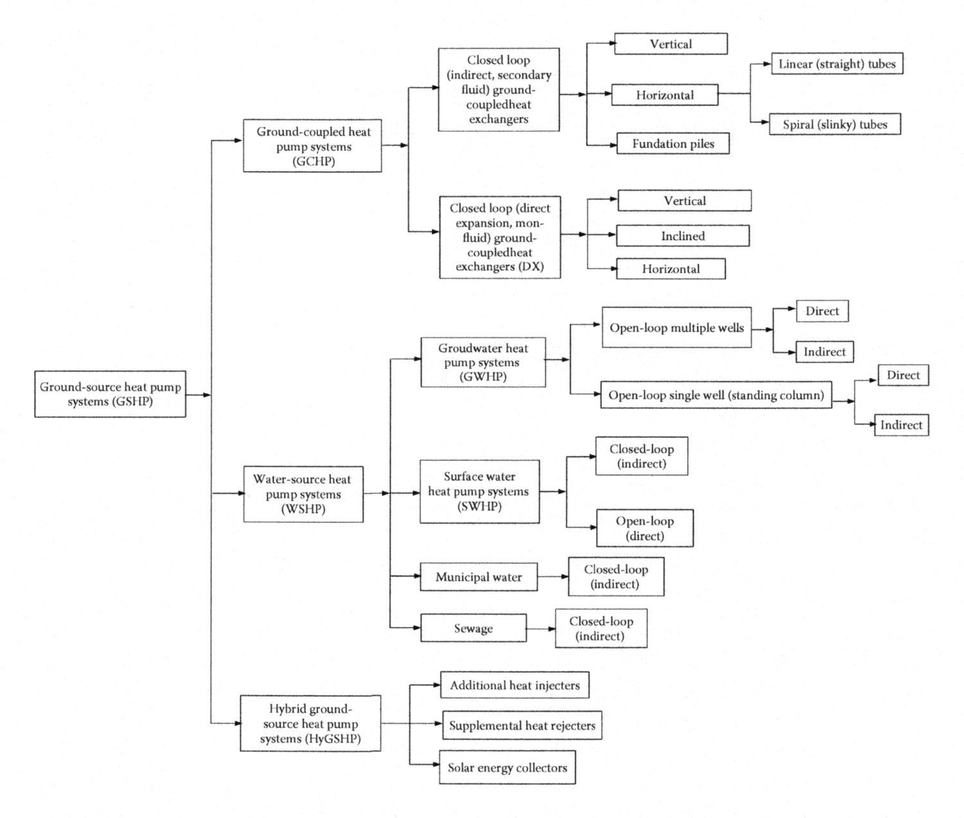

FIGURE 7.2 Classification of ground-source heat pump systems according to heat/sink sources and common configurations.

systems could be also classified as shown in Figure 7.2 (Kavanaugh 1992; DOE 1995; Kavanaugh and Rafferty 1997; McQuay 2002). As it can be seen in Figure 7.2, at least three general types of ground-source heat pump systems are available: (i) ground-coupled (GCHP), provided with closed-loop (indirect, secondary fluid) or closed-loop direct expansion (DX) vertical or horizontal ground-coupled heat exchangers placed directly in the ground/soil/rocks (or in surface waters); these systems circulate a fluid (water, brine, or refrigerant) that transports the heat between the heat/sink sources and the geothermal heat pumps; heat is extracted from or rejected to the ground/soil/rocks by means of buried pipes, through which a heat transfer fluid usually circulates water or a water/antifreeze solution. Ground/coupled heat exchangers are buried either horizontally in shallow trenches (at depths between 1.0 and 2.0 m), or vertically in boreholes up to 250 m in depth or in building foundation piles; the choice of a ground-coupled heat exchanger (vertical or horizontal) depends on available land, local ground/soil type, and thermo-physical properties, and drilling and/or excavation costs; (ii) water-source (WSHP) (groundwater, surface water, municipal water, or sewage); in groundwater and surface water open-loop systems, the water is circulated once through intermediate heat exchangers or even through geothermal heat pumps and then is returned to the same aquifer or surface water; groundwater heat pump systems are divided into open-loop multiple-, and single (standing column)-well systems where the groundwater may pass directly or indirectly (via intermediate heat exchangers) through the geothermal heat pump(s); similarly, surface water ground-source heat pump systems can use water indirectly through closed-loops containing heat exchangers immersed in the surface water body, or directly through open loops; municipal water and sewage could be used as heat/sink sources for geothermal heat pumps only indirectly via closed loops containing intermediate heat exchangers; in systems using groundwater, surface water, and sewage as heat/sink sources, the water quality is crucial since the intermediate heat exchangers and/or geothermal heat pumps can be affected by corrosion, fouling, and/or mechanical blockage; and (iii) hybrid (HyGSHP) (with additional heat injectors and/or supplemental heat rejecters) or solar energy collectors.

The selection of ground-source heat pump systems shown in Figure 7.2 is a process that depends on the local weather conditions, and the geological and thermal characteristics of the ground/soil/rocks, groundwater, or surface water. The decision to employ one of them at a given site should be based on criteria such as: (i) availability and accessibility of geothermal energy resources; (ii) size and height of buildings and their heating and cooling loads; (iii) available site (ground/soil surface) land area for vertical or horizontal ground-coupled heat exchangers; (iv) approximate drilling, trenching, and excavation costs; and (v) local regulatory requirements. Closed-loop (indirect, secondary fluid, and direct expansion) ground-coupled heat pump systems are further classified as vertical (and inclined in the case of direct expansion concepts) and horizontal (linear/straight or spiral/slinky) systems. Both vertical and horizontal indirect ground-coupled heat pump systems have the disadvantage of higher energy consumptions (up to 15%) due to brine (or water) circulation pumps. In spite of their higher initial costs, the majority of large ground-source heat pump systems today use vertical closed-loop, ground-coupled heat exchangers (DOE 1995).

REFERENCES

DOE. 1995. *Commercial/Institutional Ground-Source Heat Pump Engineering Manual.* Elaborated by Caneta Research for American Society of Heating, Refrigerating and Air Conditioning Engineers (ASHRAE), Atlanta, GA.

Kavanaugh, S.P. 1992. Ground-coupled heat pumps for commercial building. *ASHRAE Journal* 34(9):30–37.

Kavanaugh, S.P., K. Rafferty. 1997. *Ground-Source Heat Pumps. Design of Geothermal Systems for Commercial and Institutional Buildings*. American Society of Heating, Refrigerating and Air-Conditioning Engineers, Inc., Atlanta, GA.

McQuay. 2002. *Geothermal Heat Pump Design Manual* (https://www.15000inc.com/wp/wp-content/uploads/Geothermal-Heat-Pump-Design-Manual.pdf. Accessed December 2, 2019).

von Rittinger, P.R. 1867. *Lehrbuch der Aufbereitungskunde* (In German). Verlag von Ernst & Korn, Berlin, Germany.

8 Geothermal Heat Pumps

8.1 INTRODUCTION

In order to use very low temperatures of shallow ground/soil, groundwater, or surface water as heat sources for efficiently heating the buildings in cold and moderate climates, geothermal heat pumps are needed to increase their temperature levels (e.g., from 5°C to 10°C) to much higher levels (e.g., up to 35–45°C) by the addition of an external, generally, electrical work.

The geothermal heat pumps are brine (water)-to-air, brine (water)-to-water, or refrigerant-to-refrigerant machines based on mechanical vapor compression reverse Rankine cycle designed to operate over extended ranges of heat source entering temperatures (e.g., as low as –5°C in the heating mode and as high as 35–40°C in the cooling mode) (ARI 1998).

In cold and moderate climates, geothermal heat pumps are commonly available with nominal cooling capacities between 1.75 and 70 kW in a wide variety of configurations (e.g., compact or split, and vertical or horizontal), with or without desuperheaters (components used to preheat domestic/process hot water by recuperating heat from the compressor superheated discharge vapor).

8.2 THERMODYNAMIC PARAMETERS

In thermodynamic systems, the heat and work that can lead to changes in the system internal energy, are equivalent variables. The specific (mass) heat (***q***, **J/kg**) is defined as the microscopic (internal) energy naturally in transit from a high-temperature body to a lower temperature object via a heating process. On the other hand, the specific (mass) work (***w***, **J/kg**) done by the unit mass (**kg**) of a working fluid (e.g., an ideal gas) crossing the boundary of a thermodynamic system between state 1 and state 2 is defined as follows:

$$l_{12} = \int_1^2 pdv \tag{8.1}$$

where

l_{12} is the specific (mass) work (J/kg).

p is the system pressure (Pa).

dv is the infinitesimal change of the fluid specific volume (m^3/kg).

Both heat and work are not functions of state, can be measured and quantified, but not stored or conserved independently since they depend on the process. In other words, heat and work are forms of energy in transit, i.e., energy that enters a system as heat may leave as work, or vice versa.

DOI: 10.1201/9781003032540-8

The specific enthalpy of a thermodynamic system is defined as follows:

$$\boldsymbol{h = u + pv} \tag{8.2}$$

where

h is the specific enthalpy (***J*/*kg***).

u is the specific internal energy (***J*/*kg***).

p is the system pressure (***Pa***).

v is the specific volume ($\boldsymbol{m^3/kg}$).

The enthalpy is an extensive variable, thus the total enthalpy is:

$$H = m \cdot h \tag{8.3}$$

where

H is the system total enthalpy (J).

m is the mass of the thermodynamic system (kg).

h is the system specific enthalpy (J/kg).

For many substances, the enthalpy, given in property tables, diagrams, or properties software, is relative to an arbitrary reference value. For example, for water vapor, the reference state is the saturated liquid at 0°C where the enthalpy is zero. For a thermodynamic process 1–2 occurring at constant pressure (p = constant), the finite variation of enthalpy represents the heat exchanged (i.e., added to or removed from the system):

$$\Delta h = q_{12} \tag{8.4}$$

The specific heat (J/kgK) is defined as the heat required to raise the temperature of a unit mass of substance by a unit temperature. Because the changes in pressure and volume of any thermodynamic system affect the specific heat, it is expressed either at constant pressure (c_p) or at constant volume (c_v). Therefore, both constant-pressure and constant-volume specific heats are thermodynamic properties of a substance. The constant-pressure specific heat (c_p) is the amount of heat (δq) required to increase the temperature of a closed thermodynamic system of constant composition by dT when the system is heated in a reversible transformation of state at constant pressure:

$$c_p = (\delta q/dT)_{p,rev} = (dh/dT)_{p,rev} \tag{8.5}$$

where

δq and dT are differential changes in heat and temperature, respectively.

Similarly, by definition, the constant-volume specific heat (c_v) is the amount of heat required to increase the temperature by dT when the system is held at constant volume:

$$c_v = (\delta q/dT)_{v,rev} \tag{8.6}$$

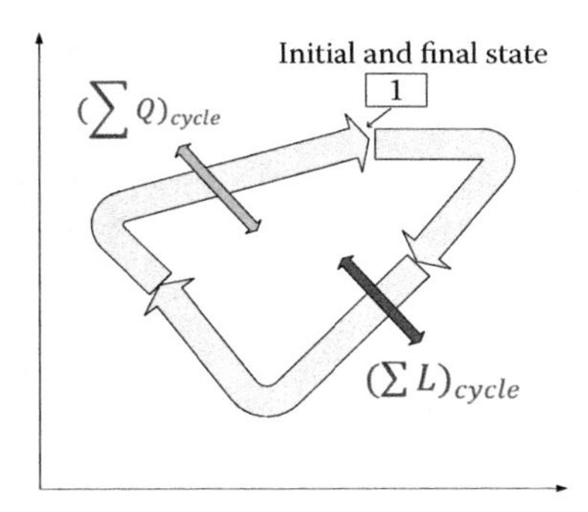

FIGURE 8.1 Thermodynamic cycle of a closed system with multiple heat and work exchanges with the surrounding medium.

The heat capacities at constant-volume and constant-pressure (J/K) are defined as, respectively:

$$C_p = m \cdot c_p \tag{8.7}$$

$$C_v = m \cdot c_v \tag{8.8}$$

The ratio of constant-pressure and constant-volume specific heats ($k = c_p/c_v$) represents the isentropic expansion factor for ideal gases or adiabatic exponent for real gases. The first law of thermodynamics (also known as the law of energy conservation) states that energy can be neither be created nor destroyed, and that it can only be transferred (or changed) from one physical body form to another. The first law of thermodynamics does not provide information about the direction of processes and does not determine the final equilibrium state, even if it is known that energy is transferred from high-temperature to low-temperature bodies.

In the case of a closed thermodynamic system undergoing a cycle by adding work (W, expressd in J) from the surrounding medium at state 1 and by transferring equivalent heat (Q, expressed in J) to the surrounding returning to state 1, the amount of the work removed (or rejected) is always proportional to heat supplied (or absorbed) (Figure 8.1).

If the cycle involves several heat and work (absorbed or rejected) quantities, the first law of thermodynamics for a closed system undergoing a cycle can be expressed as follows:

$$(\sum L)_{cycle} = (\sum Q)_{cycle} \tag{8.9}$$

Applied to open thermodynamic systems with fluid flowing in permanent regime, the first law of thermodynamics can be expressed as follows:

$$\dot{Q}_{12} - \dot{L}_{12} = \dot{m}\left[(h_2 - h_1) + \frac{1}{2}(w_2^2 - w_1^2) + g(z_2 - z_1)\right] \tag{8.10}$$

or

$$q_{12} - l_{12} = (h_2 - h_1) + \frac{1}{2}(w_2^2 - w_1^2) + g(z_2 - z_1) \quad (8.11)$$

where

$\dot{Q}_{12}$ is the total thermal power exchanged by the system with ambient medium (W).

$\dot{L}_{12}$ is the total (e.g., electrical ot thermal) power exchanged by the system with ambient medium (W).

$\dot{m}$ is the fluid mass rate (kg/s).

h_1 and h_2 are the specific enthalpy of working fluid entering and leaving the open thermodynamic system, respectively (J/kg).

w_1 and w_2 are the velocities of working fluid at the inlet and outlet of the open thermodynamic system, respectively (m/s).

z_1 and z_2 are the elevations of working fluid at the inlet and outlet of the open thermodynamic system, respectively (m).

$q_{12} = \dot{Q}_{12}/\dot{m}$ is the specific (mass) heat exchanged by the system with ambient medium (J/kg).

$l_{12} = \dot{L}_{12}/\dot{m}$ is the specific work exchanged by the system with ambient medium (J/kg).

Applied to geothermal heat pumps, the first law of thermodynamics, as expressed by Equation 8.10, states that, for transferring heat from one temperature level to a higher one, an expenditure of energy in the form of absorbed work is required, and that the amount of energy rejected at the higher temperature level (Q_{warm}) is the sum of the heat removed on the lower temperature level (Q_{cold}) plus the added energy (or equivalent work) (W) to accomplish the energy lift. Since at the end of a geothermal heat pump thermodynamic cycle, the system returns to the same state as at the beginning, the change in the system internal energy must be zero. Thus:

$$0 = |Q| + |L| \quad (8.12)$$

where

$$|Q| = |Q_{warm}| + |Q_{cold}| \quad (8.13)$$

By convention, the total heat added to the cycle ($Q_{added} = Q_{cold}$) (e.g., at the geothermal heat pump evaporator) is considered negative and, if extracted from the cycle ($Q_{extracted} = Q_{warm}$) (e.g., at the geothermal heat pump condenser), is assigned as positive. Also, the work added (or equivalent heat) (W) to a thermodynamic cycle is assigned as negative. Consequently, Equation 8.13 becomes:

$$Q_{warm} = Q_{cold} + L \quad (8.14)$$

where

Q_{cold} is the total heat extracted from the (cold) heat source (J).

L is the total work (or equivalent energy) added to the cycle (J).

Q_{warm} is the total heat removed from the cycle and transferred to the (warm) heat source (J).

The second law of thermodynamics states that heat cannot be transferred from a body at lower temperature to another body at higher temperature without the expenditure of energy. In other words, "heat will not spontaneously flow from a lower temperature to a higher temperature," or, according to Clausius (1867), "no process is possible whose sole result is the removal of heat from a reservoir at one temperature and the absorption of an equal quantity of heat by a reservoir at a higher temperature."

The second law of thermodynamics introduces the notion of entropy can be viewed as a measure of the randomness of molecules in a system. At a given absolute temperature (***T, K***), any infinitesimal heat transfer between a thermodynamic system and its surroundings will lead to an increase of the system entropy:

$$ds = \delta q / T \tag{8.15}$$

where

s is the mass (specific) entropy (J/K).

$\delta q = c_p dT$ is the infinitesimal heat transfer between a thermodynamic system and its surroundings (J).

c_p is the specific heat at constant pressure (J/kgK).

T is the temperature (C).

Enounced as the principle of the increase of entropy, the second law of thermodynamics states that in any process, whatever between two equilibrium states of a system, the sum of the increase in entropy of the thermodynamic system and the increase in entropy of its surroundings is equal to or greater than zero. In other words, entropy cannot be destroyed, but it can be created. Therefore, the entropy of any isolated system always increases, and the system spontaneously evolve towards thermal equilibrium, i.e., the state of maximum entropy. In the case of geothermal heat pumps' closed thermodynamic cycles, the second law of thermodynamics states that, for a given delivered heat ($\boldsymbol{Q_{warm}}$), better performance means a smaller input work ($\boldsymbol{W}$), but the smallest work ($\boldsymbol{W}$) will never be zero.

8.3 SUBCRITICAL MECHANICAL VAPOR COMPRESSION GEOTHERMAL HEAT PUMPS

Most geothermal heat pumps that transfer heat from low-temperature heat sources to higher-temperature heat (sink) sources work according to reversible subcritical Rankine thermodynamic cycles (see Figures 8.4c and 8.4d). To accomplish such a cycle, the geothermal heat pumps use the physical properties of volatile working fluids (refrigerants), as well as some amount of external electrical primary energy to run the compressor and auxiliary equipment as blowers and/or brine (water) circulating pumps. The selection of refrigerants depends on their thermodynamic

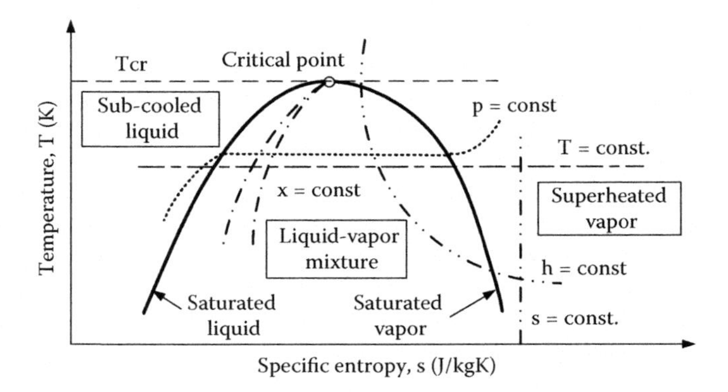

FIGURE 8.2 Schematic of the temperature – entropy diagram for pure refrigerants.

properties and environmental impacts (mainly due to leakages in the atmosphere and greenhouse gas emissions associated with electrical or thermal energy use). Refrigerants are classified as follows (http://ozone.unep.org/new_site/fr/montreal_protocol.php, accessed March 4, 2016):

- Chlorofluorocarbons (CFCs), as CFC-11 (trichlorofluorométhane) and CFC-12 (dichlorodifluoromethane), with strong destructive impact on the ozone layer if released into the atmosphere and important influence upon the global warming; the manufacture of these refrigerants was already discontinued in January 1996;
- Hydro-chlorofluorocarbons (HCFCs), as HCFC-22, with reduced impact on the ozone layer and on the greenhouse effect, have been used during the last two decades as short-term alternatives to CFCs;
- Hydro-fluorocarbons (HFCs), as HFC-134a (1,1,1,2-tetrafluoroethane) (an alternative of CFC-12 in the temperature range of –10°C to 84°C) and HFC-410A (a very efficient working fluid; it promises a higher efficiency than HCFC-22, but it requires a 4 MPa technology), are harmless to the ozone layer and with less than 1% of the global warming effect of all greenhouse gases, are used today as long-term alternatives to CFCs and HCFCs; although HFCs do not contain chlorine atoms and are very stable fluids, their global warming potential (GWP) values are relatively high (see Table 8.1); HFCs are continuously subjected to progressive phase-down in order to eliminate all adverse environmental effects;
- Natural refrigerants as carbon dioxide (CO_2, R744) and ammonia (NH_3, R717), and flammable hydrocarbons (HCs) (R-600-propane, R-600A-isobutane, ethylene, propylene); all these fluids are harmless to the ozone layer, and have less or no influence upon global warming;
- Low-GWP refrigerants as HFO-1234yf, HFO-1234ze(E), and HFO-365mfc.

ANSI/ASHRAE Standard 34 (2013) also classifies the refrigerants according to their toxicity and flammability in six safety groups coming from the least to the most hazardous.

TABLE 8.1
Some Properties of Low-Temperature Pure Refrigerants and of their Potential Replacements

Refrigerant	ODP[1]	GWP[2]	Critical temperature (°C)	Critical pressure (MPa)
Present refrigerants				
HFC-134a	0	11,300	101.08	4.0603
HFC-410A	0	11,890	72.13	4.9261
Potential replacements				
HFO-1234yf	0	4	124.69	3.382
HFO-1234ze(E)	0	6	109.35	3.632

Notes
[1] Based on ODP = 1 of CFC-11.
[2] Based on GWP = 1 of CFC-11 over 100 years.

The impact of refrigerants containing chlorine compounds on ozone destruction is estimated by using the ozone depletion potential (ODP) index which is a measure of its destructive potential on stratospheric ozone layer relative to depletion caused by an equal amount of a reference substance, generally CFC-11 of which ODP is 1.0.

On the other hand, the global warming potential (GWP) index quantifies the capability of refrigerants to absorb infrared radiation relative to carbon dioxide. It compares the amount of heat trapped by a given mass of gas over a certain time period, usually 20, 100, or 500 years, compared to that of CO_2, a natural gas assigned as a reference with GWP = 1.

Total equivalent warming impact (TEWI) is, besides a global warming potential measure used to express contributions to global warming, defined as sum of the direct (chemical) and indirect (energy use) emissions of greenhouse gases (Reiss 1996).

The TEWI index can be calculated with the following equation:

$$TEWI = n \cdot L \cdot GWP + m_{refr}(1 - \alpha_{rec})GWP + n \cdot \beta \cdot E \tag{8.16}$$

where

n is the heat pump technical life (years).

L is the annual leakage rate of the geothermal heat pump (3 − 8%).

GWP is the global warming potential of the refrigerant for a period of 100 years (-).

m_{refr} is the refrigerant charge (kg).

α_{rec} is the refrigerant recovery and/or recycling factor (0.70 − 0.85%).

β is the CO_2 emission factor (-).

E is the annual energy consumption (kWh).

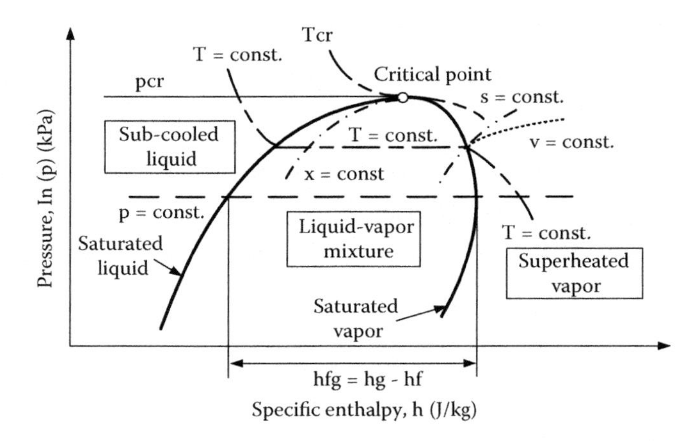

FIGURE 8.3 Schematic of the pressure-enthalpy diagram for pure refrigerants.

The thermodynamic properties (e.g., temperature, pressure, enthalpy, specific volume, and entropy) and processes of pure refrigerants are commonly represented in two-dimensional thermodynamic diagrams, as temperature-entropy (Figure 8.2) and pressure-enthalpy (Figure 8.3). In Figure 8.2, the left part of the dome curve represents the thermodynamic states of refrigerant saturated liquid and the right side, those of saturated vapor. The sub-cooled liquid region is located to the lower entropy, left side of the saturated liquid curve, while the superheated vapor region is located to the right side of the saturated vapor curve. Both saturated liquid and saturated vapor curves culminate in the critical point on top of the saturated dome in an inflection point (critical point) where the curvature of isobars changes from positive to negative and where the saturation liquid and vapor states become the same as the pressure increases. Above the critical pressure and temperature, no liquefaction will take place. In the two-phase region, i.e., under the saturation dome, the isobar lines (p = const.) are horizontal. From the saturated vapor line, the slope of constant pressure curves increases sharply into the superheated vapor region. On the other hand, the isobar curves drop quickly from the saturated liquid line into the sub-cooled liquid region.

In the pressure-enthalpy ln (p)-h diagram of a pure refrigerant (Figure 8.3), highly utilized in the heat pump technology, the ordinate shows the logarithm of the pressure, and the abscissa, the specific enthalpy. The boundary line between the subcooled liquid range and the two-phase region is the saturated liquid curve, and the boundary between the superheated vapor and the two-phase range is the saturated vapor curve. The two-phase region is found under the saturation dome, the area where the isotherms are horizontal. To the left of the two-phase region is the sub-cooled liquid region and to the right is the superheated vapor region.

The curves of saturated liquid and saturated vapor converge at the critical point. Above this point, there is no distinction between the liquid and vapor. Because both pressure and temperature are constant during the phase change processes, the isotherms are straight lines in the two-phase region located below the saturated dome. In the superheated vapor region, the isotherms (T = const.) curve quickly approaches vertical and approaches ideal gas behavior for which the enthalpy is independent of the pressure. The isentropic (constant entropy, s = const.) lines, which are important to determine compressor work, have positive slopes and have no abrupt change in the slope at the vapor saturation line. The isochores (constant specific volume, v = const.) curves, important in determining the volumetric capacity of heat pump compressors, show a much flatter slope. The thermodynamic properties of saturated liquid, as specific enthalpy, specific entropy, density, and specific volume, are respectively designed by h_f, s_f, ρ_f and v_f, and those of saturated vapor, by h_g, s_g, ρ_g, and v_g.

In the saturated region, situated under the saturation dome, the specific enthalpy, specific entropy, and specific volume are expressed as follows:

$$h = xh_g + (1 - x)h_f \tag{8.17}$$

$$s = xs_g + (1 - x)s_f \tag{8.18}$$

$$v = xv_g + (1 - x)v_f \tag{8.19}$$

where

x is the titre of saturated vapor (-).

For isobar/isotherm processes, as, for example, vaporization and condensation of pure refrigerants in the (p)-h diagram allow calculating the enthalpy (latent heat) of vaporization or condensation for a given saturation pressure as the enthalpy difference between saturated vapor and saturated liquid:

$$h_{fg} = h_g - h_f \tag{8.20}$$

where

h_g is the specific enthalpy of saturated vapor (J/kg).

h_f is the specific enthalpy of saturated liquid (J/kg).

The thermodynamic properties (e.g., enthalpy, entropy, and specific volume) of subcooled (or compressed) liquid are strongly dependent on temperature (rather than pressure) and, thus, may be approximated by the corresponding values for saturated liquid (h_f, s_f, v_f) at the existing temperature. To be acceptable for geothermal heat pump industry, future alternate refrigerants are expected to meet all, or the major part, of criteria as: (i) have properties as close as possible to those of the original HFCs, as zero ozone depleting potential (ODP) and low global warming potential (GWP); (ii) be chemically inert (i.e., non-corrosive) with construction materials and fluids, such as compressor lubricants; (iii) be chemically stable in order to prevent decomposition, e.g., during high temperature and pressure at the end of the compression process; (iv) be non-toxic for manufacturing and service personnel of refrigerants; (v) be non-flammable and without danger of explosion; (vi) provide high energetic efficiency; (vii) be inexpensive and easily available; (viii) have pressure-temperature properties that must fit with those of particular applications; and (ix) have relatively high enthalpy (latent heat) of vaporization

TABLE 8.2
Properties of Most Promising Natural Refrigerants for Geothermal Heat Pumps

			Critical point	
Refrigerant	**ODP[1]**	**GWP[2]**	**Temperature (°C)**	**Pressure (MPa)**
Propane (R-290)	0	3	96.7	4.30
Isobutane (R-600a)	0	3	134.7	3.677
Propylene (R-1270)	0	3	91.43	4.58
Carbon dioxide (CO_2, R-744)	0	1	30.98	7.474

Notes
[1] Based on ODP = 1 of CFC-11.
[2] Based on GWP = 1 of CFC-11 over 100 years.

because it directly impacts the compressor size and the system overall cost and performances.

For replacing HCFC-22 refrigerant, widely used in the past geothermal heat pump industry, considered fluids are HFC-134a (with lower vapor pressure than that of HCFC-22, resulting in a lower volumetric capacity), and HFC-410A, with 50% higher vapor pressure and better performance within ∓ 10% variation. As alternatives for replacing commonly used hydro-fluoro-carbon (HFC) low-temperature refrigerants (fluids with relatively low critical temperature) that have GWPs higher than 1,000 as, for example, HFC-134a and HFC-410a (Table 8.1), new fluids, such as low-GWP (HFO) compounds and others occurring naturally in nature, as hydrocarbons (propane and isobutane) and CO_2 (Table 8.2), can be used. Hydro-fluoro-olefins (HFOs) are synthetic refrigerants that promise to be a part of solution to the environmental problems. Among them, there are pure compound refrigerants as HFO-1234yf (2,3,3,3-tetrafluoropropene) and HFO-1234ze(E) (trans 1,3,3,3-tetrafluoropropene) as promising low-GWP substitutes to HFC-134a.

As long-term options for eliminating the influence of synthetic refrigerants on global climate change, natural refrigerants, containing neither chlorine nor fluorine molecules, such as hydrocarbons (e.g., propane R-290, propylene R-1270, and isobutane R-600a), and carbon dioxide (CO_2, R-744) (Table 8.2), must be used more and more in industrial heat pump systems as environmental-friendly alternatives. The ozone depletion potentials (ODPs) of these natural refrigerants are zero and most of them have close to zero global warming potentials (GWPs) in comparison to HFCs. Hydrocarbons, such as propane (R-290) and isobutane (R-600a) (Table 8.2), can be natural alternative refrigerants to halogenated substances, especially in direct expansion geothermal heat pump systems.

As refrigerants, the hydrocarbons present advantages such as the following: (i) available in nature at relatively low costs; (ii) have small molecular weights; (iii) have excellent thermodynamic and transport properties, similar to those of most HCFCs and HFCs; (iv) offer excellent thermal efficiency; (v) have zero ozone depletion potential ($ODP = 0$) and insignificant direct global warming potential

($GWP = 3$); (vi) are non-toxic and nearly odorless; (vii) mineral oils are miscible with hydrocarbons; (viii) compatible with most common materials; (ix) for large geothermal heat pump systems, regular monitoring and maintenance measures are required. Among the limitations of hydrocarbons which reduce their widespread use in the geothermal heat pump industry are as follows: (i) high flammability with relatively low limits; (ii) if part of the refrigerant circuit is located in an occupied spaces, the charge limit is 1.5 *kg* per system; (iii) if a geothermal heat pump is placed in an unoccupied space, the charge of hydrocarbons is limited to 5 *kg*; (iv) precautions are required to prevent ignition; (v) explosion-proof technologies are required for the development of geothermal heat pumps with hydrocarbons; (vi) today, geothermal heat pumps with hydrocarbons are available for capacities < 20 kW.

Because of its negligible impact on climate change (no ozone depletion potential and negligible direct global warming potential when used as a refrigerant in closed cycles), carbon dioxide (CO_2, R-744) is another promising natural refrigerant for use in thermosiphon ground-source heat pump systems. Carbon dioxide is a non-toxic fluid at low concentrations (but can be harmful in higher concentrations), non-flammable, colorless, odorless, non-corrosive, heavier than air, relatively inexpensive, and readily available (actually recovered from other industrial processes); compatible with most of common lubricants; and has excellent thermos-physical and transport properties (Lorentzen and Pettersen 1993).

The maximum allowable concentration of CO_2 for a workplace is 5,000 ppm or 0.5%. Immediate danger to health and life exists for CO_2-concentration over 4% (by volume) in air (40,000 ppm). Above 10% (by volume) in breathing air, CO_2 has a numbing effect and is immediately lethal above 30% (by volume).

The main components of geothermal heat pumps are: an electrical compressor that compress synthetic or natural refrigerants, an expansion device (usually, a thermostatic or electronic valve) and at least two heat exchangers (condenser and evaporator).

Real geothermal heat pumps work with pure refrigerants in the heating (Figure 8.4a) or cooling mode (Figure 8.4b) according to a non-reversible (real) subcritical Rankine reverse thermodynamic cycle. Figures 8.4a and 8.4b schematically represent a typical geothermal brine-to-air heat pump with vertical, indirect (secondary fluid), ground-coupled heat exchanger operating in heating and cooling modes, respectively. Figures 8.4c and 8.4d represent in diagrams T-s and in ln (p)-h, respectively, the real heat pump thermodynamic process of a reversible geothermal heat pump using a pure refrigerant.

In the heating mode, an anti-freeze mixture (brine) circulating through the vertical ground-coupled heat exchanger (Figure 8.4a) extracts heat from the ground/soil (acting as a heat source), while the geothermal heat pump's condenser, located inside the building, rejects it into the building's heating air acting as a heat sink medium. In the cooling mode (Figure 8.4b) the cycle is reversed, and the sensible and latent heat recovered from the building is rejected to the ground/soil. It can be seen that the entire real, subcritical thermodynamic process represented in T-s (Figure 8.4c) and in ln (p)-h (Figure 8.4d) diagrams occurs below the critical point of the refrigerant being used. Heat absorption occurs by evaporation of the refrigerant at low temperature and pressure, and heat rejection takes place by condensing the refrigerant at a higher pressure and temperature, but always below that of the refrigerant critical point.

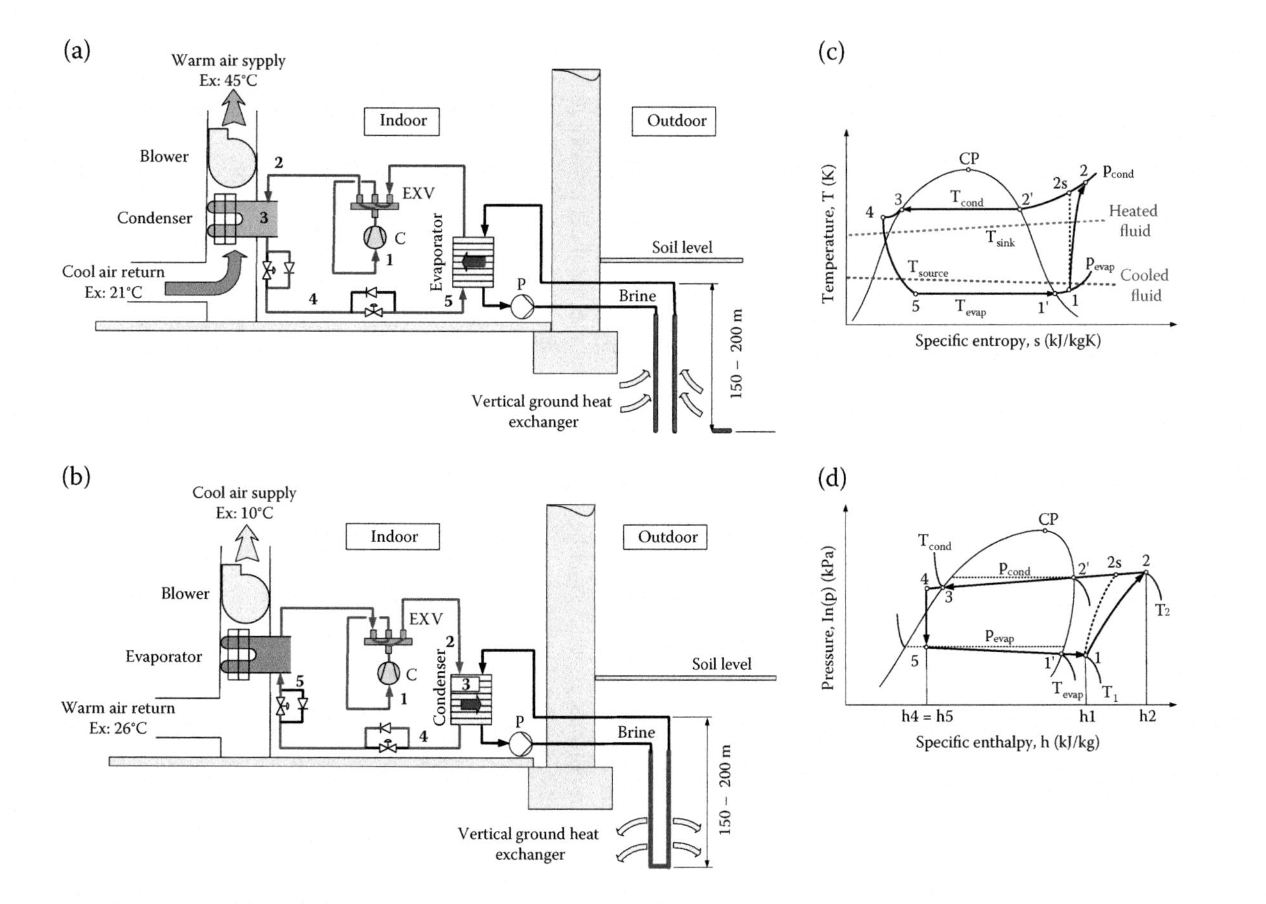

FIGURE 8.4 Reversible subcritical mechanical vapor compression geothermal heat pump with a pure refrigerant; (a) operating in heating mode; (b) operating in cooling mode; (c) thermodynamic cycle in T-s diagram; (d) thermodynamic cycle in (p)-h diagram; cond., condensing; CP, critical point; evap., evaporating; EXV, expansion valve; h, specific enthalpy, p, pressure; s, specific entropy; T, temperature.

Such a process consists of the following thermodynamic main processes: (i) the refrigerant leaving the evaporator as superheated vapor (state 1) is polytropicaly compressed up to the superheated state 2 (irreversible process 1–2); the electrical energy input is converted to shaft work to raise the pressure and temperature of the refrigerant; by increasing the vapor pressure, the condensing temperature is increased to a level higher than that of the heat source (T_{source}); at the end of the ideal (ideal, adiabatic) compression rocess ($1-2s$), the specific enthalpy (h_{2s}) of the superheated vapor is at pressure p_{cond}, but with the same entropy as the vapor before compression ($s_{2s} = s_1$); on the other hand, the final state 2 of the real (polytropic) compression process lies anywhere on the constant pressure line p_{cond}, on the right side of theoretical state point 2s; (ii) in the condenser, the superheated vapor is first desuperheated from state 2 to saturated state 2' and then undergoes a two-phase condensation at constant temperature (T_{cond}) and pressure (p_{cond}) (process 2'–3); (iii) before leaving the condenser, the saturated liquid is sub-cooled at constant pressure to a lower temperature (process 3–4) in order to reduce the risks of flashing within the expansion valve; sub-cooling of the liquid refrigerant ensures that only liquid and no vapor bubbles enter the expansion device, allowing for better flow control and smaller devices; also, if vapor bubbles were to enter the expansion device, there would be less liquid refrigerant available in the evaporator to produce cooling, while the compressor still has to compress that portion of the vapor that did not contribute to the cooling capacity; this represents a loss of capacity and efficiency; on the other hand, too much sub-cooling indicates that a considerable portion of the condenser volume is filled by single-phase liquid, thus, the area available for heat rejection from the condensing fluid is smaller than it could be; as a consequence, the saturation temperature is increased along with the saturation pressure and, as a result, the compressor work increases; the degree of sub-cooling at the condenser outlet is primarily determined by the amount of refrigerant charged to a heat pump and the setting of the expansion device primarily determines the degree of superheat at the evaporator outlet; (iv) during desuperheating ($2-2'$), condensing ($2'-3$), and sub-cooling ($3-4$) processes, heat is rejected by the condenser to the heat sink medium (gas or liquid). Within all these processes, the refrigerant exhibits pressure drops (see Figure 8.4c); (v) after the condenser, the liquefied sub-cooled refrigerant enters the expansion valve where an expansion process at constant enthalpy (i.e., drop in pressure accompanied by a drop in temperature) takes place in order to reduce the refrigerant pressure at a level corresponding to an evaporating temperature (T_{evap}) below to the heat source temperature (T_{source}) (process 4–5); the expansion valve controls the refrigerant flow into the evaporator in order to ensure its complete evaporation, maintain an optimum superheat in order to avoid the liquid refrigerant to enter the compressor and also avoid excessive superheat that may lead to overheating of the compressor; (vi) the refrigerant then enters the evaporator in a two-phase (state 5), absorbs (recover) heat from the heat source thermal carrier and undergoes change from liquid-vapor mixture to saturated vapor at constant pressure (p_{evap}) and temperature (T_{evap}); inside the evaporator, the saturated vapor (state 1') is slightly superheated up to state 1 (process $1'-1$) before entering the compressor; the superheat is desirable for two reasons.

There are pressure losses in the evaporator and condenser as well as in all connecting tubing that increase the power requirement for the cycle, and the heat transfer to and from the various components; these pressure losses are indicated by the slopes of lines representing the evaporation (5–1) desuperheating (2 – 2'), condensation (2' – 3), and sub-cooling (3 – 4) processes in the diagram ln (p)-h (Figure 8.4d); due to the pressure drop in the evaporator, the compression process (1 – 2) starts at a lower pressure and, further, into the superheated vapor region than in Figure 8.4d, which causes a double penalty for the compressor because a higher-pressure difference needs to be overcome by the compressor and the work increases due to the higher temperature of the superheated vapor inside the compressor suction line; the pressure drop of refrigerant in the suction line between the evaporator and compressor may also affect the compressor capacity because the pressure drop in the suction line increases the specific volume of the suction superheated vapor, so the mass rate of flow drops; however, this phenomenon is less penalizing at high evaporating temperatures; for example, the percentage reduction in refrigerating capacity for ammonia at two different evaporating temperatures and at drops in saturation temperature of 0.5°C and 1.0°C; at the lower saturated suction temperature caused by pressure drop in the suction line, the power requirement also decreases. However, if the compressor is already attempting to deliver full capacity, it cannot take advantage of this reduction in power to increase the refrigerating capacity. As can be seen in the T-s diagram (Figure 8.4c), the isenthalpic expansion process provides simultaneously two penalties depending on the specific heat capacity of the refrigerant, i.e., a small loss of work and loss of cooling capacity, resulting, when not extracted from the fluid, in an increased vapor quality; consequently, less liquid refrigerant is available to provide cooling, i.e., heat recovery in the evaporator; in other words, during the expansion process of refrigerants with a large specific heat capacity, a relatively large amount of refrigerant has to evaporate to cool the remaining liquid to the evaporator saturation temperature and less liquid is left to provide cooling capacity.

8.3.1 Energy Balance and Thermal Efficiency

Optimum design of evaporator and condenser heat exchangers depends on their respective thermal capacities that are function of the operating temperature ranges and on refrigerant flow rates. In heating and cooling modes, the real thermal capacities of geothermal heat pump evaporators and condensers can be calculated as functions of refrigerant flow and the refrigerant-side specific enthalpy (h, kJ/kg) changes (see Figures 8.4c and 8.4d):

$$\dot{Q}_{extracted}^{real} = \dot{m}_{refr}(h_1 - h_5) = \dot{Q}_{evap}^{real} \tag{8.21}$$

$$\dot{Q}_{supplied}^{real} = \dot{m}_{refr}(h_2 - h_4) = \dot{Q}_{cond}^{real} \tag{8.22}$$

where

$\dot{m}_{refr}$ is the refrigerant mass flow rate (kg/s).

h is the refrigerant mass enthalpy (J/kg).

The compressor real electrical power input (kW) in both heating and cooling modes can be determined by the following energy conservation expression:

$$\dot{W}^{real}_{required} = \dot{m}_{refr}(h_2 - h_1) = \dot{Q}^{real}_{cond} - \dot{Q}^{real}_{evap} \tag{8.23}$$

Geothermal heat pumps can be designed to provide all the required heat (in monovalent systems). However, because of the relatively high capital cost, it may be economic to design bivalent systems (i.e., system with geothermal heat pumps plus equipment for auxiliary heat supply) where the geothermal heat pump is designed to cover the base heating load, while an auxiliary device covers the additional peak demand.

The actual performance of the geothermal heat pump system is a function of the brine temperature produced by the ground-coupled heat exchanger (which will depend on the ground/soil temperature, pumping speed and the design of the ground-coupled heat exchanger) and the output temperature. Compared to conventional air-source heat pumps, the geothermal heat pumps can achieve higher instantaneous, overall, and seasonal coefficients of performance because the temperature of the ground, acting as a heat/sink source, is relatively constant compared to the ambient air temperatures.

The real heating coefficient of performance of subcritical mechanical vapor compression geothermal heat pumps ($COP^{real}_{heating}$) is defined as the ratio between the condenser useful (supplied) thermal power output ($\dot{Q}_{cond}$) and the electrical power input at both the compressor and blower ($\dot{W}_{compr+blower}$):

$$COP^{real}_{heating} = \frac{Usefull\ \ effect}{Net\ \ power\ \ input} = \frac{\dot{Q}^{real}_{cond}}{\dot{W}^{real}_{compr} + \dot{W}^{real}_{blower}} \approx 1 + \frac{\dot{Q}^{real}_{evap}}{\dot{W}^{real}_{compr} + \dot{W}^{real}_{blower}} < COP^{Carnot}_{heat\ pump} \tag{8.24}$$

where

$\dot{Q}^{real}_{evap}$, $\dot{Q}^{real}_{cond}$, and $\dot{W}^{real}_{compr}$ are defined by Equations 8.21, 8.22, and 8.23, respectively.

It can be noted that the air-to-air (or air-to-water) heat pumps operating in mild climates may achieve $\boldsymbol{COP^{real}_{heating}}$ up to 4.0, but at ambient temperatures below approximately $-\ 8°\boldsymbol{C}$, their $\boldsymbol{COP^{real}_{heating}}$ may drastically drop up to near 1. The average coefficient of performance of geothermal heat pump systems is at least 30% higher than that of comparable air-source heat pump to indicate how the real geothermal heat pump operation is close to the ideal optimum, the cycle efficiency can be expressed as the ratio between the ideal ($\boldsymbol{COP^{Carnot}_{heating}}$) and real ($\boldsymbol{COP^{real}_{heating}}$) coefficients of performance:

$$\eta = \frac{COP^{Carnot}_{heating}}{COP^{real}_{heating}} < 1 \tag{8.25}$$

It can be seen that $\mathrm{COP}^{\mathrm{real}}_{\mathrm{heat\ pump}}$ of a geothermal heat pump is greater than 1, i.e., a geothermal heat pump always supplies more heat than electrical energy consumed. In practice (i.e., real heat transfer systems with finite temperature differences across the heat exchangers, real working fluids, flow losses, and compressor efficiency), $COP^{real}_{heat\ pump}$ of mechanical vapor compression geothermal heat pumps varies between 40 and 60% of the maximum $COP^{Carnot}_{heat\ pump}$ and depends on the difference (temperature lift) between the condensation (or heat sink) and evaporation (or heat source) temperatures: the smaller the difference, the higher the $COP^{real}_{heat\ pump}$.

The ratio between sum of geothermal heat pump output and work input in an entire season are used to describe the seasonal heating performance factor (SHPF). In general, the SHPF for ground-source heat pump systems is equal and even higher than 3.5 compared to 2.5 for air-source heat pumps in heating mode.

In the cooling mode, the real energy efficiency ratio (EER) of the geothermal heat pump cycle is defined as follows (see Figures 8.4c and 8.4d):

$$EER^{cooling}_{subcritic} = \frac{Useful\ \ heat\ \ pump\ \ cooling\ \ effect}{Net\ \ energy\ \ input} = \frac{h_1 - h_5}{h_2 - h_1} \tag{8.26}$$

Similarly, in the cooling mode, the performance of a subcritical mechanical vapor compression geothermal heat pump can be described by the energy efficiency ratio:

$$\begin{aligned} \boldsymbol{COP^{real}_{cooling}} &= \frac{\boldsymbol{Useful\ \ heat\ \ pump\ \ cooling\ \ effect}}{\boldsymbol{Net\ \ energy\ \ input}} = \frac{\boldsymbol{h_1 - h_5}}{\boldsymbol{h_2 - h_1}} \\ &= \frac{\dot{Q}^{real}_{evap}}{\dot{W}^{real}_{compr} + \dot{W}^{real}_{blower}} \approx \frac{\dot{Q}^{real}_{evap}}{\dot{Q}^{real}_{cond} - \dot{Q}^{real}_{evap}} < COP^{Carnot}_{cooling} \end{aligned} \tag{8.26a}$$

where

$$\boldsymbol{COP^{Carnot}_{cooling}} = \frac{\boldsymbol{T_{sink}}}{\boldsymbol{T_{source} - T_{sink}}} \approx \frac{\boldsymbol{T_{evap}}}{\boldsymbol{T_{cond} - T_{evap}}} \tag{8.27}$$

represents the cooling coefficient of performance of the equivalent ideal Carnot cycle.

In United States and Canada, the heat pump's performance in the cooling mode is commonly described by instantaneous (EER) or seasonal (SEER) energy efficiency ratios, both expressed in ***Btu/Wh***.

The EER (energy efficiency ratio) is defined as the cooling energy (expressed in ***Btu***) provided by the geothermal heat pump during a given season (summer, year) divided by the total electrical energy consumed by the compressor(s), blower(s), and liquid circulation pump(s) during the same period of time, expressed in ***Wh*** (**1** ***Btu/h*** = **0.** 2931 ***W***). With existing ground-source heat pump technologies, EER ranges from 15 to 30.

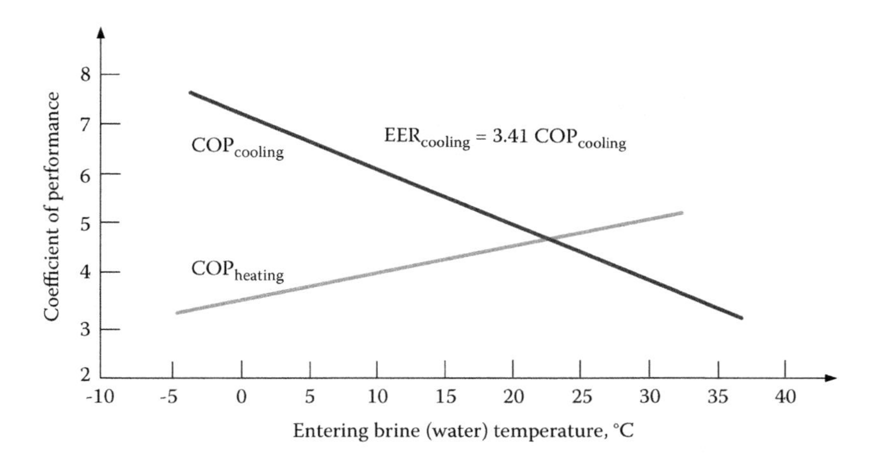

FIGURE 8.5 Typical variation tends of heat pumps' $COP_{heating}$ and $COP_{cooling}$($EER_{cooling}$) with the entering brine (or water) temperatures.

The cooling EER (energy efficiency ratio) is the ratio of Btu/h cooling capacity per watt of power input, expressed in Btu/h per watt.

The geothermal heat pumps' heating coefficients of performance (COP) and cooling EER depend on entering brine (or water) temperatures. In the cooling mode, the geothermal heat pump provides better performances as the entering brine (or water) temperature decreases, while, in the heating mode, the performances improves with higher entering fluid temperature (see Figure 8.5).

8.3.2 Exergy Analysis

Energy analysis, based on the principle of energy conservation, is often inadequate to identify other aspects of energy utilization with the majority of thermodynamic systems as, for example, the magnitudes and locations of the main sources of thermodynamic irreversibility and inefficiency, as well as the potentials for energy-efficiency improvements to minimize the entropy generation and overall energy consumption. Such a task can be accomplished by exergy analysis, a technique based on the second law of thermodynamics (Schmidt 2009).

8.3.2.1 Definitions

A working fluid can move from an initial state 1 (that isn't in thermodynamic equilibrium with the ambient medium) to a final (ambient) state *a* (where this equilibrium is attaint) by an adiabatic (1–2) ($ds = \delta q = 0$) process followed by an isothermal (2–a) ($dT = 0$) process (Vladea 1974; Kotas 1995) (Figure 8.6). A common definition of the surrounding (ambient) medium is (despite variations in climate across the world) a large thermal reservoir with a fixed temperature T_a = 25°C (or ≈ 298 K) and a fixed pressure p_a = 101.33 kPa. Two systems are in thermodynamic equilibrium when they are at the same temperature and the same pressure.

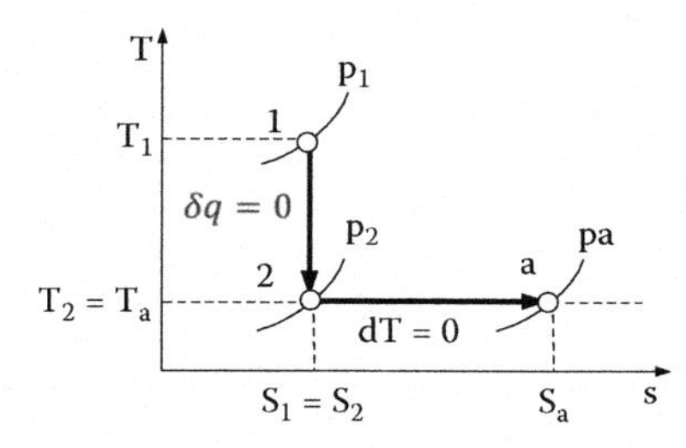

FIGURE 8.6 Adiabatic (1–2) and isotherm (2–a) successive processes to pass from state 1 to state *a* in thermodynamic equilibrium with the ambient medium represented in T-s diagram; a, ambient; p, pressure; s, specific (mass) entropy; T, temperature.

For an open system, the adiabatic (isentropic) process 1 − 2, produces the specific work:

$$l_{rev,1} = (u_1 + p_1 v_1) - (u_2 + p_2 v_2) + 1/2(w_1^2 - w_2^2) + g(z_1 - z_2) \quad (8.28)$$

where

$l_{rev,1}$ is the specific work produced in the adiabatic (isentropic) ($s_1 = s_2$) process 1 − 2 (J/kg).

u_1 is the specific (mass) internal energy of fluid at state 1 (J/kg or m^2/s^2).

u_2 is the specific (mass) internal energy of fluid at state 2 (J/kg or m^2/s^2).

p_1 is the fluid pressure at state 1 (Pa).

v_1 is the fluid specific volume at state 1 (m^3/kg).

w_1 is the fluid velocity at state 1 (m/s).

w_2 is the fluid velocity at state 2 (m/s).

g is the gravitational acceleration (m/s^2).

z_1 is the fluid elevation at state 1 (m).

z_2 is the fluid elevation at state 2 (m).

With such an isentropic transformation, the system attains the temperature $T_2 = T_a$; thus, it is in thermal equilibrium with the ambient medium, but not at pressure p_a. In order to realize the thermodynamic equilibrium with the ambient medium, it is required an additional process $2-a$ when is produced the specific work (with $w_a = 0$ because of thermodynamic equilibrium with the ambient medium):

$$l_{rev,2} = [(u_2 + p_2 v_2) - (u_a + p_a v_a) + 1/2 \cdot w_2^2 + g(z_2 - z_a)] - T_a(s_a - s_2) \quad (8.29)$$

where

$l_{rev,2}$ is the specific (mass) work produced in the isothermal ($T_2 = T_a$) process (J/kg).

s_a is the specific (mass) entropy of the fluid at state a (J/kgK).

s_2 is the specific (mass) entropy of the fluid at state 2 (J/kgK).

Because in both processes there are not internal energy variations ($du = 0$), the maximum reversible work produced is:

$$l_{rev,max} = l_{rev,1} + l_{rev,2} = (p_1 v_1 - p_a v_a) + 1/2 \cdot w_1^2 + g(z_1 - z_a) - T_a(s_a - s_1) \quad (8.30)$$

Or, because, in this case, the specific enthalpy is $h = pv$:

$$l_{rev,max} = l_{rev,1} + l_{rev,2} = (h_1 - h_a) + 1/2 \cdot w_1^2 + g(z_1 - z_a) - T_a(s_a - s_1) \quad (8.31)$$

where

$$T_a(s_a - s_1) = q_{isothermic} \quad (8.32)$$

is the heat exchanged during the isothermal process 2–a.

From Equation 8.31 it can be concluded that the kinetic ($1/2 \cdot w_1^2$) and potential [$g(z_1 - z_a)$] energies are totally transformable into mechanical work, while from the enthalpy difference ($h_1 - h_a$) only a part can be transformed into mechanical work, i.e., the part that remains after removing from it the term $q_{isothermic} = T_a(s_a - s_1)$. Since the process 1–2 is an open (non-cyclic) state transformation, the entropy difference ($s_a - s_1$) could be positive or negative; thus, the thermodynamic equilibrium with the ambient medium can be attained with both introduction or extraction of heat from/to the ambient medium. This is in conformity with second law of thermodynamics because the entropy decrease of the ambient medium (caused by the heat extraction) is compensated, within a reversible process, by an equivalent increase of entropy of the rest of the system.

In Equation 8.31, the quantity $e = (h_1 - h_a) - T_a(s_a - s_1)$ is called exergy (Rant 1956; Bosnjakovic 1963), and $T_a(s_a - s_1)$, anergy (Rant 1956). Thus, the enthalpy difference ($h_1 - h_a$) equals the sum of exergy (e) and anergy (a) (Figure 8.7). In other words, heat (as a form of energy) can be divided into one part that can produce work (its exergy), and one part that cannot be used to produce work, which is commonly referred to as anergy. For heat $q_{isothermic}$ exchanged at temperature T, the following decomposition can be made with reference to the ambient temperature T_0:

$$h_1 - h_a = \text{Energy} = \text{Exergy} + \text{Anergy} \quad (8.33)$$

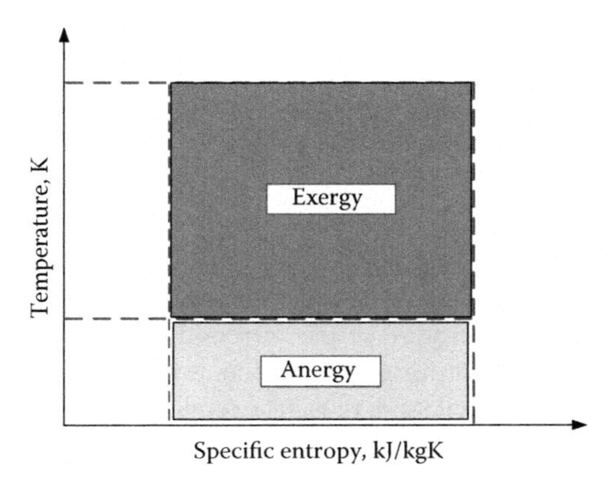

FIGURE 8.7 Decomposition of heat into exergy and anergy.

Based on the previous definitions, it can be noted that:

1. Any system not in equilibrium with its environment can produce work via an appropriate engine, and the maximum quantity of work equals the initial exergy value of the system. The surrounding medium at temperature T_a acts as a cold reservoir, while the heat at temperature T_l acts as a hot reservoir (Q_{hot}).
2. The maximal possible conversion of a given amount of heat to work depends on the temperature at which heat is available (T_l) and the temperature level at which the heat can be rejected is the temperature of the surrounding medium (T_a). The maximum work is obtained when the reversible process brings up the working fluid in thermodynamic equilibrium with the ambient medium since, at this state, it is not capable to produce more work, even in another processes. This means that exergy is the maximum work (also known as available or utilizable energy, and reversible work) that can be obtained via only reversible processes, from a hot reservoir at temperature T_l, when the cold sink is the surroundings at temperature T_a. After the system and surroundings reach the thermodynamic equilibrium, the exergy is zero.
3. Different energy forms have different quality (or different amounts of exergy) in the sense that they have different capabilities to generate work.
4. The amount of work obtained is always less than the heat input, and the ratio is termed the Carnot efficiency:

$$\eta_{Carnot} = 1 - T_{cold}/T_{hot} = 1 - T_a/T_l \quad (8.34)$$

   where
   T_l is the absolute temperature of the heat source (K).
   T_a is the absolute temperature of the surrounding (K).
   The exergy content of heat Q_{hot} is then the maximum amount of work that can be extracted:

$$Exergy = L_{max} = Q_{hot} \cdot \eta_{Carnot} = Q_{hot}\left(1 - \frac{T_{cold}}{T_{hot}}\right) = Q_{hot}(1 - T_a/T_l) \quad (8.35)$$

   where
   Exergy is the total exergy (J).
   L_{max} is the maximum (total) work (J).
   T_l is the absolute temperature of the heat source (K).
   T_a is the absolute temperature of the surrounding (K).
   Equation 8.35 shows that for a given amount of heat (Q_{hot}), the amount of work that can be produced (and, thus, its exergy or energy quality) increases with temperature, which is a quality parameter for heat.
5. There is a strong link between exergy and entropy since entropy production is equivalent to exergy loss which, in turns, is equivalent to lost work. While

exergy is the ability to produce work, entropy is the system's inability to do work. This means that exergy is an indication of energy quality.

6. Unlike energy, exergy is conserved only during ideal processes and destroyed due to irreversibility in real processes. Energy is neither created nor destroyed during a process, changing from one form to another (first law of thermodynamics). In contrast, exergy is always destroyed when a process is irreversible. This destruction is proportional to the entropy increase of the system together with its surroundings (entropy production). The destroyed exergy is called anergy, and this corresponds to the waste heat. Exergy can be used to identify the main sources of irreversibility (i.e., exergy losses which lower the mechanical work to compensate the entropy increase) and to minimize the generation of entropy in processes where the transfer of energy takes place.
7. Exergy can't be conserved as can the energy because each irreversible process destroys the exergy by the dissipative work corresponding to the respective entropy increase.
8. Exergy is a combination property of a system and its environment because it depends on the state of both. In other words, exergy is the energy that is available to be used.

The energy analysis, based on the first law of thermodynamics, remains the most common method for thermal evaluation of vapor compression geothermal heat pumps, but, without distinguishing the quality of various energy forms. This means that to fully understand all important aspects of energy utilization and to indicate the possibilities for thermodynamic improvements, energy analysis alone is insufficient. Therefore, the exergy analysis based on both first and second laws of thermodynamics may contribute to evaluate the quality of different energy flows, identify the system components where inefficiencies occur, minimize the process irreversibility, and, finally, optimize the geothermal heat pump thermodynamic cycle (Kuzgunkaya and Hepbasli 2010; Stanek et al. 2019).

The real (irreversible) thermodynamic cycle of mechanical vapor compression geothermal heat pumps consists of irreversible processes producing entropy that result in increased power consumptions compared with the ideal (Carnot) cycle. This means that the rate of entropy generation is a measure of irreversibility and, thus, it is reasonable to identify processes that contribute the most to the entropy creation and try to minimize their negative influences.

In order to illustrate a typical exergy analysis of a closed-loop (indirect, secondary fluid) ground-source heat pump system provided with reversible brine-to-water mechanical vapor compression geothermal heat pumps and vertical borehole heat exchangers, the concept shown in Figures 8.4a-d can be considered. It can be seen that the real thermodynamic cycle of mechanical vapor compression geothermal heat pumps consists of the following processes: (i) polytropic compression (1–2), (ii) condensation (2–4), (iii) isenthalpic expansion (4–5), and (iv) vaporization (5–1).

During exergy analysis of geothermal heat pump thermodynamic cycles, assumptions as the following are generally used: (i) heat transfer processes and fluid flows are steady-state with negligible potential and kinetic energy effects and no chemical reactions; (ii) heat transfer and refrigerant pressure drops in the tubing connecting the main geothermal heat pump components are neglected; (iv) compressor mechanical and compressor motor electrical efficiencies can be considered 80% and 70%, respectively; (v) circulating pump mechanical and circulating pump motor electrical efficiencies can be 90% and 86%, respectively; and (vi) fan mechanical and fan motor electrical efficiencies can be taken 40% and 70%, respectively.

The exergy losses in the main components of mechanical vapor compression geothermal heat pumps can be expressed as follows:
Compressor:

$$Ex_{compr} = T_a(s_2 - s_1) \tag{8.36}$$

Condenser:

$$Ex_{cond} = h_2 - h_4 - T_a(s_2 - s_4) \tag{8.37}$$

Expansion valve:

$$Ex_{exp} = T_a(s_4 - s_5) \tag{8.38}$$

Evaporator:

$$Ex_{evap} = h_5 - h_1 - T_a(s_5 - s_5) \tag{8.39}$$

The specific exergy (J/kg) of a refrigerant is:

$$ex = (h - h_a) - T_a(s - s_a) \tag{8.40}$$

where

h is mass enthalpy at a given state (J/kg).

s is the mass entropy at the same given state (J/kg).

h_a and s_a are the mass enthalpy and mass entropy at the ambient conditions.

T_a is the ambient temperature (C).

The exergy rates (W) of a refrigerant flowing through the geothermal heat pump components is determined as:

$$\dot{E}x = \dot{m} \cdot ex \tag{8.41}$$

where

$\dot{m}$ is the refrigerant mass flow rate (kg/s).

ex is the refrigerant specific (mass) exergy (J/kg).

The exergetic efficiency of a geothermal heat pump is expressed as the ratio of total exergy output to total exergy input:

$$\eta_{ex} = \dot{E}x_{output}/\dot{E}x_{output} = [\dot{Q}_{cond}(1 - T_a/T_{cond})]/\dot{W}_{compressor} \tag{8.42}$$

where "output" refers to "net output" and "input" refers to "driving input."

The exergy efficiency is defined as the ratio of exergy output rate (or useful exergy being produced by the system) to exergy input rate:

$$\eta_{ex} = (Useful\ \ exergy\ \ output)/(Total\ \ exergy\ \ input) \tag{8.43}$$

When applying the exergy efficiency defined by Equation 8.43 to the case of ground-source heat pump systems, the exergy efficiency is:

$$\varepsilon_{GSHP,\ system} = \frac{\dot{E}x_{in,\ cond} - \dot{E}x_{out,\ cond}}{\dot{W}_{comp,\ elec} + \dot{W}_{pump,\ elec} + W_{fans,\ elec}} \tag{8.44}$$

where

$\dot{E}x_{in,\ cond}$ is the exergy input rate into condenser (W).

$\dot{E}x_{out,\ cond}$ is the exergy output rate from the condenser (W).

$\dot{W}_{comp,\ elec}$ is the compressor electrical power input (W).

$\dot{W}_{pump,\ elec}$ is the brine circulation pump electrical power input (W).

$\dot{W}_{fans,\ elec}$ is the is the fan(s) electrical power input (W).

REFERENCES

ARI. 1998. ARI (Air-Conditioning and Refrigeration Institute) 330–98 Standard – *Ground Source Water-Source Closed-Loop Heat Pumps* (https://webstore.ansi.org/standards/ari/ari33098. Accessed August 6, 2019).

Bosnjakovic, F. 1963. Reference states of the exergy in a reacting system. *Forschung Im Ingenieurwesen Engineering Research* 20:151–152.

Clausius, R. 1867. *The Mechanical Theory of Heat*. Taylor and Francis, eBook, London, UK.

Kotas, T.J. 1995. *The Exergy Method of Thermal Plant Analysis*. Paragon Publishing, St. Paul, Minnesota.

Kuzgunkaya, E.H., A. Hepbasli. 2010. Exergetic performance assessment of a ground source heat pump drying system. In *Proceedings World Geothermal Congress*. Bali, Indonesia, 25–29 April.

Lorentzen, G., J. Pettersen. 1993. A new, efficient and environmentally benign system for car air-conditioning. *International Journal of Refrigeration* 16(1):4–12.

Rant, Z. 1956. Exergy, a new word for technical available work. *Forschung auf dem Gebiete des Ingenieurwesens* (in German) 22:36–37.

Reiss, M. 1996. Total equivalent warming impact of refrigerants. *Fuel and Energy Abstracts* 37(2):147–147(1).

Schmidt, D. 2009. Low exergy systems for high performance buildings and communities. *Energy and Buildings* 41:331–336.

Stanek, W., T. Simla, W. Gazda. 2019. Exergetic and thermo-ecological assessment of heat pump supported by electricity from renewable sources. *RenewableS Energy* 131:404–412.

Vladea, I. 1974. *Tratat de Termodinamica Tehnica si Transmiterea Caldurii* (in Romanian). Editura Didactica si Pedagogica, Bucuresti, Romania.

9 Refrigerant-to-Air Condensers

9.1 INTRODUCTION

The term *air-side* concerns the refrigerant-to-air condensers (that heat the building indoor air) (see Figure 9.1), to air-to-refrigerant evaporators that cool and dehumidify the building indoor air (see Figure 10.1) of geothermal heat pumps, all provided with blowers (fans) that circulate the air across the coil between the vertical fins and over the horizontal tubes thus intensifying the heat transfer. Refrigerant-to-air (air-cooled) condensers are heat exchangers where unmixed fluids (i.e., refrigerant and air) move roughly perpendicular (cross-flow) to each other (Figure 9.1a).

Generally, they are fin-and-tube heat exchangers that reject heat from the refrigerant circulating inside horizontal tubes to the building indoor air. When the refrigerant leaves the compressor(s) of geothermal heat pump(s), it enters the condenser as a superheated vapor and leaves as a sub-cooled liquid (see Figures 9.1a and 9.1b). The condenser can be separated into three sections according to the refrigerant thermodynamic states: superheated, saturated, and sub-cooled. The specific heat rejected can be found by evaluating the refrigerant enthalpies or temperatures at the inlets and outlets of each section. In more detailed words, the refrigerant (with the mass flow rate, $\dot{m}_{refr}$) enters the geothermal heat pump condenser coming from the compressor as superheated vapor at temperature $T_{refr,\ in}$ and specific mass enthalpy h_2. The superheated vapor is first cooled (de-superheated) to saturation ($T_{refr,sat}$) (process 2−a), condensed (with the specific heat $c_p \rightarrow \infty$) at constant temperature ($T_{refr,\ sat}$ = const.) and pressure, and then slightly sub-cooled up to temperature $T_{refr,\ out}$ and specific mass enthalpy h_3, transferring heat to the cooler airstream flowing over the outside finned tubes, while the temperature of the colder air increases from $T_{air,\ in}$ to $T_{air,\ out}$ (see Figures 9.1a and 9.1b). It can be seen that in the de-superheating and sub-cooling sections, both the refrigerant and heated air change in temperature, while during the phase change the temperature of the refrigerant is constant as it condenses. Although the majority of the heat transfer from the refrigerant to air occurs during the phase change section, up to 15–20% does occur during de-superheating and sub-cooling processes.

9.2 TYPICAL CONSTRUCTION

The refrigerant-to-air fin-and-tube condensers (with circular internally smooth or enhanced tubes and plate fins on the air-side) are generally oriented vertically (Figure 9.2).

DOI: 10.1201/9781003032540-9

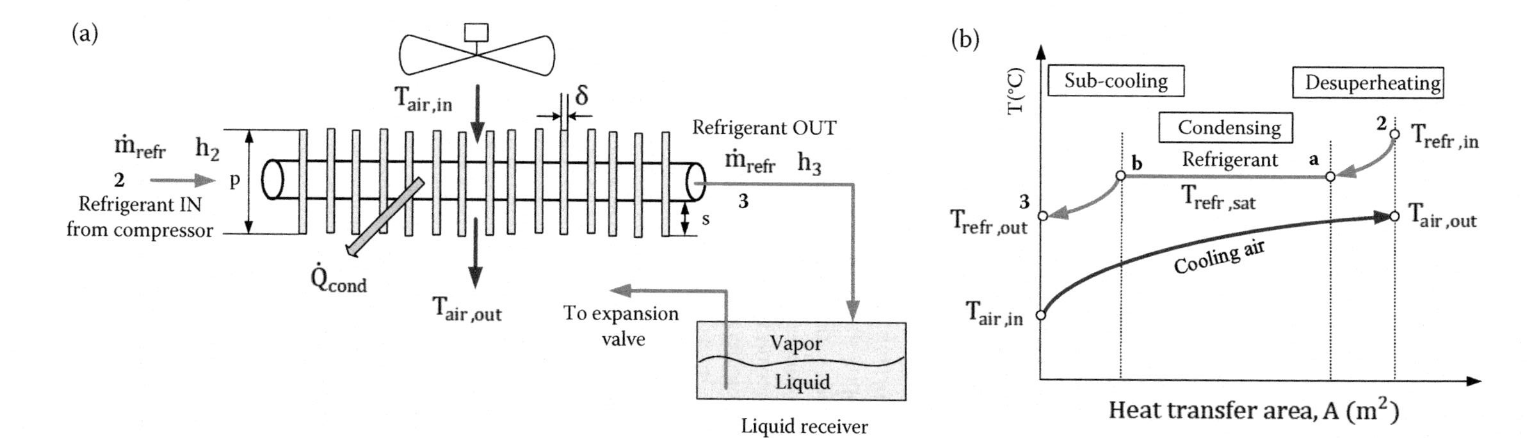

FIGURE 9.1 Cross-flow refrigerant-to-air condenser providing refrigerant vapor de-superheating and two-phase condensation, and liquid sub-cooling; (a) schematic; (b) temperature–heat exchange area diagram; h, specific enthalpy; $\dot{m}_{refr}$, refrigerant mass flow rate; $\dot{Q}_{cond}$, condenser heat transfer rate; T, temperature (Note: schematic not to scale).

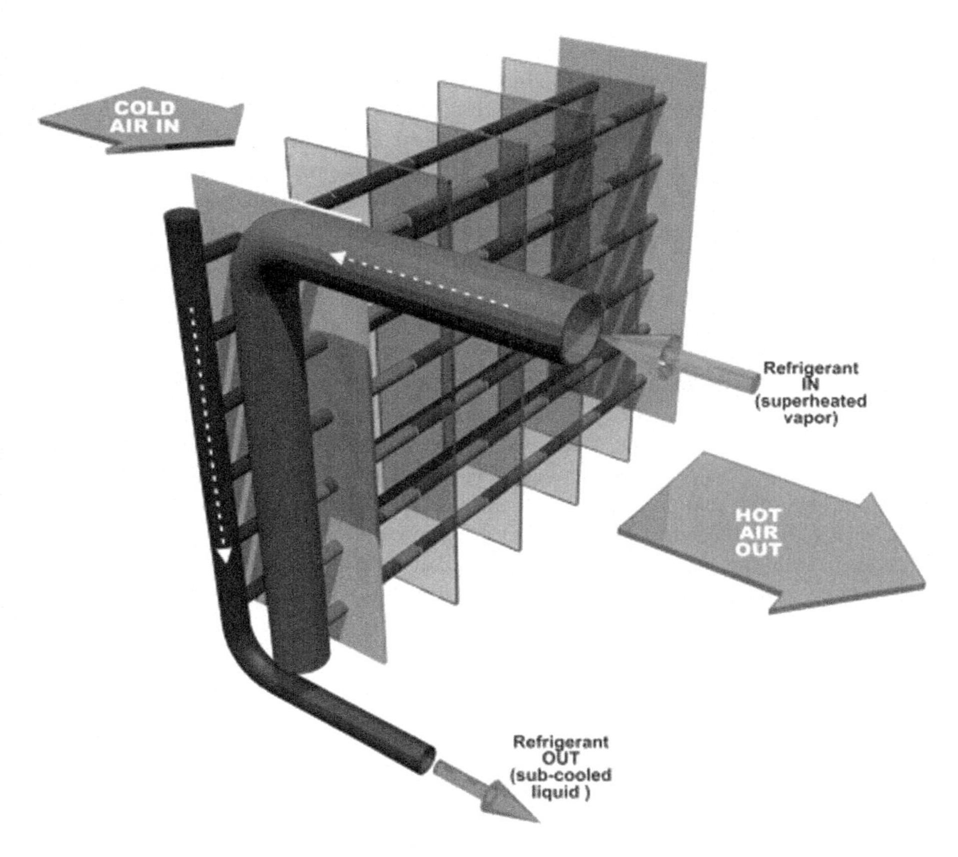

FIGURE 9.2 Vertical arrangement of an air-cooled refrigerant-to-air condenser (Notes: schematic not to scale; not all components shown).

Currently, fin-and-tube condensers are made of copper tubing with aluminum fins, both tubes and fins of aluminum, or combination of other special materials, as copper-nickel, carbon steel, or stainless steel. Standard fin thickness varies from 0.0065 to 0.016 inch (1 inch = 25.4 mm), depending on the materials. Fin spacing may include 4–20 fins per inch (1 inch = 25.4 mm), depending on the tubes' outside diameters. Usually, the core tubes penetrate directly into the headers to a depth no greater than 3.048 mm without the use of intermediate adapter tubes. The return bends must be of the same material as the core tubing, with a wall thickness no less than that of the core tube. In operation, the finned tubes and fans should be kept as clean as possible (i.e., free from dirt so that maximum airflow can take place) to maintain a maximum heat transfer efficiency. If the flow of air is blocked, the condenser thermal efficiency will be negatively affected and the geothermal heat pump compressor could fail in the medium or long term.

The refrigerant-to-air condenser coils are factory pressurized at a pressure of over 2, 758 kPa and a maximum 150°C temperature and then completely submerged in warm water containing special wetting and final cleaning agents for leak testing. At the end, the standard copper tube coils, for example, are air tested at a minimum of 2, 172 kPa pressure and then filled with nitrogen, assuring they remain leak free and clear of internal contamination prior factory- or in-field installation. The large-scale geothermal heat pump condensers drain the liquid refrigerant into high-pressure liquid receivers (see Figure 9.1a) which provide volume for storing all refrigerant charge. The piping arrangement allows the outlet pipe to not touch the bottom of the liquid receiver, thus avoiding passing solid contaminants into the system.

9.3 THERMAL DESIGN

There are two well-established methods available for the thermal heat exchanger design (McQuiston et al. 2005): (i) the log-mean temperature difference (LMTD) method; (ii) the effectiveness-number of transfer units ($\varepsilon - NTU$) approach, where ε is defined as the ratio between the actual heat transfer rate and the maximum amount that can be transferred, the $\varepsilon - NTU$ method avoids the iterations required by the *LMTD* for outlet temperature calculations.

After calculating the refrigeration-side and air-side heat transfer coefficients, the overall heat transfer coefficient may be calculated as follows:

$$\frac{1}{\bar{U}} = \frac{1}{h_{refr} * A_{refr}} + \frac{ln\,(d_{ext}/d_{int})}{2\pi * L * k_{wall}} + \frac{1}{\eta_{air} * h_{air} * A_{air}} \tag{9.1}$$

where

$\bar{U}$ is the overall heat transfer coefficient (W/m^2K).

h_{refr} is the refrigerant vaporization heat transfer coefficient (W/m^2K).

A_{refr} is the refrigerant-side heat transfer area (m^2).

d_{ext} is the external diameter of circular tubes (m).

d_{int} is the internal diameter of circular tubes (m).

L is the total length of circular tubes (m).

k_{wall} is the thermal conductivity of the tube wall material) (W/mK).

η_{air} is the air-side heat transfer efficiency (-).

h_{air} is the air-side heat transfer coefficient (W/m^2K).

A_{air} is the air-side heat transfer area (m^2).

It can be seen that, neglecting both vapor and liquid sensible heat transfer, the overall heat transfer coefficient (Equation 9.1) accounts for three thermal resistances: (i) vaporization (forced boiling) thermal resistance on the refrigerant side, (ii) conductive thermal resistance through the tube cylindrical wall, and (iii) convective thermal resistance on the air side.

The air-side fin heat transfer efficiency (that is, the ratio of actual heat flow of the fin to that which would be obtained with a fin of constant temperature uniformly equal to the base surface temperature; that is, one with infinite thermal conductivity) is defined as follows (see Figure 9.1a):

$$\eta_{air} = 1 - A_{fin}/A_{air}\,(1 - \eta_{fin}) \tag{9.2}$$

where

η_{air} is the air-side overall heat transfer efficiency (-).

A_{fin} is the fin heat transfer area (m^2).

A_{air} is the total air-side surface heat transfer area (m^2).

$\eta_{fin} = \frac{\tanh(ml)}{ml}$ is the fin efficiency (-).

m is a dimensionless function given by (1/m):

$$m = \sqrt{\frac{2 \cdot h_{air}}{k_{fin} \cdot \delta}} \left(1 + \frac{\delta}{s}\right)$$

h_{air} is the heat transfer coefficient (W/m^{2}K).

k_{fin} is the fin thermal conductivity (W/mK).

δ is the fin thickness (m).

s is the tube row depth (m).

$$l = p/2 - \delta$$

p is the fin height (m).

Based on the energy balance of the refrigerant-to-air condenser, its steady state thermal capacity (also called heat transfer rate or total heat rejection) can be calculated from both the refrigerant-side enthalpy changes and flow rate, and the air-side temperature change, mass flow rate, and specific heat as follows (see Figures 9.1a and 9.1b):

$$\dot{Q}_{cond} = \bar{\eta}_{cond} * \dot{m}_{refr} * (h_2 - h_3) = \dot{m}_{air} * \bar{c}_{p,air} * (T_{air,out} - T_{air,in}) \quad (9.3)$$

where

$\dot{Q}_{cond}$ is the refrigerant-to-air condenser thermal capacity (or heat transfer rate) (kW).

$\bar{\eta}_{cond}$ is the overall thermal efficiency of the refrigerant-to-air condenser (-).

$\dot{m}_{refr}$ is the refrigerant mass flow rate (kg/s).

h_2 is the specific enthalpy of refrigerant superheated vapor entering the condenser (J/kg).

h_4 is the specific enthalpy of refrigerant sub-cooled liquid leaving the condenser (J/kg).

$\dot{m}_{air}$ is the air mass flow rate (kg/s).

$\bar{c}_{p,air}$ is the average isobaric specific heat of air (J/kg·K).

$T_{air,in}$ is the dry-bulb temperature of air leaving the condenser (°C).

$T_{air,out}$ is the dry-bulb temperature of air entering the condenser (°C).

Even temperature profiles are complex because of vapor de-superheating and liquid sub-cooling; it can be assumed that condensing temperature prevails throughout the condenser and, thus, the condenser operates with the constant temperature equal to the refrigerant saturation temperature. If the single-phase zones of the condenser (i.e., vapor de-superheating and liquid sub-cooling are neglected), the model reduces to a one-zone heat exchanger. In this case, an average condensing temperature ($\bar{T}_{cond}$) is defined as the weighted average of the actual temperatures occurring in the three zones (single-phase desuperheating, two-phase condensation, and single-phase sub-cooling).

In this case, the steady state thermal capacity (heat transfer rate) of the refrigerant-to-air condenser can be written as follows:

$$\dot{Q}_{cond} = \bar{U} * A * (LMTD) \tag{9.4}$$

where

$\dot{Q}_{cond}$ is the thermal capacity (heat transfer rate) of the refrigerant condenser (kW).

$\bar{U}$ is the overall heat transfer coefficient expressed by Equation 9.1 (W/m^2K).

A is the condenser heat transfer area (m^2).

$LMTD = \frac{\Delta T_{max} - \Delta T_{min}}{\ln(\Delta T_{max} / \Delta T_{min})} = \frac{T_{air,out} - T_{air,in}}{\ln(T_{cond} - T_{air,in} / T_{cond} - T_{air,out})}$ is the logarithmic mean temperature difference (°C).

where

$$\Delta T_{max} = T_{refr,sat} - T_{air,in}$$

$$\Delta T_{min} = T_{refr,sat} - T_{air,out}$$

$T_{refr,\ sat}$ is the condensing temperature (°C).

From Equation 9.4, the heat transfer surfaces of refrigerant-to-air condensers can be calculated using the following equation:

$$A = \dot{Q}_{cond} / \bar{U} \cdot (LMTD) \tag{9.5}$$

where

$\dot{Q}_{cond}$ is the thermal capacity of the designed heat exchanger (kW).

$LMTD$ is the logarithmic mean temperature difference (°C).

Manufacturers of refrigerant-to-air condensers provide performance data, allowing selecting of the appropriate heat exchanger for given applications at specific design parameters. By applying some fundamentals of heat transfer, it is possible to translate catalog data to non-design conditions. The strategy in extending catalog data to non-design conditions is usually to compute the *UA* value (the product of the overall heat transfer coefficient and the heat transfer area) and for situations where the *UA* remains essentially constant, apply the *UA* value to the new set of operating conditions.

REFERENCE

McQuiston, F.C., J.D. Parker, J.D. Spitler. 2005. *Heating, Ventilating, and Air Conditioning Analysis and Design*, 6th edition, John Wiley & Sons, Inc., Somerset, NJ.

10 Air-to-Refrigerant Evaporators

10.1 INTRODUCTION

Air-to-refrigerant evaporators used by geothermal heat pumps to recover sensible and latent heat recovery from the buildings' indoor air are finned-tube heat exchangers where unmixed fluids (moist air and refrigerants) move roughly perpendicular (cross-flow) to each other.

Because a refrigerant vaporization (convective flow boiling) process occurs with specific heat, $c_{p,refr} \rightarrow \infty$, it remains at a nearly uniform saturation temperature during vaporization ($T_{refr,sat}$ = const.), while the temperature of the hot moist air decreases from $T_{air,in}$ to $T_{air,out}$ (Figures 10.1a, 10.1b, and 10.1c). As can be seen in the Mollier diagram (Figure 10.1c), the moist air is first cooled (process 1−a) and then the water vapor is condensed (process a−2) on the external finned tubes of air-to-refrigerant evaporators.

10.2 CONSTRUCTION

The augmentation of heat transfer from the building cooled/dehumidified air to vaporizing refrigerant is a challenging task because it can reduce the size of geothermal heat pump evaporators in a significant manner (Incropera and DeWitt 2002).

Better thermal performances can be achieved by using geothermal heat pump evaporators with large face area, minimum depth and fin spacing, appropriate airflow rates to avoid the moisture freezing, and optimum refrigerant superheating across the coil. Air-to-refrigerant geothermal heat pump evaporators are compact, extended surface, and direct expansion heat exchangers. They are provided with dense arrays of finned tubes (Figures 10.1a) in order to achieve as large as possible heat transfer surface area per unit volume on the air side. The major components of air-to-refrigerant evaporators are tubes and tube sheets, fins, and drain pans. Fins are used on the air side to increase the heat transfer area because the thermal resistance associated with the airflow is too high compared with that of the refrigerant. The refrigerant circuitry also plays a major role in the design, construction, and operation of air-to-refrigerant geothermal heat pump evaporators. The performances of air-to-refrigerant evaporators are affected by factors as the type of refrigerant- and air-side heat transfer surfaces, fin spacing, tube pitch, depth row pitch, refrigerant circuitry, and air velocity distribution over the frontal surface of the heat exchanger (Domanski et al. 2004; Stoecker 1998).

DOI: 10.1201/9781003032540-10

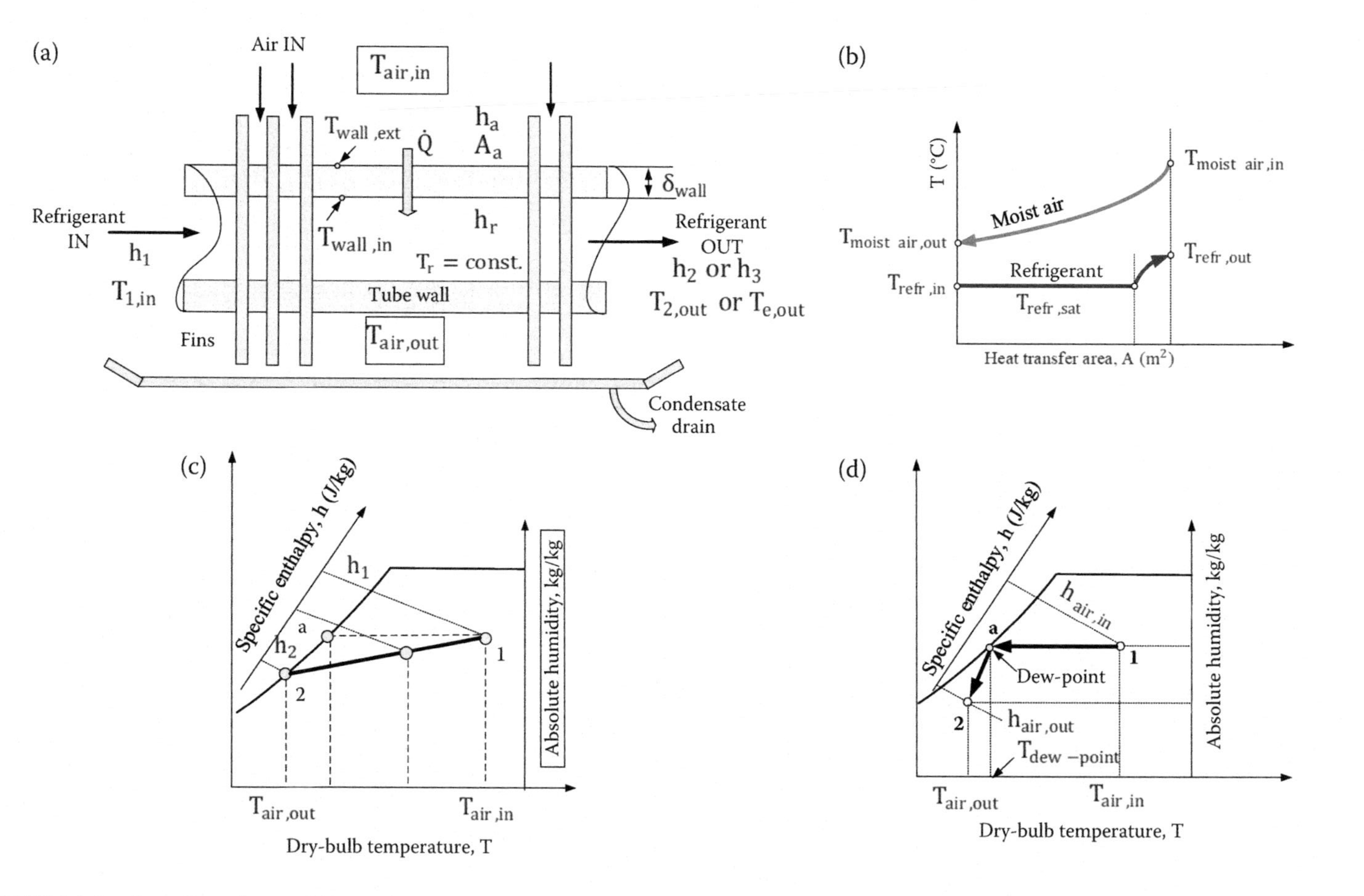

FIGURE 10.1 Cross-flow air-to-refrigerant finned-tube heat pump evaporator with phase change; (a) schematic; (b) temperature variations represented in a temperature–heat exchange area diagram; (c) and (d) moist air process represented in the Mollier diagram; h, specific enthalpy; in, inlet; out, outlet; T, temperature; refr, refrigerant; sat, saturation (Note: schematic not to scale).

10.2.1 Tubes

Under air cooling and dehumidifying conditions, the geometry of fin-and-tube air-to-refrigerant geothermal heat pump evaporators (e.g., coil configuration, number of tube rows, columns, modules and circuits, tube outer diameters, and longitudinal tube pitch) drastically affects the performances' heat transfer capability, pressure drops on both refrigerant and air sides, the refrigerant and air mass flow rates, and system overall energy consumption (Erdem 2015; Wang et al. 1999).

The tubes of air-to-refrigerant evaporators of geothermal heat pumps are generally circular and horizontally oriented. At each end of the coils, heavy plates support the tubes with holes through which the tubes pass (Figure 10.2). The pattern of these holes defines whether the tubes are in line or staggered. A coil with a staggered-fin pattern provides improved heat transfer but with a slight increase in air pressure drop (Stoecker 1998). For air-to-refrigerant geothermal heat pump evaporators with fewer number of tube rows, the surface heat transfer is more sensitive to wetting conditions at low airflow rates. Increasing in the number of rows did not always increase the heat transfer coefficients that depend on Reynolds number. However, even with a small number of tube rows, relatively high heat transfer coefficients have been obtained because of turbulent vortex pattern associated with the number of tube rows (Rich 1975). As the number of rows of

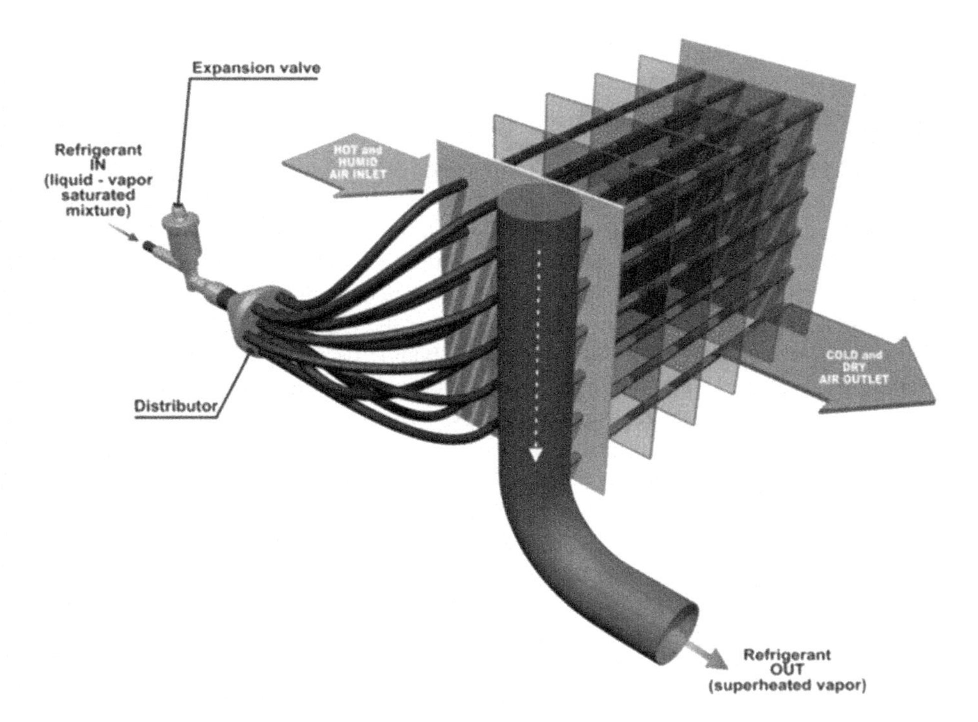

FIGURE 10.2 Schematic representation of a typical finned-tube geothermal air-to-refrigerant heat pump evaporator with one refrigerant circuit (Notes: schematic not to scale; not all components shown).

air-to-refrigerant geothermal heat pump evaporator increases, the heat transfer area also increases. In some applications, increasing the number of rows from 2 to 4 may reduce the energy consumption and cooling/dehumidification process time by 12% and 14%, respectively. However, in these cases, further increasing the number of rows does not significantly change the energy consumption or process time (Erdem 2015). Under both dry and wet surface conditions, increasing the number of tube rows decreases the value of the friction factor (Wang et al. 1999), notably when the Reynolds number is between 300 and 2,000 (Halici and Taymaz, 2005).

Also, decreasing the longitudinal tube pitch leads to an increase in heat transfer coefficient and pressure drop under both dry and wet surface conditions when the Reynolds number is greater than 500 (Halici et al. 2001; Halici and Taymaz 2005). On the other hand, as the outer diameters of tubes increase, the total cross-sectional air-side finned area decreases, which contributes to reduce the heat transfer in the evaporator (even the air velocity and the Reynolds number increase), as well as the total air-side pressure drop. As the number of columns increases, the cross-sectional area of the airflow increases and, therefore, the air velocity and Reynolds number decreases. However, because of the increase in the heat transfer area, the evaporators' heat transfer increases, as well as the air-side pressure drop. In some applications, increasing the number of columns from 2 to 4 may decrease the energy consumption and cooling/dehumidification process time by 8% and 4%, respectively. On the other hand, using evaporators with more than 4 columns does not have a significant effect on the energy consumption and/or drying time (Erdem 2015).

10.2.2 Refrigerant Distribution

Air-to-refrigerant geothermal heat pump evaporators use refrigerant expansion devices (e.g., thermostatic, electric, or electronic valves) in order to regulate the overall superheat at the evaporators' exit, commonly between 4°C and 7°C (Figure 10.2). This is achieved by controlling the flow rate of saturated refrigerant (liquid-vapor mixture) entering the coil as a function of temperature and pressure of the refrigerant vapor leaving the evaporator. After the expansion valves, refrigerant distributors are usually provided in order to blend the vapor/liquid mixture and evenly distribute it to each individual row of tubes, thus avoiding the degradation of the evaporator's heat transfer capacity. Large refrigerant distributors are provided inside with properly sized nozzles (or orifices) where the refrigerant velocity increases, providing a pressure drop and a "flash" temperature in the evaporator. The refrigerant tubes after the distributors are of same lengths and diameters, ensuring a nearly identical pressure drops between the distributor and each coil circuit. Without such distributors, the vapor-liquid stream may be not properly mixed and distributed, and then some tube circuits may receive mostly vapor, and others, mostly liquid. Inside air-to-refrigerant geothermal heat pump evaporators, the refrigerant flows through parallel refrigerant circuits, which are designed to optimize between the benefit of improved refrigerant heat transfer and the penalty of refrigerant pressure drop. Ideally, the tube circuits inside the evaporator coils are designed to be identical to each other for consistent pressure drops and flash temperatures. An extremely large number of refrigerant circuitry possibilities aim to provide maximum heat transfer rates and evaporator capacities. Optimized refrigerant circuit arrangements may reduce the needed heat transfer area by about 5% for the same evaporating capacity. The air-to-refrigerant evaporators of geothermal heat pumps perform optimally when the superheat at individual circuit exits matches the desired overall superheat in the exit manifold (Choi et al. 2003; Lee and Domanski 1997).

Refrigerant side maldistribution may lead to refrigerant dry-out in certain sections of the evaporator with significant reduction in cooling and dehumidifying capacity. However, the evaporator heat transfer rates and capacities are much more sensitive to moist air maldistribution (see Section 10.2.4) than to refrigerant maldistribution causing further degradation of the coil cooling and dehumidifying capacity. Consequently, designing optimized refrigerant circuitries is particularly difficult if the airflow is not uniformly distributed over the coil surface (Chwalowski et al. 1989; Domanski et al. 2004).

10.2.3 Fins

The use of (extended surface) fins is a way of increasing the effective heat transfer surface area without increasing the actual size of the air-to-refrigerant geothermal heat pump evaporators. The main role of plate (or circular) fins is thus to maximize the external, moist air-side heat transfer area. This is because, in the case of air-to-refrigerant heat pump evaporators, the major thermal resistance is on the air side (e.g., 0.01305 m^2K/W), while on the refrigerant side, the thermal resistance is much lower (e.g., 0.000833 m^2K/W, i.e., up to 20 times lower). This comparison suggests that, to increase the overall heat transfer coefficient, attention should be directed toward increasing the air-side heat transfer surface, not on that of the refrigerant side. That is because if the heat transfer coefficient on the refrigerant side is, for example, doubled, the overall heat transfer coefficient would increase by only 3%. Other methods to increase the overall heat transfer coefficient would be to increase the air-side average heat transfer coefficient ($\bar{h}_{moist\ air}$) by increasing the air velocity. However, by increasing the air velocity, additional electrical blower power is required that ultimately appears as a parasite cooling load. Generally, up to 10–20% of the heat removed by the evaporator from the air comes from the blower and its electrical motor. That means that the air velocity, which provides optimum air-side heat transfer coefficients, requires a reasonably sized blower and motor. On the other hand, according to one of fans' laws, the blower power requirement increases exponentially as the airflow rate increases. For example, the blower input electrical power increases more than 8 times as the airflow rate increases from 0.4 to 1 kg/s.

The air-to-refrigerant evaporators of geothermal heat pump (as shown in Figure 10.2) are equipped with fins formed from flat metal plates that are then punched on the tubes inserted in the holes. The fins must form good bonds to the tubes because, otherwise, there will be additional heat transfer resistance through air gaps. For that, when the tubes are in position, they are expanded against the collar of the fins either hydraulically or mechanically to provide good thermal contact with the fins (Rich 1973). As the fin pitch of geothermal heat pump evaporators increases, the air-side heat transfer coefficient, for a given velocity, increases only slightly. For 4-row coils with 7 mm diameter tubes, the effect of varying the fin pitch on the air-side heat transfer performance and friction characteristics is negligible, i.e., independent of fin pitch. For plain plate fin configurations ranging from 8 to 14 fins per inch (1 inch = 25.4 mm), the effect of longitudinal tube pitch on the air side is negligible for both the air-side heat transfer and pressure drop. However, the heat transfer performance increases with reduced fin pitch (Rich 1973; Wang et al. 1999). For the air side and Reynolds numbers greater than 1,000, the effect of fin pitch on the air-side friction pressure drop is negligibly small. As both fin depth and tube spacing increase, with all other variables constant, the air-side heat transfer coefficient decreases. However, by reducing the tube spacing and the tube diameter, an increase in the air-side heat transfer coefficient occurs (Shepherd 1956).

10.2.4 Air Distribution

Evaporator air-side velocities may vary due to the geometry of the heat exchanger installation, nearness of the blower, blockage of the air filter, and other factors. Non-uniform airflow can cause some circuits to have excessive superheat while others may remain at two phases at the evaporator exit. In such situations, some circuits inefficiently use the coil area when transferring heat with superheated vapor instead of two-phase refrigerant (Choi et al. 2003).

Airflow maldistribution (together with the refrigerant inside tubes) could lead to significant loss in the heat transfer capacity of the compact evaporators (Fagan 1980; Kirby et al. 1998; Kandlikar 1990a, 1990b). The evaporator capacity degradation due to both non-uniform refrigerant and airflow distribution can be as much as 30%.

10.2.5 Condense Draining

Water condensed from the building's indoor warm and moist air either forms a thin layer or appears as discrete droplets on the finned heat exchanger surfaces. The water film or droplets are then swept from the surface due to gravity and/or drag forces. The condensed water is sometimes entrained in the airstream because of carryover resulting in undesirable moisture blow-out, a situation highly undesirable that must be avoided. The effect of the retained water on the heat transfer rate is found to be small, while its effect on the pressure drop may be significant. Sometimes the air passages may become partly blocked, resulting in considerable increase in pressure drop. With the gradual reduction in the free flow area, the increased pressure drop causes a cyclic but sudden outburst of water from the blocked passages, a cyclic phenomenon. To avoid the problem, proper drainage path for condensed water must be provided (Kandlikar 1990a, 1990b). For that, the channels on the moist air side are specially designed to promote the drainage of the condensed water from the air-to-refrigerant evaporators, thereby reducing the blow-out of water droplets in the outlet airstream. Recent developments aiming to improve the water film drainage include the coating on the fin surface with thick coating of a hydrophilic film (≈1 micron). The primary effect of the coating is to reduce or even avoid the water hold-up in the evaporator, especially in compact heat exchangers with narrow passages. It may result in up to a 30% reduction in the air-side pressure drop and fan power requirements, and up to a 3% increase in the heat transfer rate. Air-to-refrigerant evaporators of a geothermal heat pump must be equipped with drain pans where the condensate is collected and drained to convenient destinations.

10.2.6 Materials

In the past, most geothermal heat pump evaporators with halocarbon refrigerants commonly used copper for both the fin and tube construction because it has a high thermal conductivity, and it was relatively inexpensive. With rising copper costs, the industry widely adopted aluminum for both tubes and plain fins. Aluminum tubes can be extruded with a wide range of passage sizes and tube expansion methods that provide good fin-tube thermal contact for a wide range of fin thicknesses. The degradation in thermal performance, due to the lower thermal conductivity of aluminum, was offset by using more of a less-expensive material.

Typical combinations of tube/fin materials are as follows (Stoecker 1998): (i) copper tube/aluminum fin or aluminum tube/aluminum fin for moisture condensers operating with halocarbon refrigerants, (ii) carbon steel tube/carbon steel fin for evaporator coils using ammonia inside the tubes, and (iv) stainless steel tube/stainless steel fin when special cleaning provisions are required on the air side.

The advances in fin-making technologies allow the use of very thin fins, but in spite of this advance, most geothermal heat pump evaporators currently rely on round-tube designs with plain fins, similar to flat plates. In residential and commercial geothermal heat pump evaporating coils with thin aluminum fins, the fin spacing may be 470 per meter.

10.2.7 Moisture Frosting

Under abnormal operating conditions, the evaporators of geothermal heat pumps can be exposed to moisture freezing. That means that the surfaces of evaporator coils may operate at temperatures below the air dew-point temperature and, thus, the moisture from the building indoor air may first condense to liquid water and then freeze to ice. In this case, frost or ice buildup may form on the evaporator external surface, partially and even totally blocking airflow through the coil while the geothermal heat pump may continue to run, drastically affecting the operating parameters and overall cooling and dehumidification performances.

The frost has detrimental effects as: (i) increased resistance to heat transfer, (ii) restriction of airflow, and (iii) irreversible damage to the heat pump's compressor.

As frost accumulates, the airflow rate decreases and the pressure drop through the evaporator increases if the fan input power is not progressively increased. The reduction of airflow rate and air velocity reduce the mean overall heat transfer coefficient, one the most serious effects. If ice formation is extreme, nearly all of the airflow across the coil is blocked and the geothermal heat pump does not cool and dehumidify the indoor air.

Moisture frost and/or ice buildup may occur on the evaporator coils due to: (i) improper refrigerant charge, (ii) uncontrolled refrigerant leakage, (iii) inadequate (maldistribution) airflow across the moisture condenser coil, (iv) incorrect sizing or inappropriate operation of thermostatic expansion valve(s), (v) other heat pump control (e.g., suction and differential pressure switches) defects, (vi) dirty air filters, (vii) coil-damaged fins, and (viii) air blower defects.

10.3 DESIGN

The heat exchanger design consists of selecting an appropriate heat exchanger type and determining the size, that is, the heat transfer surface area A, required to achieve the desired outlet temperature. For performing a heat exchanger design, both the LMTD method and the NTU approach can be used to obtain equivalent results. However, depending on the nature of the problem, the NTU approach may be easier to use (Incropera and DeWitt 2002). The use of the LMTD method is facilitated by knowledge of the air and refrigerant inlet and outlet temperatures. Typically, the fluid inlet temperatures and flow rates, as well as desired air or refrigerant outlet temperatures, are prescribed.

In finned-tube air-to-refrigerant evaporators of geothermal heat pumps operating with building warm and humid indoor air and refrigerants as a heat sink, some of the moisture present in the air condense over the outside fins and tube surfaces modifying the flow field, while inside the tubes, partial or total vaporization of the refrigerant may cause inhomogeneous distribution of the flow. For proper design of such phase-change heat exchangers, information of the heat transfer rates as functions of several variables of the system must be determined, usually from empiric (experimental) correlations (Kandlikar 1990a, 1990b; Pacheco-Vega et al. 2001). Air-to-refrigerant evaporators of geothermal heat pumps are relatively complex devices due to their various geometrical configurations, large number of variables, and operating conditions. Such refrigerant evaporators are most often selected via computer design and sizing programs that generally assume steady-state conditions with non-homogeneous, two-phase refrigerant flow inside the heat exchangers' tubes and outside air in cross-flow, and allow specifying the operating conditions, fin-and-tube parameters, refrigerant and airflows, and the type of fins for overall heat transfer enhancement (McQuiston et al. 2005).

10.3.1 Air Side

The air-to-refrigerant evaporator of a geothermal heat pump operates partially under external dry-surface conditions (i.e., during sensible heat recovery and refrigerant superheating) and, partially, under external wet surface conditions (i.e., during the moisture condensation resulting in condensing enthalpy recovery). On the air side of evaporators, both cooling and dehumidification processes consisting in heat and mass (water) transfers occur simultaneous. The state of the moist air (state 1 in Figure 10.1d) can move toward the saturation curve (100% relative humidity) of which average temperature is that of the wetted (tube and fins) surface. Typically, the air dry-bulb temperature and the absolute humidity both decrease as the air flows through the exchanger. Therefore, sensible and latent heat transfer occurs simultaneously (McQuiston et al. 2005).

When moist air at state 1 is cooled over a finned tube surface, several situations can occur (see Figure 10.1c) (i) if the air dew-point temperature (DEW) is less than the tube outer surface temperature (DEW $\leq T_{2,out}$), the process is totally dry; in the dry region, standard design methods, such as LMTD or ε – NTU, can be employed for designing the moisture condensers; (ii) if the air dew point temperature is equal to, or greater than the air temperature leaving the evaporator (DEW $\geq T_{3,out}$), the condensing process is totally wet; in other words, when the heat exchanger surface in contact with moist air is at a temperature below the air dew point, condensation of vapor will occur; and (iii) if the dew point temperature is between the final air temperature and the surface (DEW) temperature ($T_{2,out}$ > DEW > $T_{3,out}$), the condensing process is partially wet.

By neglecting the fouling and thermal efficiency of both smooth internal tube surface and finned-tube external surface, there are three series thermal resistances.

On the air side, the thermal resistance is defined as follows:

$$R_a = 1/(h_a \cdot A_a) \tag{10.1}$$

where

h_a is the heat transfer coefficient on the air side (W/m^{2}K).

A_a is the total heat transfer area (tube plus fins) on the dry air side (m^2).

The dominant thermal resistance is on the air side. To improve the overall heat transfer and reduction of the resistance in the air side, the surface of the tube is finned. On the air side of the heat exchanger, the heat, mass, and friction coefficients should be obtained from correlations based on test data, since the analogy method is unreliable (McQuiston et al. 2005).

10.3.2 Refrigerant-Side

The refrigerant side of mechanical vapor compression air-to-air heat pumps concerns a compressor (that moves the refrigerant), the evaporator (that vaporizes the refrigerant), the condenser (that condenses the refrigerant), the expansion device (that reduces the refrigerant pressures from high to low values), and sometimes the suction-line (liquid sub-cooling) internal heat exchanger. The knowledge of refrigerant-side flow patterns, pressure drops, and convective heat transfer coefficients and rates are much more complex and uncertain parameters than those of single-phase flows, and is essential to design phase-changing heat exchangers as the heat pump evaporators and condensers (Kandlikar 1990a, 1990b).

Boiling and convective (forced) evaporation are thermodynamic processes consisting of heat addition that converts a liquid into its vapor. They are the most common processes in heat pump evaporators (McQuiston et al. 2005). The boiling process occurs when vapor bubbles are formed at a heated solid surface without fluid flow, while the convective (forced) evaporation occurs when a liquid of a pure or mixed refrigerant is superheated at the surface, and vaporizes at liquid-vapor interfaces within a flowing fluid. If boiling and convective evaporation occur simultaneously in a flowing fluid, the coexisting heat transfer mechanisms are described as forced convective vaporization (or flow boiling) where the flow is due to a bulk motion of the fluid, as well as to buoyancy effects characterized by rapid changes from liquid to vapor in the flow direction (Incropera and DeWitt 2002).

The refrigerant coming from the expansion valve, as a low-quality mixture of saturated liquid and vapor enters the evaporator tubes, vaporizes during its passage through the coil and, finally, leaves it at a slightly superheated (2–5°C) state. The temperature of the refrigerant is constant during the phase change process when the heat capacity of the saturated refrigerant mixture is infinite. The process by which the refrigerant changes from a liquid to a vapor is a combination of boiling and evaporation, although a clear distinction between the two is usually not made. In boiling, vapor is produced at the solid surface and bubbles up through the liquid to the surrounding vapor, while in evaporation the liquid changes phase at the interface between the liquid and the vapor. No universal methods are available to correlate and predict the combined boiling and evaporating (i.e., vaporization) heat transfer coefficients. The actual fluid temperature has little effect on the heat transfer, and the proposed equations can all be put in a form that relates the heat transfer coefficient to the difference between the surface temperature and saturation temperature of the refrigerant at that pressure.

The thermal resistance on the refrigerant side is:

$$R_{refr} = 1/(h_{refr} \cdot A_{refr}) \tag{10.2}$$

where h_{refr} is the heat transfer coefficient on the refrigerant side (W/m^2K); in particular, for the refrigerant HCFC-22 vaporizing at temperatures from 4.4 to 26.7°C in a tube diameter of 8.7 mm and 2.4 m long, the refrigerant-side heat transfer coefficient can be determined with the following correlation:

$$Nu = (h_{refr} \cdot d_{int})/k_{refr} = 0.0225\ (Re^2 * K_f)^{0.375} \tag{10.3}$$

where

h_{refr} is the refrigerant-side heat transfer coefficient (W/m^2K).
d_{int} is the tube interior diameter (m).
k_{refr} is the refrigerant thermal conductivity (W/m·K).
Re is the Reynolds number (-).
$K_f = \Delta x \cdot h_{fg}/L \cdot g$ is the boiling number (-).
Δx is the difference of refrigerant quality (generally, between 0.2 and 1) (-).
h_{fg} is the water enthalpy (latent heat) of condensation (J/kg).
L is the tube section length (m).
g is the gravitational acceleration (m/s^2).
A_r is the internal area of the moisture condenser tube on the refrigerant side (m^2).

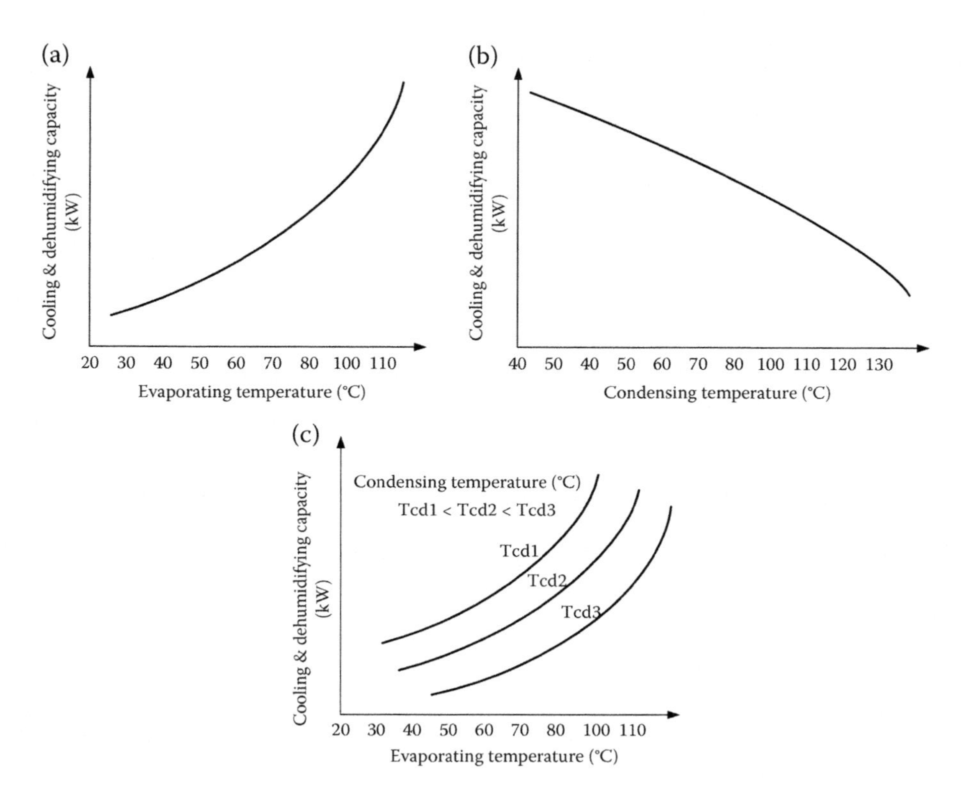

FIGURE 10.3 Cooling and dehumidifying capacity variation with evaporating and condensing temperatures.

10.3.3 Overall Heat Transfer Coefficient

The overall heat transfer coefficients (U) (also known as U – values) depend on sensible and latent heat transfers on both the air side (from hot, moist air to the external surface of the evaporator finned tubes) and refrigerant side (from the tube wall to the vaporizing refrigerant) of finned-tube moisture condensers (Figure 10.3).

The overall heat transfer coefficient ($\bar{U}$) (W/m^2K) can be derived by using the total thermal resistance as follows:

$$\bar{U} = 1/R_{tot} = \frac{1}{\frac{1}{h_{air} \cdot A_{air}} + \frac{ln(d_{ext}/d_{int})}{2\pi \cdot L \cdot k_{wall}} + \frac{1}{h_{refr} \cdot A_{refr}}} \tag{10.4}$$

The thermal resistance of the circular tube wall is:

$$R_{wall} = ln(d_{ext}/d_{int})/2\pi \cdot L \cdot k_{wall} \tag{10.5}$$

where

d_{ext} is the external diameter of circular tubes (m).

d_{int} is the internal diameter of circular tubes (m).

L is the condensing tube length (m).

k_{wall} is the tube wall thermal conductivity (W/m·K).

By applying the electrical analogy, the total thermal resistance (R_{tot}) (K·m^2/W) can be expressed as follows:

$$R_{tot} = R_{air} + R_{wall} + R_{refr} = \frac{1}{h_{air} * A_{air}} + \frac{ln\,(d_{ext}/d_{int})}{2\pi * L * k_{wall}} + \frac{1}{h_{refr} * A_{refr}} \tag{10.6}$$

During normal operation of air-to-refrigerant evaporators, surfaces are often subject to fouling by air impurities, rust formation, or other reactions between the air and the wall material. The subsequent deposition of a film or scale on the surface can greatly increase the resistance to heat transfer between the air and the evaporating refrigerant. This effect can be accounted for by introducing an additional thermal resistance, termed the *fouling factors* of which values depend on the operating temperature, fluid velocity, and length of service of the heat exchanger. Therefore, in practice, the total thermal resistances comprise the air-side, air-side fouling, tube wall, fin-to-tube contact, refrigerant-side fouling, and refrigerant-side thermal resistances. The fouling factors are variable during a heat exchanger operation (increasing from zero for a clean surface, as deposits accumulate on the surface) (Incropera and DeWitt 2002). Accordingly, with the inclusion of surface fouling and fin (extended surface) effects, the overall heat transfer coefficient must be expressed as (McQuiston and Parker 1994; Incropera and DeWitt 2002):

$$\frac{1}{U} = \frac{1}{\eta_{air} \cdot h_{air} \cdot A_{air}} + \frac{1}{\eta_{air} \cdot h_{a,f} \cdot A_{air}} + \frac{1}{R_c \cdot A_c} + \frac{1}{2\pi \cdot k_{wall} \cdot L} ln\,(d_{ext}/d_{int}) + \frac{1}{\eta_{refr} \cdot h_{r,f} \cdot A_{refr}} + \frac{1}{\eta_{refr} \cdot h_{refr} \cdot A_{refr}} \tag{10.7}$$

where

U is the overall heat transfer coefficient (W/m^{2}K).

η_{air} is the air-side overall surface efficiency of enhanced finned tubes (-); for industrial in-use coils, η_{air} generally range between 0.3 and 0.7.

h_{air} is the air-side heat transfer coefficient (W/m^{2}K).

A_{air} is the air-side heat transfer area (m^2).

$h_{a,f}$ is the air-side fouling factor (heat conductance) (W/m^{2}K); for industrial moisture condensers in operation for more than 5 years, the air-side fouling factor may be 2.84 kW/m^{2}K (Rohsenow et al. 1985).

R_c is the thermal contact conductance between the tube and fin surfaces in contact where a temperature drop exists; it is defined as the ratio between this temperature drop and the average heat transfer density across the interface (W/m^{2}K). The contact resistance between moisture condenser's tubes and fins (that can range between 10 and 16 kW/m^{2}K, with mean values of around 13 kW/m^{2}K) appears because the heat is exchanged across an interface where two surfaces are in imperfect contact.

A_c is the contact area between fins and tubes (m^2).
k_{wall} is tube wall thermal conductivity (W/m·K).
L is the total length of circular tubes (m).
d_{ext} is the external diameter of circular tubes (m).
d_{int} is the internal diameter of circular tubes (m).
η_{refr} is the refrigerant-side overall surface efficiency (-).
$h_{r,f}$ is the refrigerant-side fouling factor (conductance) (W/m^2K).
h_{refr} is the refrigerant-side heat transfer coefficient (W/m^2K).
A_{refr} is the refrigerant-side tube surface area (m^2).

Depending on operating conditions, the mean overall heat transfer coefficients can be determined by using more or less exact correlations for both h_{air} and h_{refr}.

10.3.4 Heat Transfer Rate

The geothermal heat pump cooling and dehumidifying capacity always decreases as the evaporating temperature drops (Figure 10.3a) and, also, as the condensing temperature increases (Figure 10.3b).

Compared to the influence of the evaporating temperature, each degree change in the condensing temperature affects the refrigerating capacity, however, to a lesser extent than a degree change in evaporating temperature (Figure 10.3b). The reason for this difference is that changes in the evaporating temperature exert a considerable effect on the specific volume entering the compressor, while the condensing temperature does not. At high evaporating temperatures, the increase in cooling and dehumidifying capacity is approximately 4%/°C and at low evaporating temperatures, near the maximum pressure ratios of reciprocating compressors, the decrease in refrigerating capacity is about 9%/°C. The comparison of the influences of evaporating and condensing temperatures on the refrigerating capacity also appears on the complete map of refrigeration capacity, as shown in Figure 10.3c, as controlled by the evaporating and condensing temperatures.

The design of heat pump evaporators requires data as (i) the air-side parameters, (ii) the air mass flow rate through the evaporator, and (iii) the average evaporation temperature.

The evaporator heat transfer rate can be determined based on the following conservation of energy equation (see Figure 10.3):

$$\dot{Q}_{evap} = \eta_{air}\dot{m}_{air}(h_{air,in} - h_{air,out}) = \dot{m}_{refr}(h_{refr,out} - h_{refr,in}) \tag{10.8}$$

where

$\dot{Q}_{evap}$ is the evaporator heat transfer rate (W).
η_{air} is the air-side heat transfer efficiency (-).
$\dot{m}_{air}$ is the air mass flow rate (kg/s).
$\dot{m}_{refr}$ is the refrigerant mass flow rate (kg/s).

h is the air and refrigerant enthalpy at the evaporator inlets and outlets (J/kg) (see Figure 10.3).

The heat transfer rate between the moist drying air and the evaporating refrigerant can be also expressed as:

$$\dot{Q}_{evap} \approx \bar{U} * A * LMTD \tag{10.9}$$

where

$\bar{U}$ is the average overall heat transfer coefficient (W/m^2K).

A is the average heat transfer area (m^2).

LMTD is the logarithmic mean temperature difference (C).

Based on Equation 10.9, the required heat transfer area can be calculated as follows:

$$A \approx \frac{\dot{Q}_{evap}}{\bar{U} * LMTD} \tag{10.10}$$

REFERENCES

Choi, J.M., W.V. Payne, P.A. Domanski. 2003. Effects of non-uniform refrigerant and air flow distributions on finned-tube evaporator performance. In Proceedings of the 21st International Congress of Refrigeration, Washington, DC, US.

Chwalowski, M., D.A. Didion, P.A. Domanski. 1989. Verification of evaporator computer models and analysis of performance of an evaporator coil. *ASHRAE Transactions* 95(1):1229–1236.

Domanski, P.A., D. Yashar, K.A. Kaufman, R.S. Michalski. 2004. An optimized design of finned-tube evaporators using the learnable evolution model. *HVAC&R Research* 10(2):201–211.

Erdem, S. 2015. The effects of fin-and-tube evaporator geometry on heat pump performance under dehumidifying conditions. *International Journal of Refrigeration* 57:35–45.

Fagan, T.M. 1980. The effects of air flow maldistributions on air-to-refrigerant heat exchanger performance. *ASHRAE Transactions* 86(2):699–713.

Halici, F., I. Taymaz, M. Gunduz. 2001. The effect of the number of tube rows on heat, mass and momentum transfer in flat-plate finned tube heat exchangers. *Energy* 26(11):963–972.

Halici, F., I. Taymaz. 2005. Experimental study of the air-side performance of tube row spacing in finned tube heat exchangers. *Heat and Mass Transfer* 42(9):817–822.

Incropera, F.P., D.P. DeWitt. 2002. *Fundamentals of Heat and Mass Transfer*, 5th edition, John Willey & Sons, Hoboken, NJ.

Kandlikar, S.G. 1990a. A general correlation for saturated two-phase flow boiling heat transfer inside horizontal and vertical tubes. *Journal of Heat Transfer Journal of Heat Transfer* 112:219–228.

Kandlikar, S.G. 1990b. *Thermal design theory for compact evaporators*. In R.K. Shah, A.D. Kraus, D. Metzger (Eds.), *Compact Heat Exchangers*, Hemisphere, New York, pp. 245–286.

Kirby, E.S., C.W. Bullard, W.E. Dunn. 1998 Effect of airflow non-uniformity on evaporator performance. *Transactions of the American Society of Heating, Refrigeration, and Air Conditioning Engineers* 104 (2):755–762.

Lee, J., P.A. Domanski. 1997. Impact of air and refrigerant maldistribution on the performance of finned-tube evaporators with R-22 and R-407C. *Building Environment Division of National Institute of Standards & Technology*. Project Number 665-54500, pp. 1–31, Gaithersburg, MD.

McQuiston, F.C., J.P. Parker. 1994. *Heating, Ventilating and Air Conditioning – Analysis and Design*, John Wiley & Sons, New York.

McQuiston, F.C., J.D. Parker, J.D. Spitler. 2005. *Heating, Ventilating, and Air Conditioning Analysis and Design*, 6th edition, John Wiley & Sons, Inc., Hoboken, NJ.

Pacheco-Vega, A., M. Sen, K.T. Yang, R.L. McClain. 2001. Neural network analysis of fin-tube refrigerating heat exchanger with limited experimental data. *International Journal of Heat and Mass Transfer* 44(2001):763–770.

Rich, D.G. 1973. The effect of fin spacing on the heat transfer and friction performance of multi-row, smooth plate fin tube heat exchangers. *ASHRAE Transactions* 79:135–145.

Rich, D.G. 1975. The effect of the number of tubes rows on heat transfer performance of smooth plate fin-and-tube heat exchangers. *ASHRAE Transactions* 81:307–317.

Rohsenow, W., J.P. Hartnette, E.N. Ganic. 1985. *Handbook of Heat Transfer Fundamentals*. McGraw-Hill, New York, 1413 pages.

Shepherd, D.G. 1956. Performance of one-row tube coils with thin-plate fins, low velocity forced convection. *Heating, Piping Air Cond* 28:137–144.

Stoecker, W.F. 1998. *Industrial Refrigeration Handbook*. McGraw-Hill, New York.

Wang, C.-C., Y.-J. Du, Y.-J. Chang , W.H. Tao. 1999. Airside performance of herringbone fin-and-tube heat exchangers in wet conditions. *The Canadian Journal of Chemical Engineering* 77(6):1225–1230.

11 Closed-Loop (Indirect, Secondary Fluid) Ground-Source Heat Pump Systems

11.1 INTRODUCTION

In cold and moderate climates, a ground-source heat pump system should meet between 60% and 70% of the building's peak heating or cooling demand, whichever is greater, leaving the residual 30–40% to the auxiliary heat source (e.g., electric or gas-fired boilers) or to additional heat rejection devices (e.g., cooling towers) (Ni et al. 2011; CAN/CSA 2013). Ground-source heat pump systems are generally designed for a minimum brine temperature of 0°C at the geothermal heat pumps inlets at maximum heating load conditions, and ground/soil around boreholes are allowed to freeze.

Inside large-scale buildings, several liquid (water, brine)-source geothermal heat pumps might be distributed on building indoor water (or brine) closed loops (see Figure 11.1), or installed in central mechanical rooms (see Figure 11.3). Both distributed and central systems are linked to geothermal heat/sink sources as the ground/soil (via vertical or horizontal ground-coupled heat exchangers), groundwater (via single or multiple wells), or surface water (via open or closed loops). Today, most of ground-coupled heat exchangers are inserted in individual boreholes or trenches, typically constructed of thermally fused plastic tubing in which a water-antifreeze solution (brine) is circulated, are generally connected together in parallel or series, and the individual run-outs returned to the building. Inside the building they are connected via buildings' internal water (or brine) closed loops, separated or not by intermediate heat exchangers, to several geothermal heat pumps (DOE 1995). In the distributed geothermal heat pump approach, a large number of small water-to-air geothermal heat pumps are distributed throughout the buildings. In central pumping stations, the building water (brine) closed loops are connected to the geothermal heat source/sink (vertical or horizontal ground-coupled heat exchangers, groundwater wells, or surface waters) with one large (or several small parallel single-speed or variable-speed) pumps (see Figure 11.3).

In both distributed and central systems, the location of main equipment should be based on: (i) sound criteria in the conditioned space and surrounding areas, (ii) space available, (iii) cost of installation, (iv) architectural considerations, (v) serviceability and maintenance requirements, (vi) availability of outdoor air, and (vii) building zoning.

DOI: 10.1201/9781003032540-11

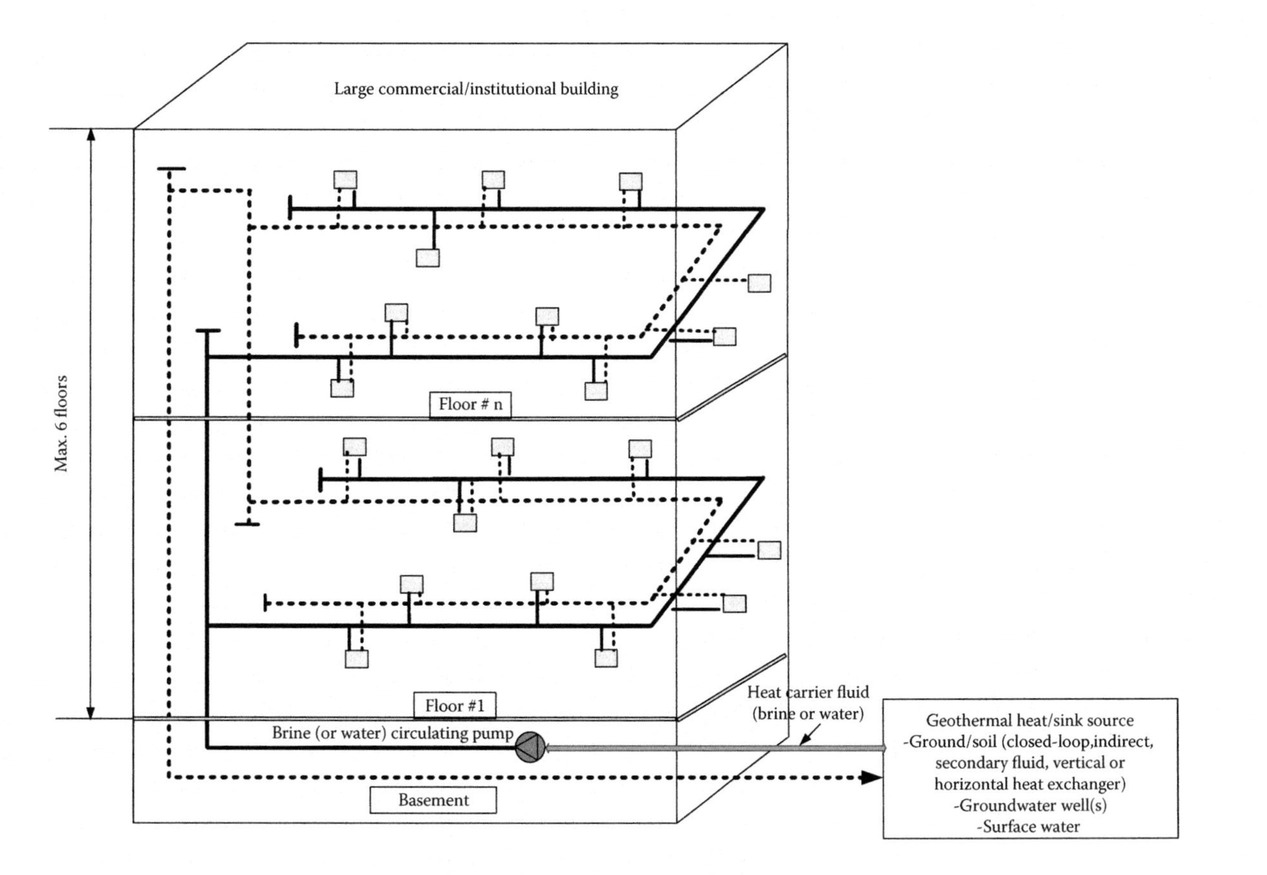

FIGURE 11.1 Schematic configuration of ground-source heat pump systems for large commercial/institutional buildings with distributed geothermal heat pumps; HP, geothermal heat pump; P, brine (water) circulating pump (Notes: schematic not to scale; not all components shown).

Among the most common applications of distributed and central ground-source heat pump systems are for: (i) office buildings that best utilize the energy-saving heat reclaim features of water-source heat pump systems; since most office buildings have significant core areas that are always cooling loads, heat can be transferred from those areas to perimeter zones when heating is required; the diversity of loads present in an office building will usually result in significantly reduced sizes of cooling towers and boilers, occasionally eliminating one or both of them entirely; (ii) hotels and motels have significant cooling load during daytime hours that can be used to offset nighttime heating loads; (iii) restaurants, swimming areas, kitchens, laundry room, and meeting rooms can be cooled during the day, and the stored heat used to warm guest rooms during nighttime occupied hours; (iv) apartments and condominiums can achieve individual tenant metering with a water source heat pump system; additionally, the first cost of a water-source heat pump system is lower than four pipe fan coil systems, and offers heating or cooling from any unit at any time; and (v) schools and universities have simultaneous periods of light and heavy loads throughout classrooms; additionally, laboratories, kitchens, and cafeterias can used to generate heat for use in perimeter classrooms.

11.2 BUILDING CLOSED-LOOPS WITH DISTRIBUTED GEOTHERMAL HEAT PUMPS

Building closed loops consist of water (or brine) circuits (supply and return) in which water (or brine) is continuously circulated throughout the building and distributes water-source geothermal heat pumps. The temperature inside the water (or brine) closed loops is maintained between a 0°C minimum and a maximum 35°C throughout the year. Distributed geothermal ground-source heat pumps can take advantage of the opportunity of internal heat recovery, most prevalent in buildings where year-round cooling is required in the core zones while the perimeter zones are more driven by the climate conditions. The supplemental heat equipment is energized when the loop temperature drops below 0°C and heat rejection equipment, when the loop temperature rises above 35°C. The closed-loop water (or brine) circuit serves as both heat source and heat sink for heat absorption and heat rejection by distributed geothermal heat pumps. Water (brine)-source geothermal heat pumps could be installed in a building's core as well as in peripheral thermal zones on reverse-return circuits. The internal (core) thermal zones require cooling only during occupied hours, or even year-round due to internal heat gains. Core thermal loads (typically, cooling loads) consist of heat gains from people, lights, equipment, and ventilation (if introduced in the core zone) in the building areas not defined as the perimeter. The heat rejected by the geothermal heat pump condensers operating in the cooling mode is recovered, instead of being rejected to the outside, and used as a heat source for geothermal heat pumps serving the perimeter spaces. This is a method that also greatly simplifies the system pressure balancing procedures by avoiding using many flow control devices. Consequently, the only balancing required will be at individual geothermal heat pumps to determine the flow rates across each distributed unit. Hence, energy can be conserved by transferring heat from warm to cold spaces in the building, whenever they coexist.

In distributed geothermal heat pump systems, individual liquid (water, brine)-source geothermal heat pumps provide heating and cooling to each thermal zone within the building. This is the case during winter months during which some units cool interior zones while others heat perimeter zones. The geothermal heat pumps are connected to a common interior building water or brine closed loops, which are used as heat and/or sink sources for the distributed geothermal heat pumps. Dedicated geothermal water (brine)-to-water heat pumps may also be connected to the loop to meet building water heating needs or to preheat or cool the outdoor fresh air intake.

The closed-loop water (brine)-source geothermal heat pump systems present features as: (i) low-cost and flexible hydronic systems (commonly using PVC pipes since the system sees only moderate temperatures and pressures) that can be utilized rather than ductwork; (ii) each unit works independently of the others, allowing heating and cooling in various areas to occur simultaneously from a single water (or brine) source; (iii) since each geothermal heat pump is able to heat or cool from a single heat/sink source, the system's layout is simplified and field-installed controls and labor minimized; (iv) water (or brine)-source geothermal heat pumps can be hidden in ceiling plenum areas; (v) because central chillers, boilers, and air handlers are eliminated, this will result in greatly reduced HVAC space requirements and maximized architectural design; (vi) failures of one or several geothermal heat pumps will affect only a small area of the building; (vii) various heat reclaim devices and solar heat collectors can be easily integrated to the system; (viii) geothermal heat pumps can be easily removed and replaced with standby units; and (ix) licensed equipment operator is not required.

Basic design steps in designing distributed ground-source heat pump systems consist of (DOE 1995): (i) determining reasonably accurate building load profile; (ii) estimating the building block and zone loads using accurate, efficient, fast, and inexpensive computerized load estimating programs (tools); selecting both perimeter and core reversible and cooling-only liquid-source geothermal heat pumps; (iii) selecting the geothermal heat pumps for each zone based on the zones' heating and/or cooling loads; (iv) generally, a single geothermal heat pump unit handles areas with similar load profiles; this means that zones from different exposures should never be combined (east and south, etc.) on a single unit; this will prevent overheating and overcooling, simplifying control arrangement and reducing equipment complications; and (v) since the heating requirements of perimeter zones rarely exceed the cooling requirements, they are most often selected based on cooling capacities.

Some of the general requirements for building water (or brine) closed-loops pipings are as follows (DOE 1995): (i) the geothermal heat pumps (available in both heating-only and heating and cooling configurations and range in heating capacity from 2.5 to 45 kW) selected for ground-source, groundwater, or surface water applications must take into account the anticipated entering brine (or water) temperature to meet the loads of residential or central and peripheral zones of commercial/institutional buildings. If this temperature is expected to be less than 10°C, extended range or low-temperature water-source geothermal heat pumps are required; (ii) for each geothermal heat pump, the heating/cooling capacity, fluid flow rate, airflow rate, external pressure, power input, and conditions of rating (i.e., voltage, air, and liquid temperatures) must be specified; (iii) geothermal heat pump distribution, materials, and equipment must comply with the standards of national and manufacturer specifications; ducted or plenum return horizontal and vertical geothermal heat pumps are generally located in ceiling spaces and closets (or mechanical rooms); (iv) in commercial/institutional buildings, console and horizontal geothermal heat pumps can be used in perimeter spaces and located in such a way to avoid noise complaints; (v) in commercial/institutional buildings' large core areas, where the heating and cooling loads are uniform, large water-to-air heat pumps (e.g., 25–35 kW nominal cooling capacity) may be located in central mechanical rooms; horizontal geothermal heat pumps shall be hung with hangers that have rubber vibration bushings and are located as shown in the drawings; (vi) vertical geothermal heat pumps shall be set on vibration absorbing pads; (vii) the locations of geothermal heat pumps in both residential and commercial/institutional buildings should consider service and maintenance access, and try to minimize the amount of connecting ductwork and piping; (viii) the runouts must be hydraulically balanced using circuit setters; (ix) individual geothermal heat pumps must be also hydraulically balanced by adjusting ball valves at their outlets; (x) limit liquid velocity to no greater than 1.22 m/s for pipes measuring less than 2 in; limit velocity to 2.44 m/s or less for larger pipe sizes; (xi) limit the head loss to 1.2192 m of water per 30.48 m of equivalent length; (xii) use PVC schedule 40 or 80 pipes of less than 203.2 mm in diameter where permitted by local codes; (xiii) use black steel or galvanized pipe longer than 101.6 mm if the plastic pipe is not allowed; use copper or black steel if less than 101.6 mm in wide; (xiv) closed-loop systems exceeding 90 kW (nominal cooling) capacity should have one diaphragm expansion tank; (xv) drain valves should be included at the base of each supply and return pipe riser; (xvi) manual air vents are needed at the top of each riser; (xvii) the electrical input power of brine circulation pump must not exceed 40–45 W per kW of system total nominal cooling capacity; (xviii) pressure and temperature ports as well as flexible hoses and ball valves on the return lines are required for each geothermal heat pump; (xix) flexible connection between the geothermal heat pump supply air discharge and ductwork are required; (xx) distribution piping must be insulated where minimum anticipated operating temperature is at 10°C or less; (xxi) distribution piping and all fittings should be insulated with impermeable vapor barriers to prevent condensation; (xxii) condensate collection and disposal system from geothermal heat pumps in accordance with the requirements of local regulations; (xxiii) condensate drains on geothermal heat pumps should be properly trapped and sloped with provisions for cleaning; (xxiv) air filters should be provided for all geothermal heat pump units; (xxv) geothermal heat pumps should be properly located to attenuate sound; and (xxvi) compressor sound shroud and discharge muffler may be needed to attenuate sound from geothermal heat pumps in critical locations.

As can be seen in Figure 11.1, one vertical reverse-return riser may serve multiple runouts that are horizontal two-pipe (supply and return) loops serving multiple geothermal heat pumps extracting or adding heat to the building closed loop, on multiple floors (up to six) for the entire building. The number of runouts depends on the zone occupancy and control requirements, as well as on economic considerations. In very large-scale buildings, multiple risers and runouts are recommended to limit the pipe size and first costs. In addition, each floor can have one perimeter runout.

In primary reverse loops with single, large circulating pumps (Figure 11.2a), the recommended flow rates vary from 0.16 to 0.19 L/s per 3.5 kW of installed nominal cooling capacity. When not correctly sized and controlled, the energy consumption of such brine (or water) circulation pumps may represent up to 25% of the total annual energy consumption of ground-source heat pump systems. However, up to 30% of the annual energy consumption could be saved by optimizing the geothermal fluid flow rates. An optimum sizing of circulating pumps corresponds to about 5.6 electric kW per 3.5 kW of nominal cooling capacity. High-efficiency motors also can save energy. For example, increasing by 4.5% the fluid circulation pump motor efficiency may result in 5% energy savings.

Pumping stations for primary reverse loops (where single or multiple parallel, variable-speed brine (water) circulating pumps could also offer the possibility of reducing pumping power during periods when building load is lower than the design values) distributed geothermal heat pump systems should be designed based on guidelines as follows (DOE 1995): (i) both the primary and standby circulation pumps, usually piped in parallel, are sized to the total fluid flow rate; (ii) each pump shall be tested to ensure its capability to provide the design flow rate at the design head; (iii) suction and discharge piping shall be adequately supported in order to not to impose loads that are too high on the pump casing; (iv) pumps shall be installed with suction reducers eccentric located at the pump suction, while discharge increasers shall be concentric with the pump discharge; (v) the suction side of the fluid circulation pumps should have an air separator with a strainer, a blow-down drain valve, and an air vent that prevents air introduced to the system during unit maintenance from getting into the ground-coupled heat exchanger; the air separation shall be at the pump suction, with its top connected to the makeup water line and a manual air vent; (vi) the fill (makeup) liquid line must have a back-flow preventer, a shut-off isolation valve, a pressure reducer, and, for larger systems, a diaphragm expansion tank located on the makeup water line between the pressure reducer and the top of the air separator; (vii) the discharge of the circulation fluid pumps should incorporate a shut-off valve, a check valve, a shut-off/isolation valve, a balancing valve/orifice with an internal flow-measuring orifice, with external ports for differential pressure measurement; on large-scale systems, a pressure safety valve should be added to prevent overpressurization; (viii) on closed-loop ground-source systems with antifreeze, the fill liquid line should be designed to be disconnected after the system is charged to prevent significant automatic makeup to dilute the antifreeze and expose the system to freeze damages; an alarm system should be provided to signal significant fluid leaks; (ix) determine maximum variation in the fluid volume due to temperature changes in the loop and size the diaphragm expansion tank accordingly; (x) verify if the pumping station equipment, pipes, and fittings can withstand the anticipated hydrostatic heads.

In large ground-source heat pump systems, most of the time, a number of geothermal heat pumps do not operate due to diversity of loads, occupancy, and other factors. For buildings with varying occupancy patterns and different internal

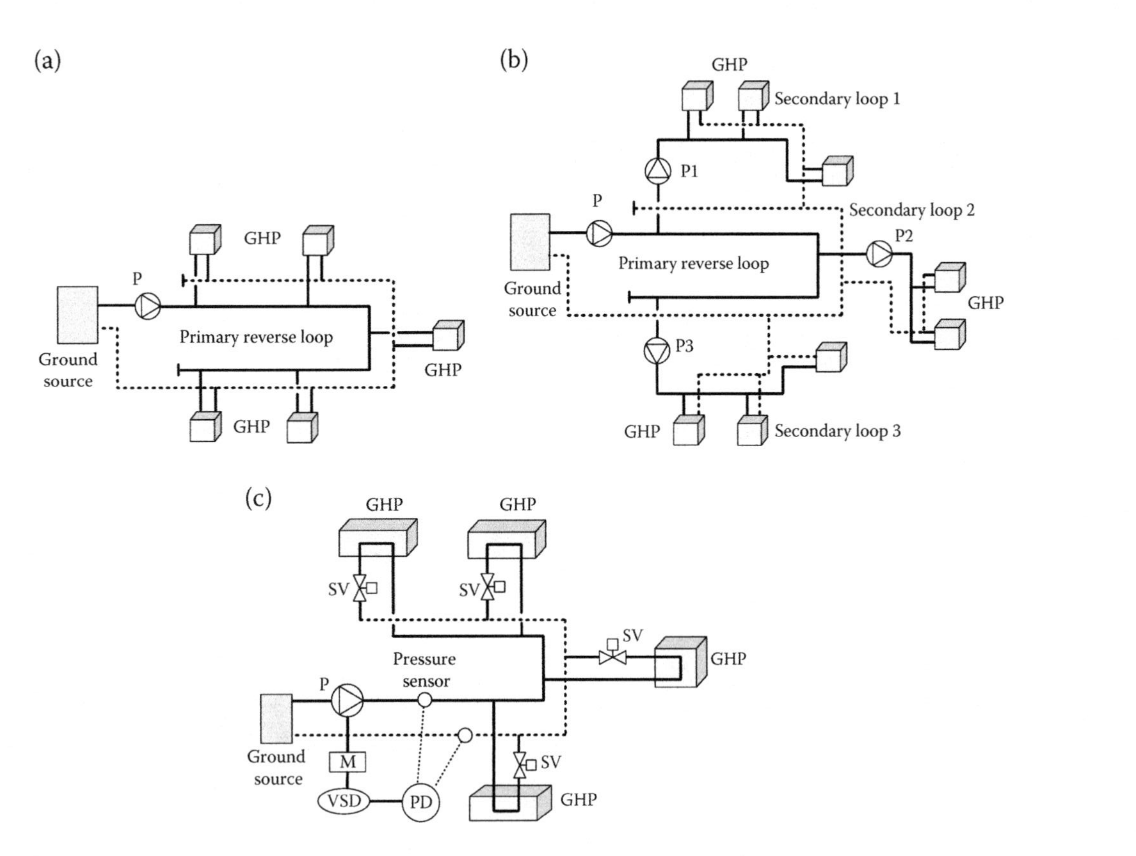

FIGURE 11.2 Ground-source heat pump systems with: (a) primary (central) constant speed pump; (b) primary and secondary constant speed pumps; (c) primary variable speed circulating pump. GHP, geothermal heat pump; P, circulating pump; SV, solenoid valve; VSD, variable speed drive; PD, pressure differential sensor.

loads, such as schools and hotels, buildings that require space heating and cooling in different zones at different times of the day, it could be advantageous to incorporate primary and several secondary pumping loops (Figure 11.2b). In this case, secondary circulating pumps provide brine flow in zones with similar load profiles.

In primary-secondary pumping arrangements (Figure 11.2b), secondary pumps are selected to provide the design constant flow rates through the geothermal heat pumps from the common pipe between the supply and return distribution mains under fixed pressure drops regardless of the thermal loads. Variable-speed pumps draw flow from the primary loop and distribute to the zone geothermal heat pumps. The speeds of the zone distribution pumps are determined by controllers measuring zone differential pressure across supply-return mains.

Another method to reduce the total flow rate consists of using variable-speed pumps on primary reverse loops because they offer substantial energy savings (Figure 11.2c). In this case, when several geothermal heat pumps are not operating, the geothermal fluid connections are closed with solenoid valves, and the circulating pump modulates to provide only the required flow rates to running geothermal heat pumps. The speeds of distributed circulation pumps are determined by controllers that measure the differential pressures across the supply-return mains. A variable-speed drive (VSD), two loop pressure sensors, and one solenoid valve (SV) at each geothermal heat pump are required. The pressure differential sensor (PD) continuously reads the difference between the supply and return sections of the loop and, then, activates the VSD in order to modulate the fluid flow rate. The solenoid valve installed on the return side of each geothermal heat pump opens when the compressor starts, and closes when the compressor shuts down. The compressors are not operating when the solenoid valves are closed. As the number of closed solenoid valves increases, the pump's speed is reduced to maintain a constant differential pressure. Such variable-speed systems may save up to 85% in pump energy consumption. However, the addition of solenoid valves, pressure sensors, and VSDs increase the initial costs. Therefore, sometimes a single-speed pump seems the better economical choice (see Figure 11.2a). For ground-source heat pump systems, some studies have looked at a variable-speed circulation pump control (Karlsson and Fahlen, 2007) and variable-speed compressor compared to fixed speed (Fahlen and Karlsson, 2005). For variable speed, the efficiency was shown to increase by 4–11% when operating at part load with less significant savings at a higher load.

Among the advantages of conventional building water (brine) closed-loop systems with distributed geothermal heat pumps versus conventional HVAC systems, the following can be mentioned (DOE 1995):

1. Smaller size buildings with improved aesthetics because most equipment is not visible, and cleaner roofs with less risks of water or brine leaks.
2. Buildings with increased architectural flexibility since water (brine)-to-air and water (brine)-to-water geothermal heat pump terminal units are available in many configurations (generally, covered by national certification programs) with longer technical life (25 years or even more since they are installed inside the buildings only when the tenants' or owners' occupancy is imminent).

3. Smaller mechanical rooms (e.g., 50% or less in new construction, which means more leasable space available) since no central heating and cooling mechanical rooms are required.
4. Minimal ductwork and shaft spaces that may increase the usable spaces and, also, reduce the number of fire dampers.
5. The geothermal heat pump systems are often installed in above dropped ceiling plenums, under windows, or in closets, which lower the building heights; in this case, the outside air ducts can be put near each unit to improve the building's indoor air quality.
6. The distributed geothermal heat pumps can be subdivided or expanded into new zones to fit building remodeling or additions.
7. Allow having instantaneous all-year heating and cooling availability in every core and peripheral thermal zone of the buildings.
8. When the building water (or brine) closed loops are thermally balanced, no net heat/sink sources are required from the geothermal energy sources.
9. Offer individual zone control being able to accommodate changes in location and sizes of geothermal heat pumps, and can be metered and billed individually by tenants for their own electrical consumption.
10. Provide permanent operating diversity and odd-hour flexibility since such systems could run only when needed.
11. Lower initial investments since water closed loops and accessories can be installed first, and geothermal heat pump units installed later as the construction of the building progresses.
12. Lower energy consumptions for geothermal heat pumps compared to those of conventional central HVAC built-up systems, mainly because the heat recovered from building interior zones can be transferred to the perimeter zones, while the geothermal heat pumps located in a building's unoccupied areas can be isolated and stopped down. In addition, the failure of a geothermal heat pump does not affect the operation of others.
13. Maintenance costs are low mainly because there is much less outdoor equipment, and no roof penetrations.
14. Relatively constant electrical demand year-around, which means better annual thermal load factors.
15. The system can be all-electric, which eliminates the need for multiple utility service entrances.
16. Skilled operators and piping insulation are not required.

The limitations of building water closed-loop heat pump systems are as follows (DOE/EE-0258. 2018. www.ecw.org, accessed January 14, 2020): (i) additional complexity compared to both conventional HVAC and ground-source heat pump–only systems, by integrating supplementary heat injection and/or auxiliary heat rejection devices; (ii) some additional noise could be generated since the geothermal heat pumps' compressors and fans are commonly located close to the occupied spaces; however, the noise can be minimized by placing the units away from the occupied space and ducting the supply air to the zone; accessibility to terminal units, particularly units located above the ceiling can be difficult;

(iii) water (or brine) loop circulating pump(s) must operate continuously with relatively high annual energy consumption and pumping costs; (iv) in buildings with infrequently used public spaces (e.g., hotels and apartment buildings), the core areas may not require many hours of cooling during the heating seasons; (v) require higher maintenance costs for the heat pumps' compressors and blowers (fans) located throughout the building; (vi) closed-loop water (or brine) piping as well as geothermal heat pumps are installed throughout the building with the possibility of water (brine) or refrigerant leaks; filter changing is required in thermal zones; (vii) increased internal electrical distribution network to serve each geothermal heat pumps; and (viii) require a separate ventilation system to supply fresh air to the building core and peripheral thermal zones.

To efficiently design the building internal water (brine) closed loops, rules such as the following should be adopted: (i) weigh advantages of central versus multiple loops; (ii) route and size piping system for low-pressure loss; (iii) provide on/off flow control through heat pumps and isolation valves; and (iv) specify material – indoor piping, insulation, antifreeze, and inhibitors.

11.3 CENTRAL GEOTHERMAL HEAT PUMPS

In central mechanical rooms, several geothermal heat pumps connected to a four-pipe system to supply hot and chilled water to the building's HVAC distributed systems can be installed (Figure 11.3).

Because of higher costs and, perhaps, higher-energy electrical consumptions, this concept typically does not result in the same level of energy efficiency as distributed geothermal heat pump systems.

11.4 MATERIALS

Because they don't need corrosion inhibitors, high-density thermally fused polyethylene pipes could be used for building interior water (brine) closed loops. These pipes use standard dimension ratios (SDRs) based on the outside diameter. An advantage of SDR ratings is the pressure rating is consistent, regardless of pipe diameter. For instance, SDR 11 is rated at 1,100 kPa and SDR 9 is rated at 1,380 kPa. However, disadvantages of these materials consist of high linear expansion coefficients, and restrictions due to local fire codes.

In central water (brine) pumping stations, the commonly used pipes are made of steel or copper. Where minimum anticipated operating temperatures are 10°C or less, all pipes and heat exchangers must be insulated to prevent vapor condensation. The insulation thicknesses depends on the pipe diameter, the type of insulation material, the lowest loop temperature anticipated, and the coincident dew point of the indoor air dry temperature.

Other materials are fittings (end caps, tees, reducing tees, elbows, etc.), isolation valves, drain valves, air vents at the top of each riser leg in high-rise buildings, pressure and temperature ports, flexible hoses, ball valves on the return lines for each geothermal heat pump, and in systems with installed capacities higher than 90 kW, diaphragm expansion tanks.

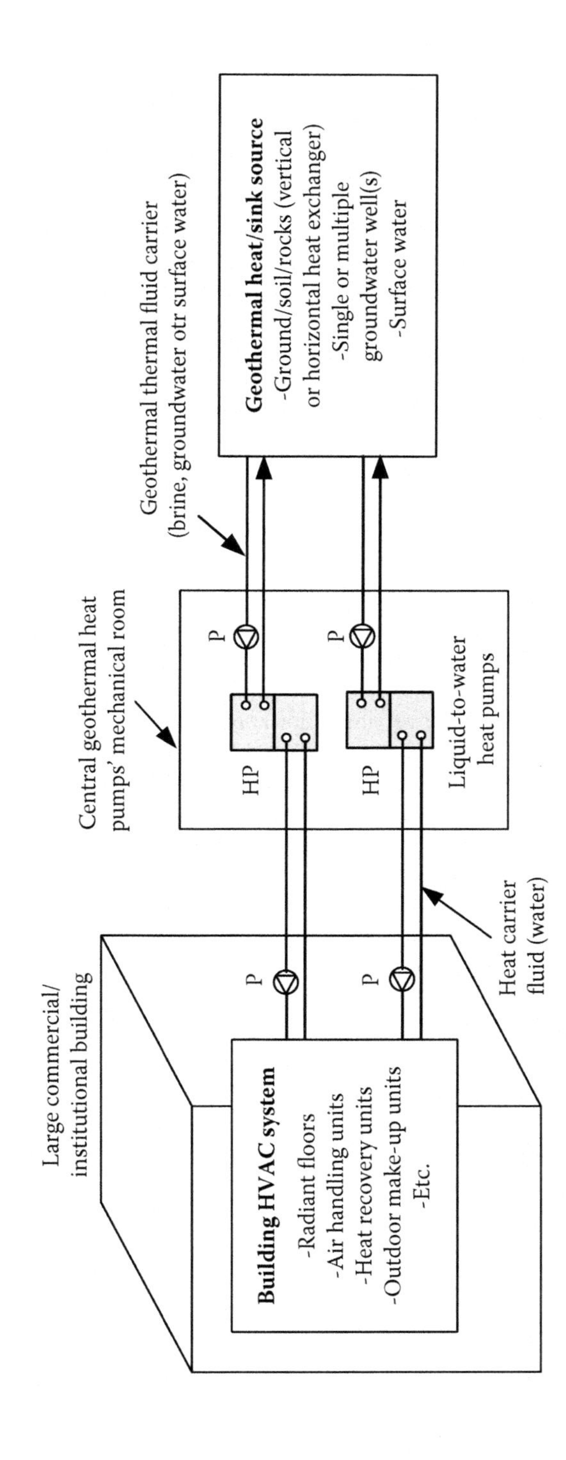

FIGURE 11.3 Schematic configuration of ground-source heat pump systems for large commercial/institutional buildings with central geothermal heat pumps' mechanical room. BP, brine circulation pump; WP, water or brine circulation pump.

11.5 BRINE AND WATER PUMPING

Usually closed-loop (indirect, secondary fluid) ground-coupled heat exchangers coupled with building internal fluid distribution networks use centrifugal circulating pumps in order to distribute and recirculate thermal carrier fluids, most commonly water/antifreeze mixtures (also called brines).

Compared with water, the brine density is slightly higher, the specific heat capacity lower, and the viscosity higher, with a tendency to increase significantly at decreasing temperatures. Therefore, the circulation pumps have to be sized for the lowest temperature that may occur in the system. Generally, systems relying on single central circulating pumps with variable-speed drives perform poorly. In larger systems with multiple distributed geothermal heat pumps, it may not be possible to closely control the flow rate to each geothermal heat pump. It is common to control the pump speed to maintain a constant pressure differential (Δp) across the geothermal heat pumps; in this case, increasing the Δp setting will increase the flow of all geothermal heat pumps.

11.5.1 CENTRIFUGAL PUMPS

Most centrifugal pumps are single-stage devices driven by electric motors with single suction impellers mounted on motor shaft extensions. Single-stage, single end-suction, frame-mounted circulating pumps and motors (flexible-coupled to the pump shafts), usually horizontally placed on common, rigid-base plate mountings requiring solid concrete pads, have single horizontal inlets and vertical discharges. Depending on the parts in contact with the liquid being pumped, centrifugal pumps are generally available with an impeller in bronze, a shaft sleeve in stainless steel or bronze, and iron casing and iron-fitted construction with mechanical seals.

In constant-speed centrifugal pumps, electric motors rotate the impellers at given rated speeds. Energy added to the fluid by centrifugal and rotational forces increases the fluid velocity that is converted into pressure energy. As most large-scale buildings operate during a substantial number of low part-load hours, a significant amount of energy can be saved by using a variable speed (e.g., with operating speeds of motors between 600 and 3,600 rotations per minute) in order to are able to move the fluids between, let's say 30%, and the maximum required flow rate.

Circulating pumps should be selected based on parameters and criteria such as: (i) nature of pumping fluid; (ii) minimum and maximum fluid design flow rate(s) that are able to ensure turbulent flow through the ground-coupled heat exchangers; (iii) pressure drop required for the most hydraulically resistant pipe circuit; (iv) pump pressure heads at maximum and minimum flow rates; (v) type of control valve (two- or three-way); (vi) constant (continuous) or variable fluid flows; (vii) pump environment; (viii) number of running and standby pumps; (ix) available electric power and current supply (usually, the pumping electrical power should not exceed 50 W per kW of nominal cooling capacity of installed geothermal heat pumps); (x) power and current starting limitations; (xi) motor quality versus service life; (xii) use motors protected against the possibility of internal condensation; (xiii) water treatment, water conditions, and material selection; (xiv) lowest temperature which may occur during the system operation (e.g., –10°C) by having their electrical motor protected against the possibility of internal condensation; and (xvi) ensure turbulent flow through the ground-coupled heat exchangers.

11.5.2 System and Pump Curves

In hydronic systems, the system curve defines the total pressure required (that must overcome the pipe, fittings, and valves friction losses) to supply a given flow rate for brine or water. Since the pressure drop is proportional to the square of the fluid flow rate, the shape of this curve is parabolic. On the other hand, the pump curve describes the pump's ability to supply a flow rate against a certain pressure head. If the system and pump curves are plotted on the same graph, their intersection is the system operating point, i.e., the point where the pump's total pressure matches the system's pressure drop (Figure 11.4). It can be seen that, if the brine or water flow rate increases, the system total discharge pressure increases. In practice, the brine or water flow rates vary and, thus, the operating point moves on both system and pump curves. To match the system variable fluid flow rates, two-way control valves are typically used in hydronic systems with constant-speed circulation pumps.

The operating point of a pump should be considered when the system includes two-way control valves. For example, at a full (100%) design flow rate, the two-way valve is wide open, and the system follows curve A (Figure 11.5). As the flow rate drops at, for example, 50% (part load), the two-way valves begin closing to match the load, which increases the friction losses and the system curve gradually changes to curve B. Point 1 shows the pump operating at the design flow rate at the calculated design pressure loss of the system. Typically, the actual system curve is slightly different than the design curve. As a result, the pump operates at point 2 and produces a flow rate higher than the design. To reduce the actual flow to the design flow at point 1, a balancing valve downstream from the pump can be adjusted while all the terminal valves are in a wide-open position. This pump discharge balancing

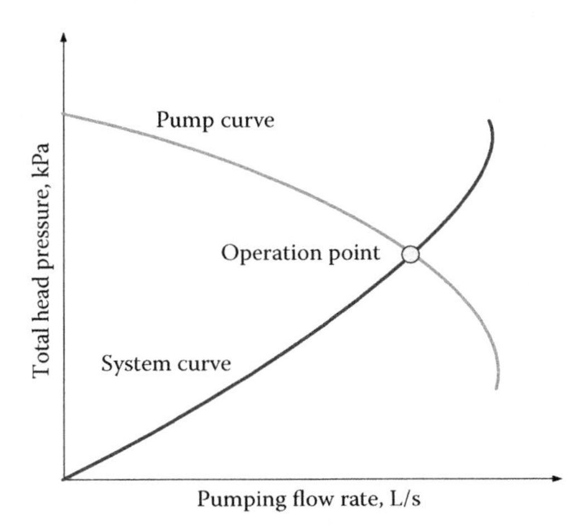

FIGURE 11.4 Typical system and pump curves.

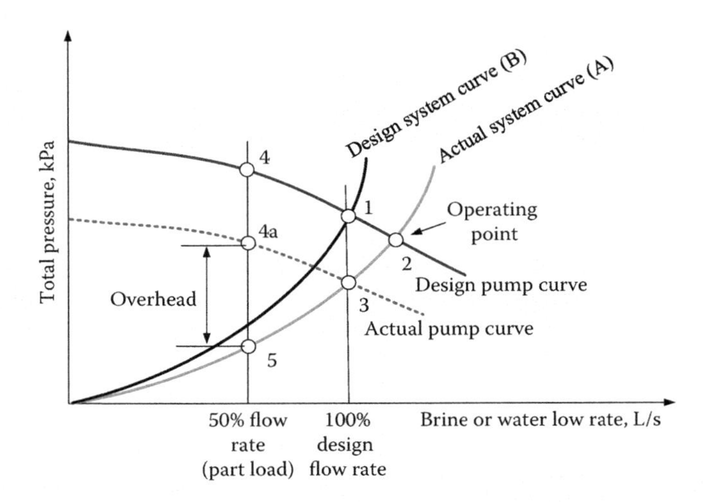

FIGURE 11.5 Moving pump operating point.

valve imposes a pressure drop equal to the pressure difference between point 1 and point 3. The manufacturer's pump curve shows that the capacity may be reduced by substituting a new impeller with a smaller diameter or by trimming the existing pump impeller. After trimming, reopening the balancing valve in the pump discharge then eliminates the artificial drop and the pump operates at point 3. Points 3 and 4a demonstrate the effect a trimmed impeller has on reducing flow (Figure 11.5).

11.5.3 Friction Losses and Pressure Drops

The pressure head loss along a pipe is incurred both by frictional resistance at the wall along the run of the pipe and at fittings such as bends or valves. For long pipes with few fittings, the overall loss is dominated by wall friction. If, however, the pipe is short and there are numerous fittings, then the principal losses are those that are produced by disturbances caused by the fittings. The friction losses in uniform, straight pipes, are caused by the effects of fluid viscosity, the movement of fluid molecules against each other or against the (possibly rough) pipes' walls. They are affected by whether the flow is laminar (Re < 2,300), transient (2,300 < Re < 4,000), or turbulent (Re > 4,000). As the velocity gradually increases, the fluid mixes with the surrounding fluid, indicating a change to transition then to turbulent motion.

In laminar flow, friction losses are proportional to fluid velocity (w) that varies smoothly between the bulk of the fluid and the pipe wall, where it is zero. The roughness of the pipe surface influences neither the fluid flow nor the friction loss.

In turbulent flow (where the velocity distribution is much flatter over most of the pipe cross section), friction losses are proportional to the square of the fluid velocity (w^2). As the Reynolds number increases, the profile becomes increasingly flat and the ratio of maximum to mean velocity reduces slightly.

In order to use a simple pump with moderate pumping power, the fluid should have as low as possible viscosity, giving a small pressure drop for the system fluid flow, especially with laminar flow. The friction pressure losses, due to the shear stresses between the pipe surface and the fluid flowing within, depends on the conditions of flow and the physical properties of the system, and can be estimated with the following Darcy-Weisbach equation:

$$\Delta p_f = (f/2)\rho w^2 \cdot L/d \quad (11.1)$$

where

Δp_f is the pressure drop due to friction (Pa = kg/m·s^2).
f is the friction coefficient (-).
ρ is the fluid density (kg/m^3).
w is the fluid velocity (m/s).
L is the total tube length (m).
d is the inner tube diameter (m).

The friction coefficient for laminar flow ($Re_d \leq 2300$) is:

$$f = 64/Re \quad (11.2)$$

where

Re is Reynolds number of the fluid in motion (-):

$$Re = w \cdot d/\nu = \rho \cdot w \cdot d/\mu \quad (11.3)$$

w is the fluid velocity (m/s).
d is the pipe interior diameter (m).
$\nu = \mu/\rho$ is the kinematic viscosity of the fluid (m^2/s).
ρ is the density of the fluid (kg/m^3).
μ is the dynamic (absolute) viscosity of the fluid (Pa·s = kg/m·s).

A commonly used friction coefficient equation for turbulent flow (valid for 4,000 < Re < 10,000) is the Blasius relation:

$$f = 0.316/Re^{1/4} \quad (11.4)$$

Generally, water (or brine) closed-loop piping should be designed with a pressure drop limitation of 1.21 mH_2O per 30.48 equivalent meters of pipe, while maximum water velocities of 3.048 m/s should not be exceeded.

Fittings (such as bends and couplings) and valves also induce friction losses in pipelines. In practice, for the purposes of calculating the total friction loss of a system, the sources of form friction are sometimes reduced to an equivalent length of pipe. This means that piping friction losses can be calculated by measuring lengths of pipe and adding the pressure friction losses in valves and fittings. Where actual piping layouts have not been determined, a rough estimate of equivalent length can be obtained by measuring the length of the circuit to the unit with the highest pressure drop and multiplying that length by 1.3 to obtain an estimation of total system pressure drop.

11.5.4 Pumping Power

Excessive electrical energy consumption of brine (water) circulating pumps is generally attributed to: (i) non-optimal or overly high flow rate; (ii) using U-tubes of too small a diameter; (iii) using unnecessary control valves (or choosing a design that requires control valves when an alternative is available); and (iv) operating a central circulating pump on a 24/7 basis instead of shutting it off when it is not needed.

The energy used for circulating fluid between the ground-coupled heat exchangers and the geothermal heat pumps is often erroneously considered to be negligible. While it can represent a fairly small portion of the total energy consumed, too many ground-source heat pump systems have been found to have excessive pumping energy.

The pump's input power can be calculated from the system pressure drop, overall pump efficiency, motor efficiency, and water (or brine) flow rate and thermophysical characteristics.

Pump input power (also called pump shaft power) is the mechanical power delivered to the pump shaft (kW), calculated as follows (see Figure 11.7):

$$P_{input} = P_{shaft} = \dot{m} \cdot g / \eta_{pump} = \rho \cdot \dot{V} \cdot g / \eta_{pump} = \dot{m} \cdot \Delta p / \rho \cdot \eta_{pump} \tag{11.5}$$

where

$P_{input} = P_{shaft}$ is the pump input (or shaft) power (kW).

$\dot{m}$ is the mass flow rate of the fluid (brine or water) (kg/s).

g is the gravitational acceleration (m/s^2).

η_{pump} is the pump efficiency (-).

ρ is the brine or water average mass density (kg/m^3).

$\dot{V}$ is the brine or water volumetric flow rate (m^3/s).

Δp is the required pressure increase based on a summation of the pressure drops calculated individually across each component in the brine or water closed loop, most of which is caused by pipe friction losses due to flow through the ground-coupled heat exchanger (Pa = kg/m·s^2).

ρ is the brine or water average mass density (kg/m^3).

Pump output power (also called hydraulic power) is the useful power delivered by the pump, usually expressed as follows (Figure 11.6):

$$P_{output} = P_{hydraulic} = \dot{m} \cdot \Delta p / \rho \tag{11.6}$$

where

$P_{output} = P_{hydraulic}$ is the pump output (or hydraulic) power (kW).

$\dot{m}$ is the mass flow rate of the fluid (brine or water) (kg/s).

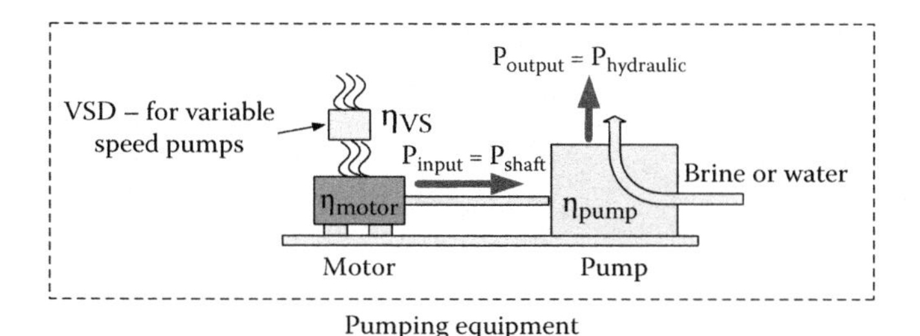

FIGURE 11.6 Schematic representation of pump input and output powers.

Δp is the required pressure increase based on a summation of the pressure drops calculated individually across each component in the brine or water closed loop, most of which is caused by pipe friction losses due to flow through the ground-coupled heat exchanger (Pa = kg/m·s^2).

ρ is the brine or water average mass density (kg/m^3).

The total power, P_{pump}, required to operate the pump is determined by the manufacturer's test of an actual pump running under standard conditions to produce the required flow and pressure, as shown in Figure 11.7. The usual installed pumping powers vary from 0.0085 to 0.045 kWelect/kWtherm of heat pump power. This represents 4–21% of the total demand of typical ground-source heat pump systems and up to 50% of the total energy for some pump control schemes.

By assuming all the pumping power is added into loop except the motor/drive losses, the exiting fluid temperature of pump is calculated:

$$T_{outlet} = T_{inlet} + \frac{\eta_{\text{motor}} * Power}{\dot{\text{m}}_{\text{w}} * c_{p,w}} \tag{11.7}$$

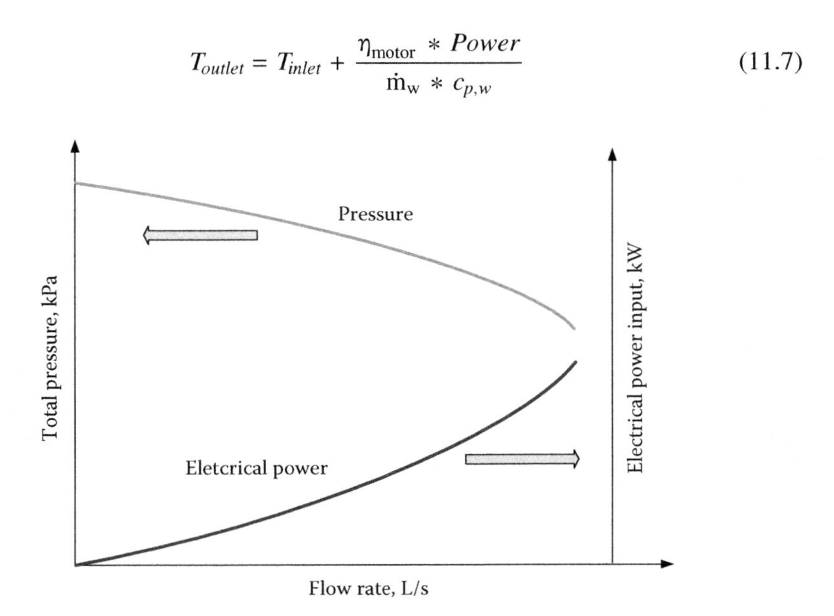

FIGURE 11.7 Typical pump electrical power increase with fluid (brine or water) flow rate.

In order to avoid having pumping energy requirements higher than desirable, borefield pressure losses should be no greater than 75 kPa and the total system pressure losses should be no more than 150 kPa.

11.5.5 Pump Efficiency

Pump efficiency is defined as the ratio of the pump output (P_{output}) (hydraulic) power (kW) and the pump input (P_{input}) (shaft) power (kW):

$$\eta_{pump} = \frac{P_{output}}{P_{input}} \cdot 100\% \tag{11.8}$$

Figures 11.8a and 11.8b show typical pump efficiency curves versus brine or water flow rates and brine or water flow rates and impeller size, respectively.

For electric-driven, constant-speed pumps, the equipment efficiency is:

$$\eta_{equip} = \eta_{pump} \cdot \eta_{motor} \tag{11.9}$$

where

η_{equip} is the equipment efficiency (0–1) (-).

η_{motor} is the motor efficiency (0–1) (-).

η_{pump} is the pump efficiency (0–1) (-).

For variable speed pumps, the equipment efficiency of the variable-speed drive efficiency must be included:

$$\eta_{equip} = \eta_{motor} \cdot \eta_{pump} \cdot \eta_{VSD} \tag{11.10}$$

where

η_{VSD} is the variable-speed drive efficiency (0–1) (-).

11.5.6 Affinity Laws

The centrifugal pump, which converts the velocity energy to static (pressure) energy, can be categorized by a set of relationships called affinity laws (Table 11.1). The affinity laws are useful for estimating pump performances at different impeller diameters (*d*) or rotating speeds (*N*) based on a pump with known characteristics. They assume that the system curve is known and that pressure varies as the square of the flow rate. Because the affinity laws are used to calculate a new condition due to a flow rate or pressure change (e.g., reduced impeller diameter or pump speed), this new condition also follows the same system curve.

The first pump affinity law shows that the pump flow rate (capacity) varies directly with the impeller diameter. The second pump affinity law shows that pressure varies as the square of the impeller diameter. The third affinity law states that pump power varies as the cube of the impeller diameter. By changing the impeller diameter and

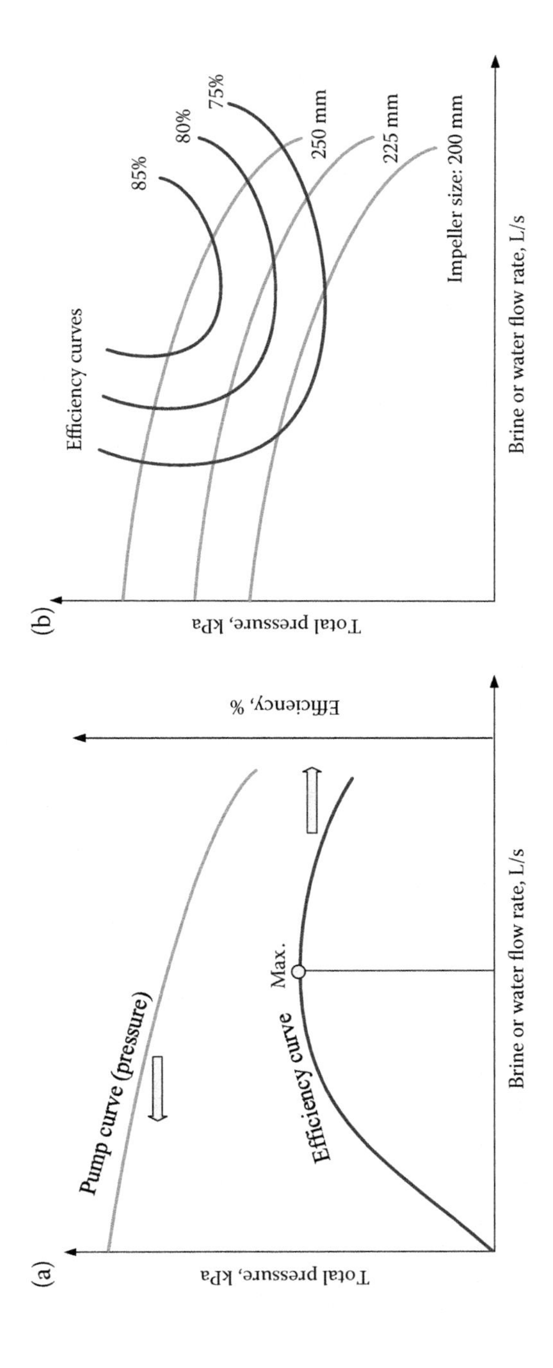

FIGURE 11.8 (a) Pump efficiency curve versus brine or water flow rate; (b) pump efficiency versus total pressure and impeller size.

TABLE 11.1
Pump Affinity Laws

Function	Speed change	Impeller diameter change
Flow rate	$\dot{m}_2 = \dot{m}_2(N_2/N_1)$	$\dot{m}_2 = \dot{m}_2(d_2/d_1)$
Head pressure	$p_2 = p_1(N_2/N_1)^2$	$p_2 = p_1(d_2/d_1)^2$
Input electrical power	$W_2 = W_2(N_2/N_1)^3$	$W_2 = W_2(d_2/d_1)^3$

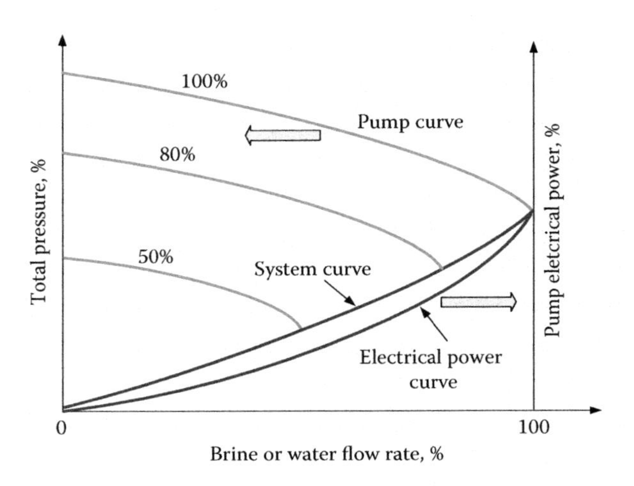

FIGURE 11.9 Pumping electrical power, pressure, and flow rate versus pump speed.

maintaining constant speed, the pump efficiency for a diffuser pump is not affected if the impeller diameter is changed by less than 5%. However, efficiency changes if the impeller size is reduced enough to affect the clearance between the casing and the periphery of the impeller. Figure 11.9 shows the relationship of pump flow rate, total pressure, and electric power as expressed by the affinity laws.

11.5.7 Pump Arrangements

In large capacity ground-source heat pump systems, a single pump may not be able to satisfy the full design flow rates and yet provide both economical operation at partial loads and a system backup. The following alternative pumping arrangements and control scenarios should be considered: (i) use multiple pumps connected in parallel or series; (ii) install standby pumps; (iii) use pumps with two-speed motors; (iv) use primary-secondary pumping arrangements; (v) use variable-speed pumping; and (vi) adopt a distributed pumping concept.

11.5.7.1 Parallel

Ground-source heat pump systems usually employ a two-circulating-pump arrangement: a main and a standby pump. The pumps are piped in parallel and designed to assure water flow to the loop system. When the pumps are arranged in parallel, each pump operates at the same pressure and provides its share of the system flow at that pressure. The piping of parallel pumps (Figure 11.10) should permit running either pump. A check valve is required in each pump's discharge to prevent backflow when one pump is shut down. Hand valves and a strainer allow one pump to be serviced while the other is operating. A strainer protects a pump by preventing foreign material from entering the pump. Gages or a common gage with a trumpet valve, which include several valves as one unit, or pressure taps permit checking pump operation. Generally, pumps of equal size are recommended, and the parallel pump curve is established by doubling the flow of the single pump curve.

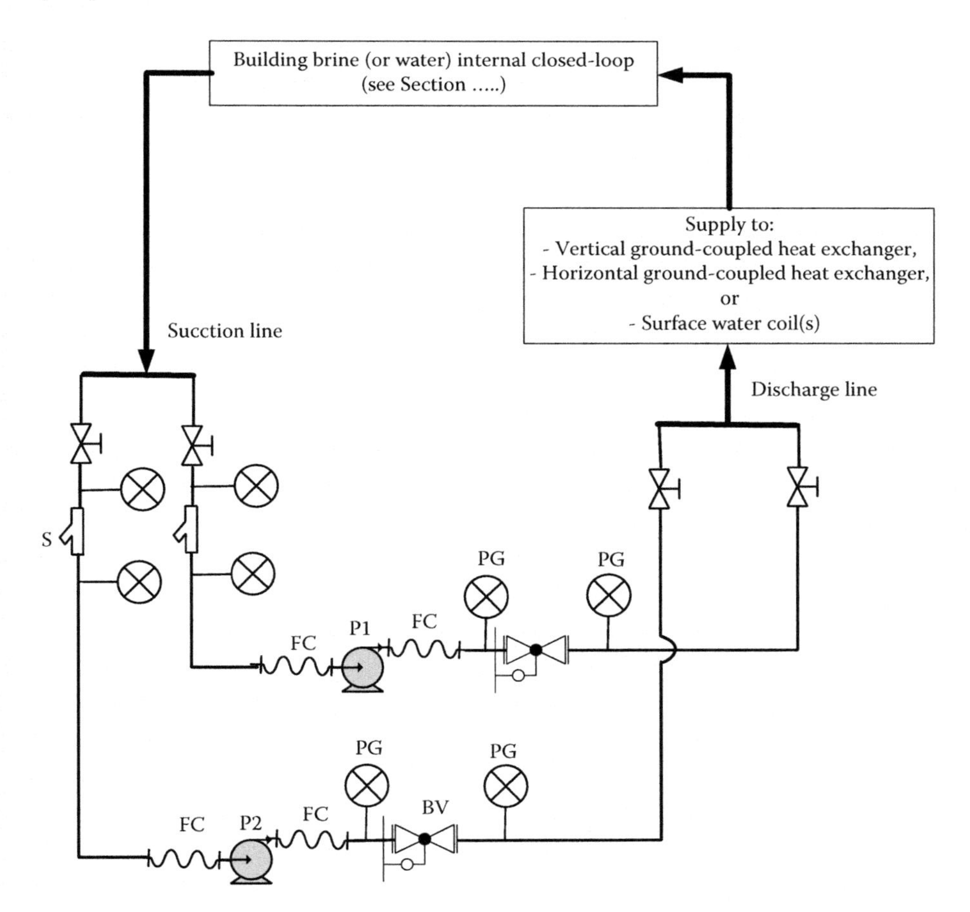

FIGURE 11.10 Typical piping for parallel pumps; BV, balancing valve; FC, flexible connection; P, circulation pump; PG, pressure gage; S, strainer.

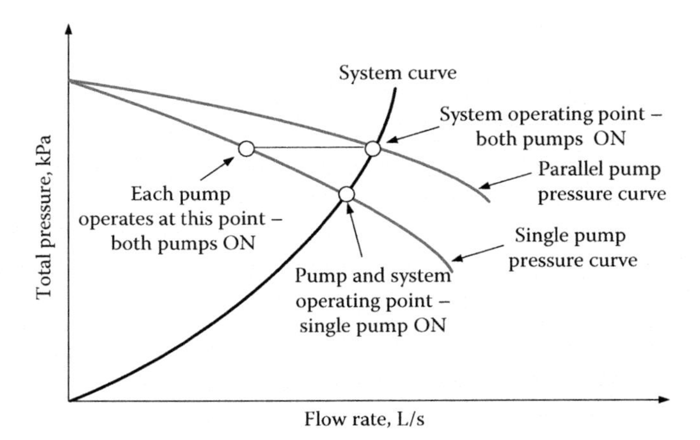

FIGURE 11.11 Operating conditions for parallel pump arrangement.

Plotting a system curve across the parallel pump curve shows the operating points for both single and parallel pump operation (Figure 11.11). Note that single pump operation does not yield 50% flow rates. The system curve crosses the single pump curve considerably to the right of its operating point when both pumps are running. This leads to two concerns: (i) the motor must be selected to prevent overloading during operation of a single pump and (ii) a single pump can provide standby service for up to 80% of the design flow, the actual amount depending on the specific pump curve and system curve.

11.5.7.2 Series

When pumps are applied in series, each pump operates at the same flow rate and provides its share of the total pressure at that flow. A system curve plot shows the operating points for both single and series pump operation (Figure 11.12). Note that

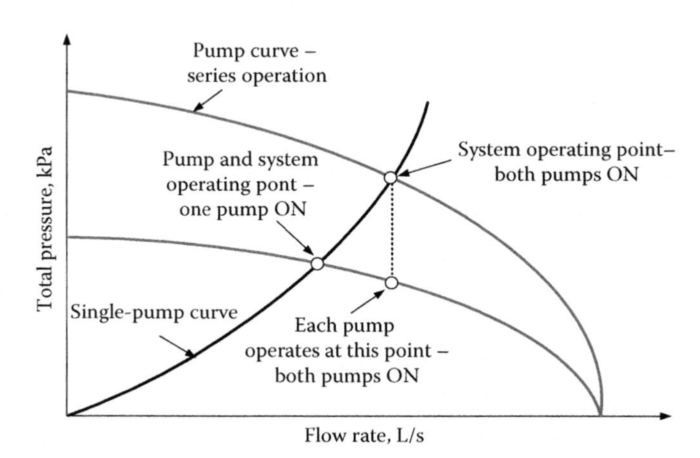

FIGURE 11.12 Operating conditions for pump series arrangement.

the single pump can provide up to 80% flow for standby and at a lower power requirement.

As with parallel pumps, piping for series pumps should permit running either pump (Figure 11.12). A bypass with a hand valve permits servicing one pump while the other is in operation. Operation and flow can be checked the same way as for parallel pumps. A strainer prevents foreign material from entering the pumps. As a safety measure, standby fluid circulating pumps of equal capacity should be used, piped in parallel, utilizing a pump sequencer, manual or automatic. A backup or standby pump of equal capacity and pressure installed in parallel to the main pump is recommended to operate during an emergency or to ensure continuous operation when a pump is taken out of operation for routine service.

REFERENCES

ASHRAE. 2016. *Standard 62.1. Ventilation for Acceptable Indoor Air Quality* (https://www.ashrae.org/technical-resources/bookstore/standards-62-1-62-2. Accessed August 6, 2019).

CAN/CSA. 2013. *Design and Installation of Earth Energy Systems*. Can/CSA (Canadian Standard Association)-C448 series.

DOE. 1995. *Commercial/Institutional Ground-Source Heat Pump Engineering Manual*. CANETA Research Inc. for U.S. Department of Defense, U.S. Department of Energy, Oak Ridge National Laboratory, Oak Ridge, TN, US.

Fahlen, P., F. Karlsson. 2005 Optimizing and Controlling Media Flows In Heat Pump Systems. In Proceedings of the 8th IEA (International Energy Agency) Heat Pump Conference, Las Vegas, US, May 30–June 2.

Karlsson, F., P. Fahlen. 2007 Capacity-controlled ground source heat pumps in hydronic heating systems. *Building Services Engineering* 30(2):221–229.

Minea, V. 2009. Improvements of large institutional ground-source heat pump systems. *IEA (International Energy Agency) Heat Pump Centre Newsletter* 27(1):24–38.

Ni, L., W. Song, F. Zeng, Y. Yao. 2011. Energy saving and economic analyses of design heating load ratio of ground source heat pump with gas boiler as auxiliary heat source. In *Proceedings of International Conference on Electric Technology and Civil Engineering (ICETCE), Institute of Electrical and Electronics Engineers (IEEE)*, Lushan (China), April 22–24, pp. 1197–1200.

12 Vertical Closed-Loop (Indirect, Secondary Fluid) Ground-Source Heat Pump Systems

12.1 INTRODUCTION

In practice, for buildings with less than 6–8 stores, vertical borehole heat exchangers are usually preferred to horizontal arrangements due to reasons such as: (i) reduced needs for land area/surface to install the borehole field and (ii) more stable (and even increasing with depth) far-field ground/soil/rock temperatures. However, their feasibility depends on the availability of local reliable drilling equipment at less expensive drilling costs, and on availability of efficient sizing/designing tools.

12.2 RESIDENTIAL AND SMALL COMMERCIAL/INSTITUTIONAL BUILDINGS

Ground-source heat pump systems with vertical ground-coupled heat exchangers for residential and small commercial/institutional buildings contain components such as the following (Figure 12.1) (Bose et al. 1985; IGSHPA 1988; DOE 1995; Muraya et al. 1996; Rottmayer et al. 1997; Kavanaugh and Rafferty 1997; Yavuzturk et al. 1999; Shonder and Beck 1999; Berntsson 2002; Bose et al. 2002): (i) at least one vertical ground-coupled heat exchanger consisting of a sealed, high-density polyethylene (a material immune to corrosion, oxidation, and galvanic or electrolytic attack, typically having thermal conductivity of 0.40 W/mK) U-shaped 19–50 mm ID diameter tube (socket or butt fused with heat fused joints), and inserted into a 45–200 m deep drilled borehole (depending on building thermal loads, soil/ground/rock conditions, and drilling equipment available, with typical diameters of 76–150 mm; the U-shaper tube, relatively inexpensive and easy to install, consists of two similar cylindrical pipes thermally fused at the bottom of the borehole to a close return (cross connector) U-bend; after the U-tube insertion, the borehole is backfilled and grouted with an enhanced material (e.g., bentonite-based grout) with a thermal conductivity of 0.69 W/m·K or higher that provides a good thermal contact between the U-tube and the surrounding soil/ground/rock, and protects groundwater from possible contamination; it can be noted that, since the general range of thermal conductivity values for geologic materials is 0.5–3.8 W/m·K for ground/soils and 1.0–6.9 W/m·K for rocks, it is evident that both grout and U-tubes provide thermal insulating effects to the borehole global heat transfer; (ii) at least one reversible brine-to-air or brine-to-water

DOI: 10.1201/9781003032540-12

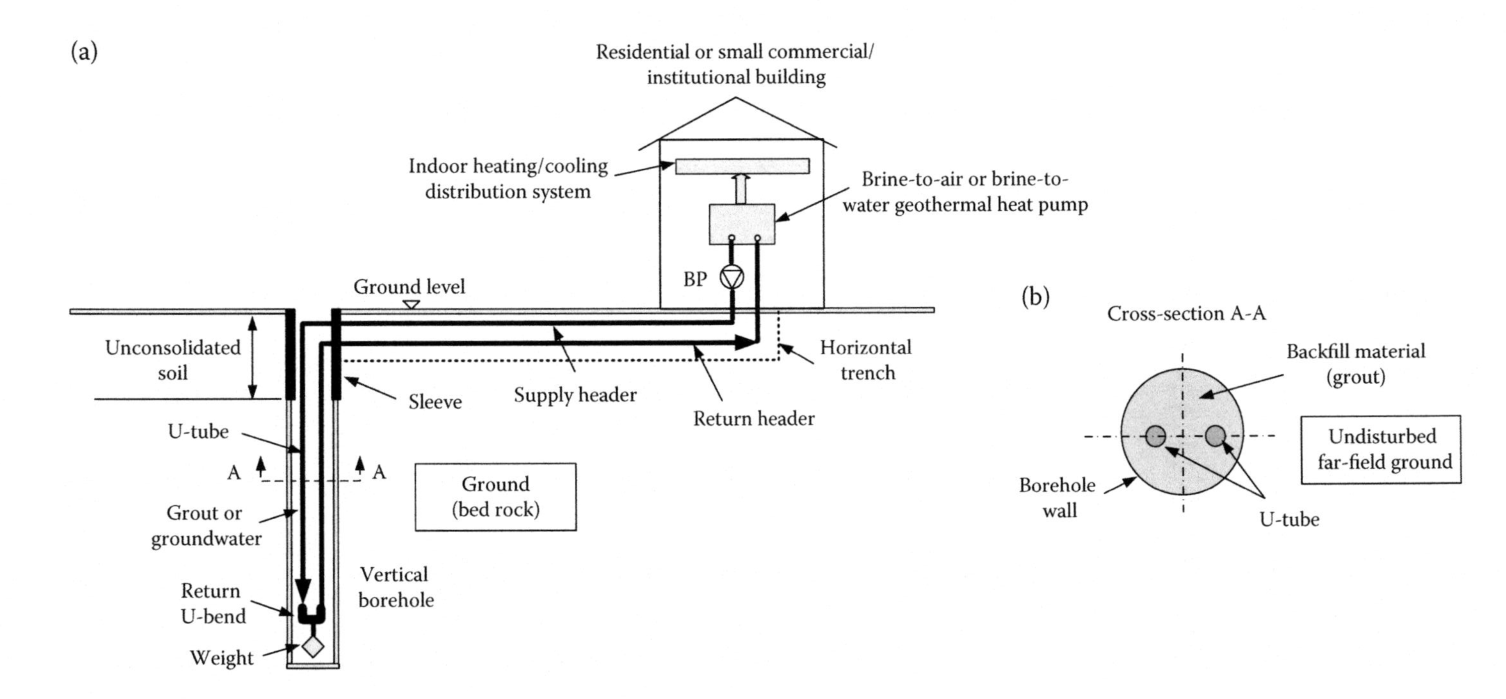

FIGURE 12.1 Typical ground-source heat pump system with vertical ground-coupled heat exchanger for residential and small commercial/institutional buildings; (a) schematic of vertical borehole and horizontal trench; (b) cross-section of the single borehole with vertical U-tube heat exchanger; BP, brine pump.

geothermal heat pump with or without desuperheater for domestic hot water preheating, typically ranging in size from 3.5 kW to 35 kW of nominal cooling capacity; (iii) a brine circulating pump; (iv) an internal building heating/cooling distribution system that delivers space heating and cooling via, for example, air-handling units, underfloor heating pipes, or radiators.

For the most effective performance, the heat transferred from the source via geothermal heat pump(s) used to heat liquids inside radiant floor (hydronic) systems (i.e., there is a relatively low temperature difference between the heat source and sink).

12.2.1 Operating Modes

In ground-source heat pump systems, the ground/soil is utilized as either a heat source (during winter) or sink (during the summer) source. In winter months, heat is extracted from the surrounding ground/soil by a thermal carrier fluid (brine) circulating through the geothermal heat pump(s) where it is used to heat the building indoor air or water.

In the heating mode (Figure 12.2), the antifreeze mixture (brine) circulating through the vertical ground-coupled heat exchanger extracts heat from the ground/soil/rock (acting as a heat source), while the geothermal heat pump condenser, located inside the building, rejects it into the building's indoor air (or water) distribution system acting as a heat sink medium. The heat absorbed from the ground/soil is added to the compressor equivalent electrical consumed energy to meet the space requirements. The brine circulation pump (BP) circulates the brine through the vertical ground-coupled heat exchanger at very low temperatures (up to –5°C) and delivers heat to the building indoor air via air handling units, or to hot water serving for space heating (e.g., through underfloor heating) at temperatures up to 30–45°C, and/or to hot water tank(s) for use as domestic tap water, while the geothermal heat pump's evaporator where the refrigerant vaporizes by extracting heat from the heat carrier fluid. The supply fluid temperature should be –6.66°C to –1.11°C warmer than the undisturbed ground/soil temperature for cooling and –12.22°C to –6.66°C colder for heating.

In the cooling mode, the cycle is reversed (Figure 12.3), the building excess sensible and latent heat being dissipated (rejected) in the ground/soil/rock. In the cooling mode, the geothermal heat pump(s) reject the heat collected from the space and another 25% of the heat from the compressor. In areas with suitable earth structure and thermal conditions, it could be possible to partially store the rejected cooling energy in the ground/soil/rock to be used during the next heating season.

When the geothermal heat pump operates in the cooling mode, there are potentially two practical issues: (i) because the heat rejection capacity at cooling operation is usually higher than the heat extraction capacity at heating operation, the length of ground-coupled heat exchanger required cooling operation will be larger than that required for heating operation; (ii) because the heat rejection during the cooling mode, the ground/soil is heated up and its moisture migrates away from the ground-coupled heat exchanger; this process results in a drop of the thermal conductivity of the surrounding ground/soil structure and, partly, in an increase of the thermal resistance between the ground/soil and the heat exchanger.

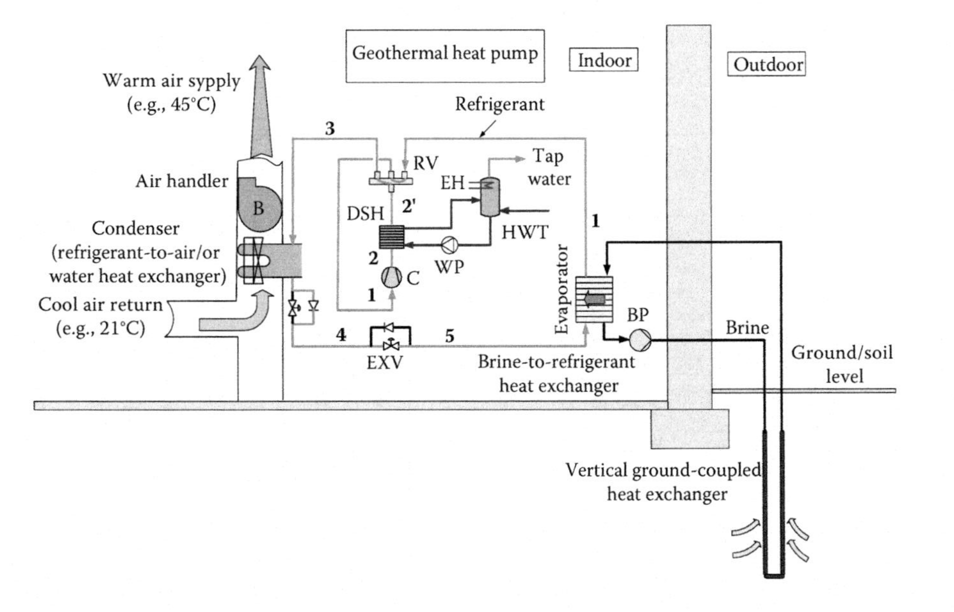

FIGURE 12.2 Schematic representation of a geothermal heat pump with vertical closed-loop (indirect, secondary fluid) ground-coupled heat exchanger operating in the heating mode (Note: not all components shown); B, air handler blower; BP, brine pump; C, compressor; DSH, desuperheater; EH, electric heater; EXV, expansion valve; HWT, domestic hot water storage tank; RV, four-way reversible valve; WP, water pump.

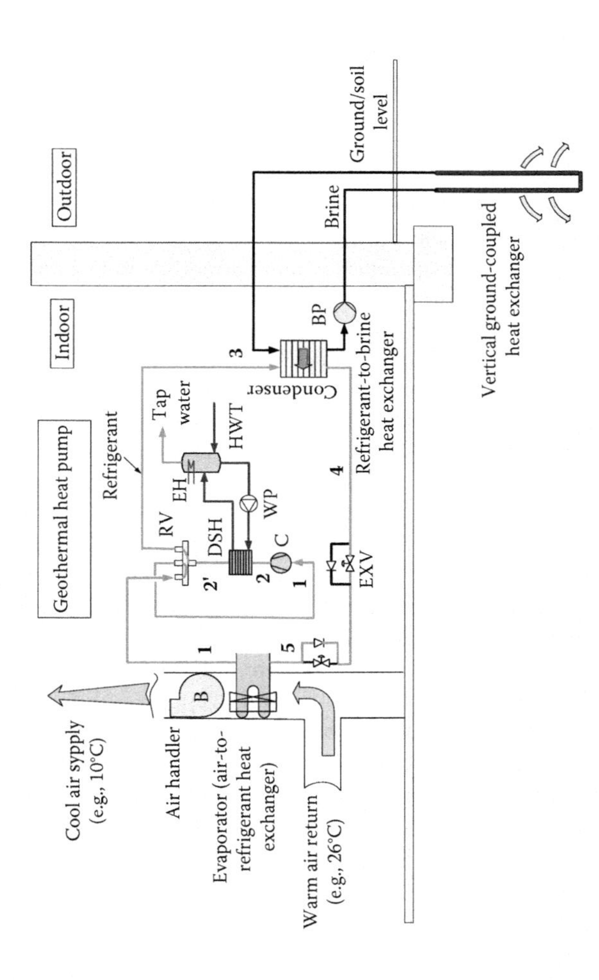

FIGURE 12.3 Schematic representation of a geothermal heat pump with vertical closed-loop (indirect, secondary fluid) ground-coupled heat exchanger operating in the cooling mode (Notes: not all components shown; for legend, see Figure 12.2).

During the summer months, heat is collected from the air inside the building using a heat pump and stored in a fluid. This fluid is then passed through either a closed or open loop of circulation tubes embedded in ground. Heat is partially dissipated from the circulation fluid to the ground/soil surrounding the circulation tubes. In other words, in cold and moderate climates, by circulating thermal carrier fluids (generally, antifreeze mixtures called brines), through vertical closed-loop (indirect, secondary fluid) ground-coupled heat exchangers, heat extraction (in the heating mode), and heat rejection (in the cooling mode) are accomplished via geothermal heat pumps. Inside the geothermal heat pump, the thermal carrier fluid circulates through a heat exchanger that acts as a refrigerant evaporator (brine-to-refrigerant heat exchanger in the heating mode) (Figure 12.2) and as a refrigerant condenser (refrigerant-to-brine heat exchanger in the cooling mode) (Figure 12.3). The temperature of thermal carrier fluid entering the geothermal heat pump may vary over extended temperature ranges going from about –5°C (in the heating/winter mode) to up to 35°C (in the cooling/summer mode).

Alternate heating and cooling operation is an ideal utilization of the ground/soil thermal capacity. In such a way, heat is extracted (in the heating mode) and rejected (in the cooling mode) from and to the ground/soil, respectively. For both operating modes ground/soil temperatures are excellent, especially at the beginning of the heating season (with high temperatures resulting from the heat rejection during summertime) and at the beginning of the cooling season (with low temperatures resulting from the heat extraction during the heating season).

The thermodynamic process of reversible geothermal heat pumps consists of the following thermodynamic main processes (see Figures 12.2 and 12.3): (i) the refrigerant leaving the evaporator as superheated vapor (state 1) is polytropicaly compressed through the compressor C up to the superheated state 2; the electrical energy input in the compressor is converted to shaft work to raise the pressure and temperature of the refrigerant; by increasing the vapor pressure, the condensing temperature is increased to a level higher than that of the heat source; (ii) in the desuperheater (DSH), the superheated refrigerant vapor is slightly cooled by transferring heat to cold water in order to preheat and store it (in a hot water storage tank HWT) prior to being distributed to domestic hot water consumers; (iii) in the condenser, the superheated vapor is cooled, and then undergoes a two-phase condensation at constant temperature and pressure and, finally, the saturated liquid is sub-cooled at constant pressure to a lower temperature in order to reduce the risks of flashing within the expansion valve; sub-cooling of the liquid refrigerant ensures that only liquid and no vapor bubbles enter the four-way expansion valve, allowing for better flow control and smaller devices; also, if vapor bubbles were to enter the expansion device, there would be less liquid refrigerant available in the evaporator to produce cooling, while the compressor still has to compress that portion of the vapor that did not contribute to the cooling capacity; this represents a loss of capacity and efficiency; on the other hand, too much sub-cooling indicates that a considerable portion of the condenser volume is filled by single-phase liquid; thus, the area available for heat rejection from the condensing fluid is smaller than it could be; as a consequence, the saturation temperature is increased along with the saturation pressure and, as a result, the compressor work increases; the degree of sub-cooling at the condenser outlet is primarily determined by the amount of refrigerant charged to a heat pump and the setting of the expansion device primarily determines the degree of superheat at the evaporator outlet; (iv) after the condenser, the liquefied sub-cooled refrigerant enters the four-way expansion valve where an expansion process at constant enthalpy (i.e., drop in pressure accompanied by a drop in temperature) takes place in order to reduce the refrigerant pressure at a level corresponding to an evaporating temperature below to the heat source temperature (process 4–5); the expansion valve controls the refrigerant flow into the evaporator in order to ensure its complete evaporation, maintain an optimum superheat in order to avoid the liquid refrigerant to enter the compressor, and also avoid excessive superheat that may lead to overheating of the compressor; and (v) the refrigerant then enters the evaporator in a two-phase state (state 5), absorbs (recover) heat from the heat source (brine), and undergoes change from liquid-vapor mixture to saturated vapor at constant pressure and temperature (process 5–1); inside the evaporator, the saturated vapor is slightly superheated up to state 1 before entering the compressor; a reasonable vapor superheat is desirable to ensure that all the refrigerant flow is evaporated in order to achieve full evaporating capacity and, also, to supply to the compressor only refrigerant vapor without any liquid entrained that could damage the compressor; however, excessive superheat may lead to overheating of the compressor; at this point, the heat pump real thermodynamic cycle restarts.

The desuperheater (DSH), used in situations where the building cooling loads exist or are dominate, is a refrigerant superheated vapor-to-water heat recovery heat exchanger installed between the compressor and the four-way reversing valve. It is operating only when the geothermal heat pump is running with relatively small thermal power outputs (about 10–15% of the total geothermal heat pump thermal power output). In practice, the cost benefits of using desuperheaters need to be carefully assessed according to the geothermal heat pump thermodynamic cycle and properties of the refrigerant used. In practice, even if it contributes to improve the seasonal performance factor for the geothermal heat pumps, the maximum output temperatures of cold water leaving the geothermal heat pump's desuperheater is around 35–40°C, much lower than the required domestic hot water supply temperature (up to 60–65°C to reduce the risk of Legionella). With relatively low water-flow rates, output temperatures up to about 70°C can be achieved. Therefore, a hot water storage tank (HWT) with an inserted auxiliary electric heater(s) (EH) is required (as shown in Figures 12.2 and 12.3) to further increase the temperature of the domestic hot water before delivery.

In the heating mode (Figure 12.1a), the brine pump (BP) circulates the antifreeze mixture (brine) through the horizontal-coupled heat exchanger that extracts heat from the ground/soil/rock (acting as a heat source), and through the geothermal heat pump's evaporator where the refrigerant vaporizes by extracting heat from the heat carrier fluid, while the geothermal heat pump condenser, located inside the building, supplies heat and ejects it into the building's indoor air (or water) distribution system acting as a heat sink medium. The brine-to-air or brine-to-water geothermal heat pump recovers heat from the thermal carrier fluid (brine) at very low temperatures (ideally, between –5°C and 0°C, while the temperature of the far-field undisturbed ground/soil could rest at 5–10°C and even higher) and deliver heat at temperatures up to 35–45°C to the house indoor air (via warm air-handling units) or to hot water serving for space heating (e.g., through underfloor heating systems in the new, low-energy, highly insulated houses).

In the cooling mode, the cycle is reversed (Figure 12.1b). The building excess sensible and latent heat is dissipated (rejected) in the ground/soil/rock via the horizontal ground-coupled heat exchanger. In areas with suitable earth structure and favorable thermal conditions, it could be possible to partially store the rejected cooling energy in the ground/soil/rock to be used during the next heating season. During summer cooling-dominated periods, and depending on the size of building, it is possible to provide free (passive) cooling by by-passing the geothermal heat pump(s) and thus circulating the brine directly from the ground-coupled heat exchanger through, for example, one or several fan convectors.

When the geothermal heat pumps operate in the cooling mode, there are two potential issues: (i) because the quantity of heat rejection is usually higher than that of heat extraction during heating operation, the length of ground-coupled heat exchanger required for cooling operation will be larger than that required for heating operation; this means that in horizontal ground-coupled heat exchangers, the ground-coupled heat exchanger should be sized for cooling operation; (ii) because of heat rejection, the ground/soil is heated up and its moisture migrates away from the ground-coupled heat exchanger; this process results in a drop of the thermal conductivity of the surrounding ground/soil structure and, partly, in an increase of the thermal resistance between the ground/soil and the horizontal ground-coupled heat exchanger; these conditions may remain even at the beginning of the next heating season, lowering both heat transfer rates and the geothermal heat pump(s) heating coefficients of performance; moisture migration could be avoided if the ground/soil heat exchangers are installed below the groundwater table or in regions with abundant rain and/or snow precipitations. In residential applications, usually, on/off and/or programmable electric/electronic thermostats are used to efficiently control the operation of geothermal heat pump(s). Most geothermal systems are designed with 0.0647 L/s (3.0 US gpm/ton; 1 US gpm = 0.075768 L/s; 1 ton = 3.51 kW) flow rates. Lower flow rates will reduce the brine circulation pump size, but they will also decrease the performance of the geothermal heat pump.

12.3 LARGE-SCALE COMMERCIAL/INSTITUTIONAL BUILDINGS

In large-scale commercial/institutional buildings (limited however at six to eight stories to stay within the pressure rating of the pipe used, unless stronger pipe, more expensive and difficult to work, must be used), the ground-coupled heat exchangers consist of one thermally fused high-density polyethylene U-shaped tube buried (at depths varying from 60 to 180 m, and in place grouted) inserted in multiple (tens or even hundreds) vertical borehole arrays, normally spaced at between at least 4.5 and 10 m (to reduce thermal interference between individual adjacent boreholes and/or long-term heat buildup between boreholes, depending on grid pattern, building annual heating load versus cooling load, and groundwater movement), and in which a heat carrier fluid circulates. Usually, such borehole fields are connected in reverse-return piping layouts and returned to the buildings where they are linked to internal

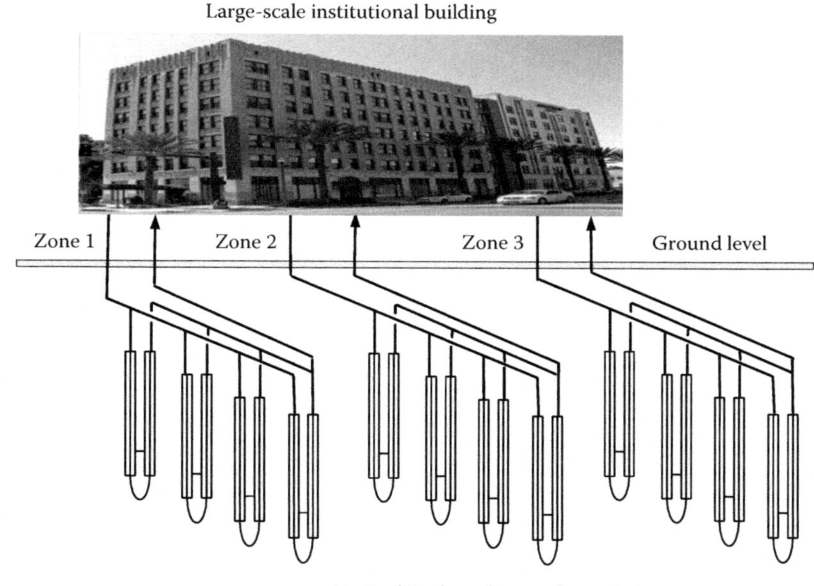

FIGURE 12.4 Schematic of a multiple vertical U-tube borehole ground-coupled heat pump system for large-scale institutional/commercial buildings (building view retrieved from: https://www.nps.gov/tps/how-to-preserve/briefs/14-exterior-additions.htm, accessed March 9, 2020) (Notes: drawing not to scale; not all components shown).

water closed loops separated or not by intermediate heat exchangers to the geothermal heat pumps (Figure 12.4). Large-scale commercial/institutional buildings must have relatively large areas suited to installation of ground-coupled heat exchangers, such as grass or parking lots, playing fields, or even beneath the buildings. The vertical ground-coupled heat exchangers may act as short- (daily) and long-term (seasonal, annual) heat storage mediums that may help to improve the overall thermal performances of ground-source heat pump systems.

Among the most common applications of ground-source heat pump systems provided with distributed on building indoor water (or brine) closed loops and central mechanical rooms' geothermal heat pumps (Figure 12.5) are for: (i) office buildings that best utilize the energy-saving heat reclaim features; since such buildings with significant load diversity have core areas that even in winter may require cooling that can be transferred to perimeter zones where simultaneously space heating is required; (ii) hotels and motels with significant cooling loads during daytime hours that can be used to offset nighttime heating loads; the restaurants, swimming areas, kitchens, laundry rooms, and meeting rooms can be cooled during the day, and the stored heat used to warm guest rooms during nighttime occupied hours; (iii) apartments and condominiums that offer heating or cooling from any geothermal heat pump at any time and achieve individual

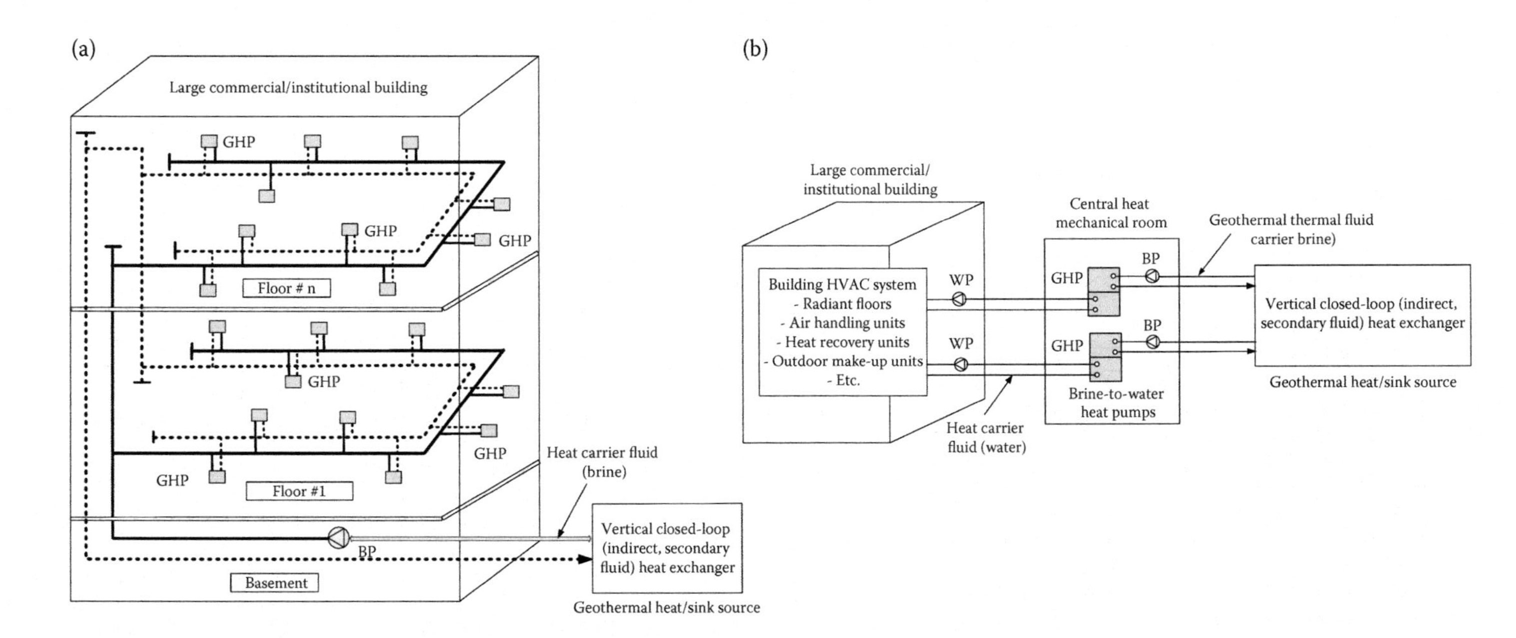

FIGURE 12.5 Schematics of geothermal heat pumps arrangements in large commercial/institutional buildings: (a) distributed and (b) central. BP, brine circulation pump; GHP, geothermal heat pump; HVAC, heating, ventilating and air conditioning; WP, water circulation pump (Note: not all components shown).

metering of consumed thermal energy; (iv) schools and universities that have simultaneous periods of light and heavy loads throughout classroom; additionally, core-located laboratories, kitchens, and cafeterias can be used to generate heat for use in perimeter classroom zones.

REFERENCES

Berntsson, T. 2002. Heat sources-technology, economy and environment. *International Journal of Refrigeration* 25:428–438.

Bose, J.E., J.D. Parker, F.C. McQuiston. 1985. *Design/Data Manual for Closed-Loop Ground-Coupled Heat Pump Systems*. Oklahoma State University. American Society of Heating, Refrigerating and Air-Conditioning Engineers, Inc., Tullie Circle, NE, Atlanta.

Bose, J.E., M.D. Smith, J.D. Spitler. 2002. Advances in ground source heat pump system – An international overview. In Proceedings of the 7th International Energy Agency (IEA) Heat Pump Conference. Volume 1, pp. 313–324. Beijing, May 19–22.

DOE. 1995. *Commercial/Institutional Ground-Source Heat Pump Engineering Manual*. Prepared by CANETA Research Inc. for U.S. Department of Defense, U.S. Department of Energy, Oak Ridge National Laboratory, Oak Ridge, TN, and American Society of Heating, Refrigerating and Air Conditioning Engineers (ASHRAE), Atlanta, LA.

IGSHPA. 1988. *Closed-loop/ground-source heat pump systems, installation guide* (National Rural Electric Cooperative Association [NRECA] Research Project 86-1). International Ground Source Heat Pump Association, Oklahoma State University, Stillwater, Okla.

Kavanaugh, S.P., K. Rafferty. 1997. *Ground-Source Heat Pumps. Design of Geothermal Systems for Commercial and Institutional Buildings*. American Society of Heating, Refrigerating and Air-Conditioning Engineers, Inc, Atlanta.

Muraya, N.K., D.L. O'Neal, W.M. Heffington. 1996. Thermal interference of adjacent legs in a vertical Utube heat exchanger for a ground-coupled heat pump. *ASHRAE Transactions* 102(2):12–21.

Shonder J.A., I.V. Beck. 1999. Determining effective soil formation thermal properties from field data using parameter estimation technique. *ASHRAE Transactions* 105(1).

Rottmayer, S.P., W.A. Beckman, J.W. Mitchell. 1997. Simulation of a single vertical U-tube ground heat exchanger in an infinite medium. *ASHRAE Transactions* 103(2):651–658.

Yavuzturk, C., J.D. Spitler, S.J. Rees. 1999. A transient two-dimensional finite volume model for the simulation of vertical U- Tube ground heat exchangers. *ASHRAE Transactions* 105(2):465–474.

13 Heat Transfer

13.1 INTRODUCTION

The design of vertical (indirect, secondary fluid) U-shaped ground-coupled heat exchangers is based on two unsteady (transient) heat transfer processes that occur concurrently in two separate regions over wide range of spatial and time scales: (i) heat transfer inside the borehole, from borehole wall through the grout material and the high-density polyethylene pipe up to the heat carrier fluid (brine) circulating inside the U-tube (and vice versa); (ii) heat transfer outside single or multiple boreholes from far-field dry or wet surrounding the ground/soil/rock formations to the borehole wall(s) (and vice versa).

To analyze the heat transfer inside and outside borehole(s), simplifying assumptions as the followings are usually employed: (i) brine temperature (or thermal flux) is considered uniform at vertical any pipe cross section; (ii) thermal effects of ground/soil/rocks, backfill grout, and heat transfer fluid are accounted for as separately interacting entities; (iii) thermophysical properties (as thermal resistance) of plastic U-tubes are usually ignored; (iv) average physical properties of ground/soil/rocks are constant; (v) effective (average) thermal conductivity, density, and specific heat of ground/soil/rocks are considered because they have a great impact on the heat transfer; (vi) if considered, the influence of groundwater flow on heat transfer depends on its velocity (the larger the velocity the more the positive influence); and (vii) ground/soil/rocks temperature at the far-field boundary remains constant over time.

The mathematic models that have been developed for predicting the transient heat transfer inside and outside vertical U-shaped ground-coupled heat exchangers can be classified into analytical (e.g., infinite and finite line-source) and numerical (more complex, normally implemented in computer-based simulation tools) models.

13.2 HEAT TRANSFER INSIDE BOREHOLES

In vertical ground-coupled heat exchangers, consisting of sealed, U-shaped tubes inserted in boreholes (used for extracting and/or rejecting heat from or into the ground/soil/rocks) and connected to brine-to-air and/or brine-to-water geothermal heat pumps (used for rejecting and/or extracting heat to or from the buildings), heat is transferred by circulation an antifreeze (brine) heat transfer medium. In both heating and cooling operating modes, the heat transfer inside such borehole heat exchangers depends on their geometry and non-homogeneous composition, including the pipes' arrangement, thermo-physical properties of adjacent surrounding ground/soil/rocks, backfill materials and heat carrier fluids, being more or less sensitive to short-, medium-, and long-term changes in actual extraction and

DOI: 10.1201/9781003032540-13

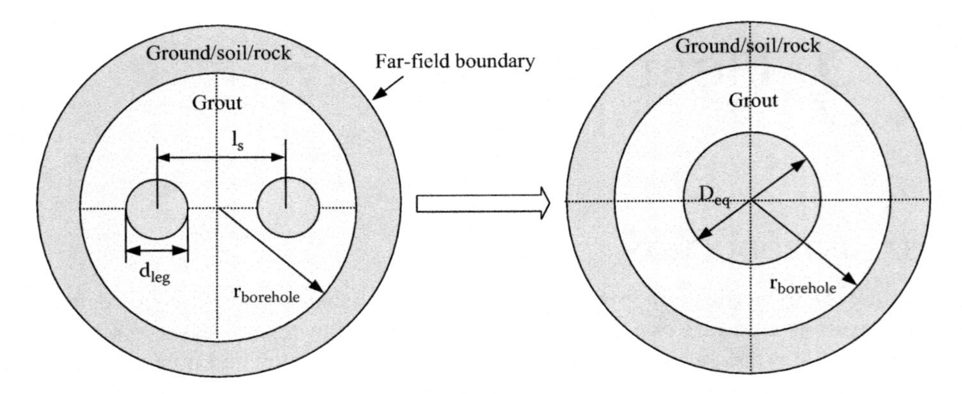

FIGURE 13.1 Equivalent diameter of a borehole heat exchanger with single U-tube.

injection heat rates. Thermal processes between the circulating brines and the surrounding ground/soil/rocks are classified into parts as the following: (i) convective heat transfer between the circulating fluid and the pipes' interior walls; (ii) conductive heat transfer through the pipes' walls; and (iii) conductive heat transfer through the grouting materials and the ground/soil/rocks.

13.2.1 Borehole Equivalent Diameter

Any U-shaped, two-leg pipe inserted within a vertical borehole can be represented as a single tube concentric with the borehole having an "equivalent" diameter defined as follows (Figure 13.1) (Gu and O'Neal 1998):

$$D_{eq} = \sqrt{2 \cdot d_{leg} \cdot l_s} \tag{13.1}$$

where

D_{eq} is the U-tube tube equivalent diameter (m).

d_{leg} is the diameter of each leg (m).

l_s is the center-to-center (shrink) distance between the two legs of the U-tube (m); ($d_{leg} \leq l_s \leq r_{borehole}$).

$r_{borehole}$ is the borehole radius (m).

13.2.2 Heat Transfer Structure

The relatively complex heat transfer phenomenon occurring between the geothermal thermal carrier (brine) and the surrounding ground/soil/rock (and vice versa) takes place through multiple, concentric cylinders composed of (Figure 13.2): (i) the equivalent vertical tube of thermal conductivity, k_{tube}; (ii) borehole filling material (grout) of thermal conductivity, k_{grout} , and thermal diffusivity, α_{grout}; and (iii) surrounding infinite homogeneous ground/soil/rocks of thermal conductivity, k_{ground}, and thermal diffusivity, α_{ground}. Neglecting the heat transfer in axial direction and assuming constant temperatures along to cylinders' walls, a constant linear heat flux ($\dot{q}_L$, W/m) is transferred in the radial direction from the brine (flowing at average temperature, $\bar{T}_{brine}$) through the interior surface of the equivalent tube with diameter D_{eq} ($\dot{q}_{surf,L}$), through the equivalent tube wall ($\dot{q}_{tube,L}$), through the grout ($\dot{q}_{grout,L}$), and, finally, through the surrounding ground/soil/rock ($\dot{q}_{groud,L}$).

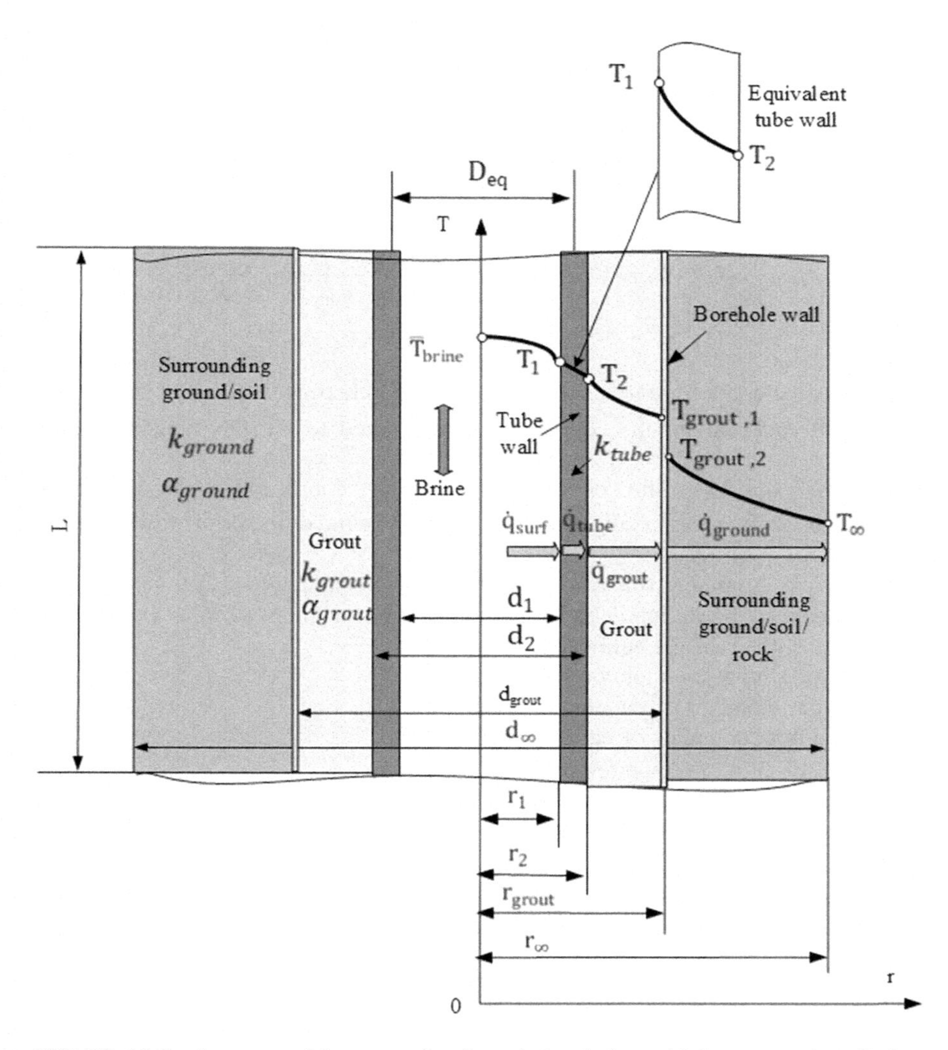

FIGURE 13.2 Structure of heat transfer through borehole multiple concentric cylinders (Notes: heat transfer is shown in cooling operating mode of geothermal heat pumps; schematic not to scale); d_1, tube interior diameter; d_2, tube exterior diameter; k, thermal conductivity; $\dot{q}$, thermal flux per borehole meter; r_1, tube interior radius; r_2, tube exterior radius; r_{grout}, grout radius; r_∞, far-field ground radius; T, temperature; α, thermal diffusivity.

The main heat transfer modes are as follows: (i) forced convection between the brine and the inner surface of the equivalent tube ($\dot{q}_{surf,L}$); (ii) conduction through the wall of the equivalent tube ($\dot{q}_{tube,L}$); (iii) conduction through the grout ($\dot{q}_{grout,L}$); and (iv) conduction through the surrounding ground/coil/rocks ($\dot{q}_{ground,L}$).

Additional assumptions as the following are usually used: (i) thermal contact resistances among the equivalent U-tube wall, backfill material, and, sometimes, borehole wall, are neglected; (ii) ground/soil/rock surface is assumed to be an adiabatic boundary; and (iii) due to the reduced thermal mass and the high aspect ratio of typical U-shaped ground-coupled heat exchangers, the internal energy variations are negligible with respect to the heat exchange at the ground/soil/rock interface.

The forced convective heat flux between the geothermal thermal carrier (brine) and the internal surface of the equivalent tube, reported to the tube unitary length (L = 1 m), is (see Figure 13.2):

$$\dot{q}_{surf,L} = h_{surf}\cdot 2\pi\cdot r_1\cdot 1\cdot(\bar{T}_{brine} - T_1) = \frac{\bar{T}_{brine} - T_1}{(1/2\pi r_1 h_{surf})} = \frac{\bar{T}_{brine} - T_1}{R_{surf}} \tag{13.2}$$

where

$\dot{q}_{surf,L}$ is the convective heat flux between the geothermal thermal carrier (brine) and the internal wall of the equivalent tube reported to the pipe unitary length (W/m).

$h_{surf} = Nu\cdot k_{brine}/d_1$ is the convective heat transfer coefficient at the equivalent tube interior surface, essentially influenced by the regime of the brine flow (laminar or turbulent) (W/m$^2\cdot$K).

Nu is Nusselt number (the ratio of convective to conductive heat transfer at the brine boundary), a criterion describing the heat transfer intensity from the brine to the interior surface of the equivalent tube (-).

k_{brine} is the brine thermal conductivity (Wm·K).

d_1 is inner diameter of the equivalent tube (m).

r_1 is the interior radius of equivalent tube (m).

$\bar{T}_{brine}$ is the brine average temperature (°C).

T_1 is the temperature of the equivalent tube interior surface (°C).

$R_{surf} = 1/2\pi r_1 h_{surf}$ is the thermal resistance of the equivalent tube interior surface (m·K/W).

The brine convective heat transfer coefficient can be determined by using the Dittus-Boelter correlation (Holman 1997; Incropera et al. 2007):

$$h_{surf} = 0.023\cdot Re_{brine}^{0.8} Pr_{brine}^{n} \tag{13.3}$$

where

h_{surf} is the convective heat transfer coefficient (W/m^{2}K).

$Re_{brine} = wd_1/\nu$ is the brine Reynolds number (the ratio of brine inertia and viscous forces) (-).

$Pr_{brine} = \nu/\alpha$ is the brine Prandtl number (the ratio of viscous diffusion rate and thermal diffusion rate) (-).

w is the brine velocity (m/s).

d_1 is the inner diameter of the equivalent tube (m).

ν is the brine kinematic viscosity (m^2/s).

$\alpha = k/\rho c_p$ is the brine thermal diffusivity (m^2/s).

k is the brine thermal conductivity (W/mK).

ρ is the brine mass density (kg/m^3).

c_p is the brine specific heat (J/kgK).

n = 0.4 for heating and n = 0.3 for cooling.

Because the temperatures, T_1 and T_2, of the interior and exterior surfaces of the equivalent tube wall are considered constants, the Laplace equation of heat conduction equation through the wall of the equivalent tube is described in cylindrical coordinates as follows:

$$\frac{d^2T}{dr^2} + \frac{1}{r}\frac{dT}{dr} = 0 \tag{13.4}$$

where

T is the temperature (°C).

r is the radial distance (m).

Equation 13.4 describes the heat transfer process through the wall of the equivalent cylindrical tube with the following first species boundary conditions (see Figure 13.2):

$$\begin{aligned} r = r_1, \quad T = T_1 \\ r = r_2, \quad T = T_2 \end{aligned} \tag{13.5}$$

The logarithmic variation of the temperature inside the wall of the cylindrical tube results by integrating Equation 13.4 with boundary conditions (Equation 13.5):

$$T = T_1 - (T_1 - T_2)[\ln(r/r_1)/\ln(r_2/r_1)] \tag{13.6}$$

From Equation 13.2, the brine average temperature is:

$$\bar{T}_{brine} - T_1 = \dot{q}_{surf,L}/(2\pi r_1 \cdot h_{surf}) \tag{13.7}$$

Fourier's law of conduction ($\dot{q} = -k \cdot dT/dr$) determines the density of the heat flux through the wall of the equivalent tube at any radial distance reported to the tube unitary length (L = 1m) (see Figure 13.2):

$$\dot{q}_{tube} = k_{tube}(T_1 - T_2)/r \cdot \ln(r_2/r_1) = \frac{T_1 - T_2}{R_{tube}} \tag{13.8}$$

where

$\dot{q}_{tube}$ is the heat flux density through the cylindrical tube (W/m^2).

k_{tube} is the tube thermal conductivity (W/m·K).

r_1 is the tube interior radius (m).

r_2 is the tube exterior radius (m).

T_1 is the temperature of the tube interior wall (°C).

T_2 is the temperature of the tube exterior wall (°C).

$R_{tube} = (r/k_{tube})\cdot \ln(r_2/r_1)$ is the thermal resistance of the equivalent tube wall (m·K/W).

The total heat flux transferred through the entire surface of the cylindrical pipe ($A = 2\pi r\cdot L$) is constant and can by expressed as follows:

$$\dot{Q} = \dot{q}_{tube}\cdot A = [2\pi\cdot k_{tube}\cdot L\cdot(T_1 - T_2)/ \ln(r_2/r_1)] \quad (13.9)$$

where

$\dot{Q}$ is the total conductive heat flux transferred through the entire surface of the cylindrical pipe (W).

k_{pipe} is the thermal conductivity of the equivalent tube (W/m·K).

From Equation 13.9, the following results:

$$T_1 - T_2 = (\dot{Q}/2\pi\cdot L\cdot k_{tube})\cdot \ln(r_2/r_1) \quad (13.10)$$

The density of the linear conductive heat flux through the grout (cylinder) at any radial distance is:

$$\dot{q}_{grout,\ L} = [(T_2 - T_{grout,1})]/[(r/k_{grout})\cdot ln(r_{grout}/r_2)] = (T_2 - T_{grout,1})/R_{grout} \quad (13.11)$$

where

$T_{grout,1}$ is the borehole-side grout temperature (°C).

k_{grout} is the grout thermal conductivity (W/mK).

$R_{grout} = (r/k_{grout})\cdot ln(r_{grout}/r_2)$ is the grout thermal resistance (mK/W).

Thus:

$$T_2 - T_{grout,1} = (r\cdot\dot{q}_{grout,L}/2\pi\cdot L\cdot k_{grout})\cdot ln(r_{grout,1}/r_2) \quad (13.12)$$

As can be seen in Figure 13.2, at the interface between the surfaces of grout and ground/soil/rocks in contact, a temperature drop occurs, which implies a contact thermal resistance defined as the ratio between the temperature drop and the average heat flow across the interface. When two surfaces as the grout and surrounding ground/soil/rocks are in contact with each other, the actual area of contact is much smaller than the apparent area of contact because the asperities of one surface are in contact with the asperities of the other surface and some fluid exists in the interstitial spaces. The amount of actual contact area is also dependent on the physical properties of the contacting materials (grout and ground/soil). The thermal contact heat transfer coefficient ($h_{contact}$) (defined in terms of the temperature drop across the interface and the heat flux through the interface) is influenced by factors as surface contact pressure, roughness, flatness, as well as interstitial imperfections and plastic or elastic deformations, and thermal conductivity.

Similarly, the density of the linear conductive heat flux through the ground/soil at any radial distance r is:

$$\dot{q}_{ground} = (T_{grout,2} - T_{\infty})/[(r/k_{ground})\cdot ln\,(r_{\infty}/r_{grout,2})] = (T_{grout,2} - T_{\infty})/R_{ground} \tag{13.13}$$

where

$\dot{q}_{ground}$ is the density of the linear conductive heat flux through the ground/soil (W/m^2).

$T_{grout,2}$ is the surrounding ground/soil-side grout temperature (°C).

R_{ground} is the thermal resistance of the ground cylindrical layer (m·K/W).

From Equation 13.13, the following results:

$$T_{grout,2} - T_{\infty} = \frac{r\dot{q}_{ground}}{2\pi\cdot k_{ground}} \ln(r_{\infty}/r_{grout,2}) \tag{13.14}$$

The sum of Equations 13.7, 13.10, 13.12, and 13.14 gives:

$$\bar{T}_{brine} - T_{\infty} = \frac{\dot{q}_{surf,\,L}}{2\pi r_1}\cdot\frac{1}{h_{surf}} + \frac{\dot{q}_{tube}}{2\pi\cdot k_{tube}}\ln(r_2/r_1) + \frac{r\cdot\dot{q}_{grout,\,L}}{2\pi\cdot L\cdot k_{grout}}\,ln\,(r_{grout,1}/r_2) + \frac{r\dot{q}_{ground}}{2\pi\cdot k_{ground}}\ln(r_{\infty}/r_{grout,2}) \tag{13.15}$$

Also, the sum of Equations 13.7, 13.10, and 13.12 gives:

$$\bar{T}_{brine} - T_{grout,1} = \frac{\dot{q}_{surf,L}}{2\pi r_1}\cdot\frac{1}{h_{surf}} + \frac{\dot{q}_{tube}}{2\pi\cdot k_{tube}}\ln(r_2/r_1) + \frac{r\cdot\dot{q}_{grout,\,L}}{2\pi\cdot L\cdot k_{grout}}\,ln\,(r_{grout,1}/r_2) \tag{13.16}$$

The linear density of heat flux transferred from the brine through the wall of the equivalent tube, grout, and ground/soil (W/m) can be expressed as follows:

$$\dot{q}_L = \dot{q}_{surf,L} = q_{tube,L} = \dot{q}_{grout,L} = \dot{q}_{ground,L} = \frac{\bar{T}_{brine} - T_{\infty}}{\frac{1}{2\pi r_1 h_{surface}} + \frac{1}{2\pi k_{tube}}\ln(r_2/r_1) + \frac{1}{2\pi k_{grout}}ln\,(r_{grout}/r_2) + \frac{1}{2\pi k_{ground}}\ln(r_{\infty}/r_{grout})} \tag{13.17}$$

The average temperature of the borehole wall ($T_{grout,1}$) (see Equation 13.19) can be determined from the following equation:

$$\dot{q}_L = \frac{\bar{T}_{brine} - T_\infty}{\frac{1}{2\pi r_1 h_{surf}} + \frac{1}{2\pi k_{tube}} \ln(r_2/r_1) + \frac{1}{2\pi k_{grout}} ln r_{grout/r_2}} \tag{13.18}$$

$$\bar{T}_{grout,1} = \bar{T}_{brine} - \dot{q}_L \left[\frac{1}{2\pi h_{wall} r_1} + \frac{1}{2\pi k_{pipe}} \ln(r_2/r_1) + \frac{1}{2\pi k_{grout}} \ln(r_{grout}/r_2) \right] \tag{13.19}$$

13.2.3 Borehole Thermal Resistance

With the steady-state assumptions, the global brine-to-ground thermal resistance that depends on brine flow rate and tube, grout, and ground/soil/rock thermal properties is the sum of convective thermal resistance between the brine and the inner side of the equivalent tube and the conductive thermal resistances of the equivalent tube wall, grout, and surrounding ground/soil/rocks:

$$\begin{aligned} R_{global} &= (R_{surf} + R_{tube} + R_{grout}) + R_{ground} = R_{borehole} + R_{ground} \\ &= \frac{1}{2\pi r_1 h_{surface}} + \frac{ln(r_2 / r_1)}{2\pi k_{tube}} + \\ &\frac{\ln(r_{grout} / r_2)}{2\pi k_{grout}} + \frac{\ln(r_\infty / r_{grout})}{2\pi k_{ground}} \end{aligned} \tag{13.20}$$

where

R_{global} is the global brine-to-ground (borehole) thermal resistance (m·K/W).

$R_{tube\ surface} = 1/(2\pi \cdot r_1 \cdot h_{surf})$ is the convective resistance of the brine within the equivalent tube (m·K/W).

$R_{tube\ wall} = [ln(r_2/r_1)/2\pi \cdot k_{tube}]$ is the conductive resistance of the equivalent tube (m·K/W).

$R_{grout} = [\ln(r_{grout}/r_2)/2\pi \cdot k_{grout}]$ is the grout conductive thermal resistance (m·K/W).

$R_{ground} = [ln(r_\infty/r_{grout})/2\pi k_{ground}]$ is the grout conductive thermal resistance (m·K/W).

$R_{tube\ surface}$ is the convective thermal resistance within the equivalent tube (m·K/W).

$h_{tube\ surface}$ is the brine convective heat transfer coefficient (W/m^{2}K).

$R_{tube\ wall}$ is the tube conductive thermal resistance (W/m·K).

r_2 is the pipe outside radii of the equivalent tube (m).

r_1 is the pipe inside radii of the equivalent tube (m).

k_{tube} is the equivalent tube thermal conductivity (W/m·K).

k_{grout} is the grout thermal conductivity (W/m·K).

From Equation 13.20 and the previous definitions, it can be seen that the borehole thermal resistance depends on: (i) diameters of equivalent tube, grout, and borehole; (ii) thermophysical properties of borehole materials; (iii) brine mass flow rate and flow regime (laminar or turbulent); and (iv) U-tube legs' spacing and thermal interference.

Based on Equation 13.20, the borehole-only overall heat transfer coefficient is:

$$U_{borehole} = \frac{1}{\frac{1}{2\pi h_{surf} \cdot r_1} + \frac{1}{2\pi k_{tube}} \ln(r_2/r_1) + \frac{1}{2\pi k_{grout}} lnr_{grout/r_2}} \tag{13.21}$$

where

$U_{borehole} = 1/R_{borehole}$ is the borehole-only overall heat transfer coefficient (W/m^{2}K).

For a borehole heat exchanger equivalent to a typical U-shaped tube, the heat transfer equation can be expressed as follows:

$$\dot{Q} = \dot{q}_L \cdot L_{borehole} = \dot{m}_{brine} \cdot \bar{c}_{brine} \cdot (T_{brine,in} - T_{brine,out}) = U_{borehole} \cdot A \cdot (\bar{T}_{brine} - \bar{T}_{borehole\ wall,1}) \tag{13.22}$$

where

$\dot{Q}$ is the total heat transfer rate (W).

$\dot{q}_{borehole}$ is the linear heat flux per unit length (W/m).

$L_{borehole}$ is the borehole depth (m).

$\dot{m}_{brine}$ is the brine total flow rate entering the borehole heat exchanger (kg/s).

$\bar{c}_{brine}$ is the brine average specific heat (j/kgK).

$T_{brine,in}$ is the brine inlet temperature (°C).

UA is the overall heat transfer coefficient (W/K).

$\bar{T}_{brine}$ is the brine mean temperature (°C).

$\bar{T}_{borehole\ wall,1}$ is the average temperature of the borehole wall (°C).

13.3 HEAT TRANSFER OUTSIDE BOREHOLES

In closed-loop (indirect, secondary fluid) ground-source heat pump systems, the relatively complex, transient heat transfer outside vertical U-shaped tubes from and to the ground/soil/rocks, that mainly occurs by conduction and, sometimes, by buoyancy-driven natural convection due to horizontal groundwater flow through fractured bedrock (advection heat transfer), varies continuously due to changing building's thermal loads and subsequent short- and long-time fluctuations (ranging from hours, days, or months, to years) in the supply and return temperatures of heat carrier fluids. All of these phenomena disturb the natural thermal behaviour of the surrounding ground/soil/rock formations. Such variations have direct impacts on the instantaneous and seasonal coefficients of performance of geothermal heat pumps and on the system overall energy efficiency.

The heat transfer outside the boreholes depends on parameters such as the following (Fan and Ma 2006; Fossa 2011): (i) ground/soil/rocks geologic/hydrologic composition as well as thermal and physical properties (e.g., effective thermal conductivity, density, and specific heat); (ii) ground/soil/rocks undisturbed temperature distribution and seasonal changes; (iii) natural groundwater movement (the larger the flow velocity, the larger the influence of the groundwater); (iv) ground/soil seepage amount and velocity; (v) ground/soil and/or groundwater freezing and thawing around the borehole; (vi) borehole depths and spacing; thermal interference between adjacent boreholes can occur in the case of large-scale buildings where tens of boreholes may be used; therefore, reasonable separation distances between boreholes should be selected to minimize required land area without increasing the thermal interference and temperature penalty in the design process.

To model and design the vertical ground-coupled heat exchangers, one-, two-, and three-dimensional methods (considering or not boreholes' interferences and/or groundwater movement) can be classified as follows: (i) analytical, as line-source (Kelvin 1882; Ingersoll and Plass 1948a, 1948b; Ingersoll et al. 1954), and cylindrical-source solutions (Carslaw and Jaeger 1959; Ingersoll et al. 1948, 1954); and (ii) numerical (Eskilson 1987).

The main objectives of these modeling methods include the following: (i) predict the heat transfer inside the boreholes; (ii) predict the conductive heat transfer around the boreholes, a slow, long-term process that influences the ground/soil/rock thermal recharge from one heating season to another with potential effects on the system overall operating costs and energy performances; (iii) evaluate the ground/soil/rocks temperatures as functions of the borehole heat exchanger's heat flux input/output; (iv) calculate the temperature difference between the undisturbed (far-field) ground/soil temperature (T_∞), and the borehole wall temperature (see Figure 13.2); and (v) determine the brine supply and return average temperatures from/to vertical boreholes.

Either analytical or numerical analysis of heat transfer outside borehole heat exchangers is usually performed considering simplifying assumptions as (Ingersoll et al., 1948; Carslaw and Jaeger 1959; Eskilson 1987; Rottmayer et al. 1997; Zeng et al. 2003a, 2003b): (i) ground/soil/rocks considered as homogeneous, semi-infinite porous medium with uniform thermal and physical properties and constant initial far-field temperature; (ii) ground/soil/rocks and groundwater are in thermal equilibrium; (iii) heat is transferred at constant rates per unit length of boreholes; and (iv) heat radiation and viscosity dissipation are neglected.

13.3.1 Heat Flux and Temperature Profile

According to Fourier's law (Fourier 1878), the conductive heat flux density through an arbitrary area, A, of a given ground/soil volume during the time, t, is:

$$\overrightarrow{q_{cond}} = Q/A \cdot t = \dot{Q}/A = -\bar{k}_{ground}\left(\frac{\partial T}{\partial x} + \frac{\partial T}{\partial y} + \frac{\partial T}{\partial z}\right) = -\bar{k}_{ground} \cdot \overrightarrow{gradT} \quad (13.23)$$

where

$\overrightarrow{q_{cond}}$ is the density of the conductive heat transfer (W/m^2).

Q is the amount of total heat transferred through the ground/soil over the heat transfer area A (J).

A is the total heat transfer area (m^2).

t is the time (s).

$\dot{Q}$ is the total conductive heat transfer rate (W).

$\bar{k}_{ground}$ is the ground/soil average (effective) thermal conductivity (W/m·K).

T is the temperature (°C).

$\overrightarrow{grad}\,T = \nabla T = \left(\frac{\partial T}{\partial x} + \frac{\partial T}{\partial y} + \frac{\partial T}{\partial z}\right)$ is the gradient of the ground/soil scalar temperature field T(x, y, z) (i.e., the derivative of T in each direction) (K/m) (Note: the gradient of a scalar temperature field is a vector field).

x, y, z are the three-dimensional spatial coordinates (m).

If the ground/soil/rocks' average (effective) thermal conductivity ($\bar{k}_{ground}$) over the heat transfer area and the temperature gradient of the earth formation in the normal direction (x) are constant, the total conductive heat transfer rate can be expressed for vertical boreholes of radii, r, and lengths (depths) L as follows (see Figure 13.3a):

$$\dot{Q} = 2\pi r \cdot L \cdot \bar{k}_{ground} \cdot dT/dx = \bar{k}_{ground} \cdot A \cdot dt/dx \tag{13.24}$$

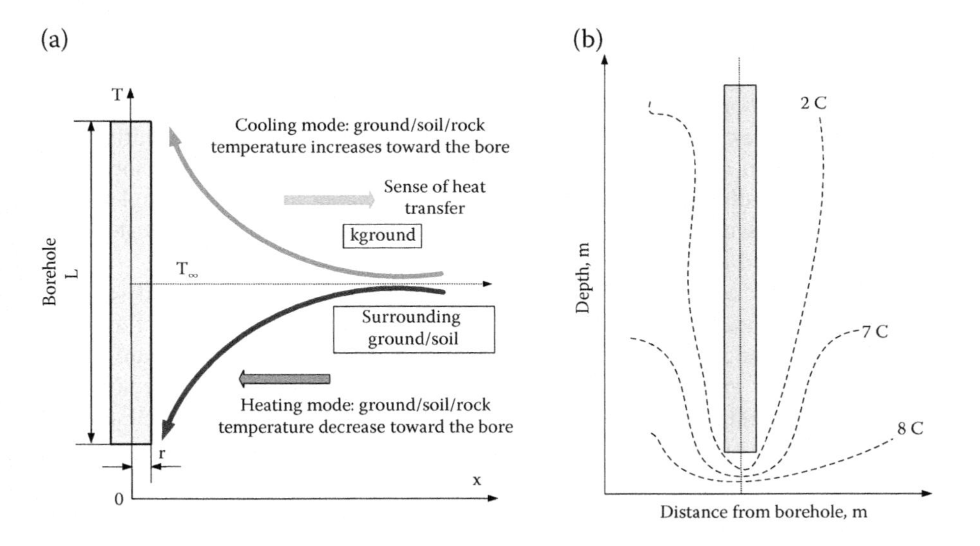

FIGURE 13.3 (a) Typical temperature variations in the ground/soil/rocks around a vertical borehole heat exchanger in heating and cooling modes; (b) temperatures around a (borehole) ground-coupled heat exchanger operating in heat extraction mode (Note: schematics not to scale).

where

$\dot{Q}$ is the total steady-state thermal flux (W).
r is the borehole radius (m).
L is the borehole depth (m).
$\bar{k}_{ground}$ is the ground/soil average (effective) thermal conductivity (W/m·K).
T is the temperature (K).
x is the vertical coordinate (m).
A is the borehole heat transfer area (m).
dt/dx is the temperature gradient (K/m).

When heat is extracted from/to the ground/soil for building heating or injected for building cooling, within a certain period of time, t, the temperature around the ground-coupled heat exchanger decreases or increases, respectively (Figure 13.3a). The temperature profiles around borehole heat exchangers strongly depend on the heat extracted from and dissipated to the surrounding ground/soil/rock formation (Figure 13.3b).

13.3.2 Analytical Models

Several one- (in the radial direction) and two-dimensional (in both radial and axial directions) analytical (e.g., line- or cylindrical-source) models have been developed to simulate and design the steady-state thermal response of vertical, single, and multiple closed-loop (indirect, secondary fluid) ground-coupled (borehole) heat exchangers embedded in semi- or infinite homogeneous ground/soil/rock formations, all being subject to constant heat transfer rates per unit borehole depth (also called heat pulses) over short- (e.g., minutes, hours or days) and/or long-time scales (e.g., years), accounting or not for end and/or leg-to-leg thermal short-circuiting effects.

Some of common assumptions used with one-dimensional analytical models are: (i) ground/soil/rock is an infinite isotropic, homogenous medium with initial uniform and constant far-field temperature; (ii) heat transfer occurs to/from the borehole in the radial direction; (iii) vertical heat transfer along the ground-coupled heat exchanger axis is neglected; and (iv) conduction is the unique heat transfer mechanism described by Fourier's law (Fourier 1878).

According to the technique employed to solve the energy equation to determine the radial temperature distribution in homogeneous ground/soil/rocks as functions of time and distance (radius) from heat source axis, analytical models are either based on infinite and finite line-source (Kelvin 1882; Allen 1920; Ingersoll and Plass 1948a, 1948b; Ingersoll et al. 1954; Ingersoll et al. 1948, 1954; Eskilson 1987) theories, or on infinite and finite cylindrical-source (Carslaw and Jaeger 1959; Ingersoll et al. 1954) concepts.

The use of analytical models is attractive because they: (i) offer better flexibility and computational times that are much less compared to those required for numerical models, and, with consistent assumptions (e.g., ignoring the geometry and thermal capacity of the borehole components, are suitable to be used for both analysis of in-situ ground/soil thermal response test data and to the design of borehole heat exchangers; (ii) can be applied to develop standards and simulation algorithms for any classical configurations of ground-coupled (borehole) heat exchangers embedded in homogeneous ground/soil/rock media; (iii) combined with temporal superposition technique, can be used efficiently to obtain time-varying (transient) solutions describing the variable thermal loads from/to the ground/soil/rocks over long-time periods (e.g., during several successive seasons or years).

On the other hand, it is more difficult to model, simulate, and design the ground-coupled heat exchangers with analytical models because: (i) in reality, the pipes are not co-axial with the borehole, and several different materials are involved; (ii) usually, the variation of brine temperature along the borehole depth is neglected; (iii) due to the complexity of borehole layout, most analytical solutions rarely consider the heat interference between the two adjacent legs of U-tubes; (iv) on long time scales, the heat transfer is transient (i.e., with time-varying heat transfer rates) within boreholes and surrounding ground/soil/rocks; thus, analytical models are less suited when one would like to take account of time varying heat transfer rates or the influence of surrounding boreholes over long time scales; (v) assuming the building load profiles on monthly basis, the accuracy of accounting for variations of average brine temperatures over short time periods decreases dramatically; (vi) in other words, since the borehole geometry is ignored, the analytical methods have limited application for short time analysis of heat transfer process; (vii) are difficult to integrate with building simulation software; and (viii) are less accurate and less able to analyze multiple borehole fields.

13.3.2.1 Infinite Line-Source Model

The infinite line-source model (Kelvin 1882) considers the two legs of U-shaped (borehole) ground-coupled heat exchangers as single, imaginary, straight infinite lines co-axial with the boreholes embedded in infinite, uniform, homogeneous ground/soil/rocks formations with constant and isotropic properties, and constant initial temperature equal to the far-field temperature (Figure 13.4). Heat extraction and heat rejection from/to ground/soil/rocks occur with constant densities of heat fluxes ($\dot{q}$) under constant mean temperature along the borehole.

The relatively simple line-source model is generally applied with assumptions, such as: (i) geometry of the borehole components is ignored; (ii) heat transfer in the axial direction is ignored, i.e., no vertical temperature gradient exists; (iii) heat transfer between the line source and ground/soil/rocks is carried out by pure, constant, radial thermal conduction; (iv) conductive heat transfer through the ground/soil/rocks is approximated as one-dimensional depending on the earth formation properties, and borehole thermal resistance; (v) heat transfer in both up and down flowing directions of brine in borehole heat exchanger is neglected; (vi) heat transfer across the upper ground/soil surface does not occur (i.e., the variation of ground/soil surface temperature is neglected it being considered adiabatic); (vii) heat transfer occurring below borehole (i.e., end thermal effect) is neglected; therefore, the infinite line-source solution is not suitable for multi-annual simulations of multiple borehole heat exchangers (e.g., over 5, 10, or 20 years where the end thermal effects may become important and, thus, need to be taken into account).

Due to such assumptions, some limitations of the line-source models exist when they are used in the analysis of ground/soil/rock thermal response tests, mainly

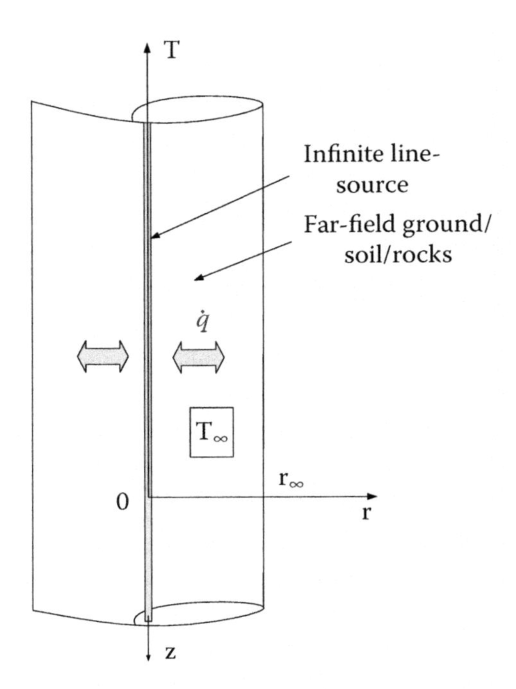

FIGURE 13.4 Schematic representation of the infinite line source; $\dot{q}$, heat flux density; r, radius; r_∞, ground/soil/rocks far-field radius; T, temperature; T_∞, ground/soil/rocks far-field temperature; z, vertical axis.

because: (i) geometry of the borehole, as well as the thermo-physical properties of the geothermal fluid (brine), U-tube, and grout are ignored; (ii) temperature of ground/soil/rock medium is initially (i.e., when a constant heat rate is switched on at time zero) uniform; (iii) perfect contact between the ground/soil/rock and the grout is presumed; and (iv) groundwater movement is generally not considered.

Because of its simplicity and efficiency in terms of computing efforts, the concept of infinite line-source solution is frequently used to determine: (i) heat transfer and performances of single borehole heat exchangers; (ii) borehole and ground/soil/rocks thermal responses; (iii) ground-coupled heat exchangers' required lengths for both heating and cooling modes of operation; (iv) temperature at the borehole wall; (v) difference between the brine temperatures and the borehole wall based on the borehole thermal resistance; (vi) temperature distribution around the line source; (vii) brine temperatures entering and leaving the geothermal heat pump (s); and (viii) overall energy efficiency of ground-source heat pump system.

To estimate the continuous time-dependent heat transfer between the infinite line-source and the surrounding ground/soil/rocks, and calculate the temperature of the undisturbed earth, the concept of far-field radius, beyond which the ground/soil/rocks temperature remains at a constant value, has been introduced. It defines the region around the infinite line source where the heat exchange between the ground/soil/rocks formation and the line-source occurs. The arbitrary magnitude of the far-field radius that depends on the line-source operating time and on the ground/soil/rock thermal diffusivity, is defined as follows (Hart and Couvillion, 1986):

$$\mathrm{r}_\infty = 4\sqrt{\alpha_g * \mathrm{t}} \tag{13.25}$$

where

r_∞ is the far-field radius (m).

α_g is the ground/soil/rocks thermal diffusivity (m^2/s).

t is the time (s).

The ground/soil/rocks formation at a distance greater than r_∞ is assumed to be at the constant undisturbed far-field temperature (T_∞) (see Figure 13.4). If, in the case of multiple borehole configurations, the r_∞ value is less than the actual minimum distance between two adjacent boreholes (in practice, of about 5 m), no thermal interference between boreholes is assumed to occur.

If groundwater movement is neglected, the heat transfer outside boreholes by pure conduction is governed by Fourier's law (Equation 13.26, expressed in cylindrical coordinates) with the following and boundary conditions (Equations 13.26a, 13.26b, and 13.26c):

$$\alpha_g\left(\frac{\partial^2 T_g}{\partial r^2} + \frac{1}{r}\frac{\partial T_g}{\partial r}\right) = \frac{\partial T_g}{\partial t} \tag{13.26}$$

$$T_g(r, t = 0) = T_\infty \tag{13.26a}$$

$$T_g(r \to \infty, \quad t) = T_\infty \tag{13.26b}$$

$$\dot{q}_L(r \to 0, t) = -(2\pi r)\cdot k_g\cdot\left(\frac{\partial T_g}{\partial r}\right)_{r\to 0} = \dot{q}_{borehole} \tag{13.26c}$$

where

α_g is the ground/soil/rocks thermal diffusivity (m^2/s).

T_g is the ground/soil/rocks temperature (°C).

r is the radial coordinate (m).

t is the time (s).

T_∞ is the (far-field) initial ground/soil/rocks temperature (°C).

$\dot{q}_L$ is the (borehole) heat exchanger heat flux per unit length (W/m).

k_g is the ground/soil/rock effective thermal conductivity (W/m·K).

To achieve errors not exceeding 2% for operating times longer than one day, the following dimensionless criterion must be satisfied (Ingersoll et al. 1954):

$$\alpha_g T_g / r_b^2 \geq 20 \tag{13.27}$$

where

α_g is the ground/soil/rocks thermal diffusivity (m^2/s).

T_g is the ground/soil/rocks temperature (°C).

r_b is the borehole diameter (m).

Assuming uniform initial temperature of ground/soil/rock formation (at $t = 0$) and constant heat flux per unit depth at $r = 0$, the ground/soil/rocks time-dependent undisturbed temperature at any point of radius (r) around the line-source with respect to far-field temperature (T_∞) (in other words, the temperature response in the ground/soil/rock due to a constant heat rate pulse), can be calculated with the following equation applicable to single and multiple vertical ground-coupled heat exchangers with diameters in the range of 50 mm or less (Kelvin 1882; Ingersoll and Plass 1948a, 1948b; Ingersoll et al. 1954; Bose et al. 1985; Hart and Couvillion 1986; Eskilson 1987):

$$T(r, t) - T_\infty = \frac{\dot{q}_L}{4\pi \bar{k}_g} \int_X^\infty \frac{e^{-\beta^2}}{\beta} d\beta \equiv \frac{\dot{q}_L}{4\pi \bar{k}_g} I(X) \tag{13.28}$$

where

$T(r, t)$ is the ground/soil/rock temperature at any selected distance, r, from the infinite line source (Note: selecting a distance equal to the borehole diameter represents the borehole surface temperature) (°C).

T_∞ is the ground/soil/rock initial uniform temperature (or far-field undisturbed temperature) (°C).

$\dot{q}_L$ is the constant linear heat transfer rate applied on unit length of borehole; when the ground/soil/rock acts as a heat source, $\dot{q}_L > 0$ (W/m).

$\bar{k}_g$ is the ground/soil/rock"effective" (average) thermal conductivity (W/m·K).

β is the integration variable (-).

$$X = r/2\sqrt{\alpha_g \cdot \Delta t}$$

r is the radius from the line heat (or sink) source (or the distance from center line of the borehole) (m); when the distance r equals to the pipe radius, the temperature, T, represents the pipe wall temperature.

$\alpha_g = \bar{k}_g/\bar{\rho}_g \cdot \bar{c}_{p,g}$ is the ground/soil/rocks thermal diffusivity (m^2/s).

$\bar{\rho}_g$ is the ground/soil/rocks average density (kg/m^3).

$\bar{c}_{p,g}$ is the ground/soil/rocks average specific heat (J/kg·K).

Δt is the time since the start of the operation (s).

$I(X) = \int_X^\infty \frac{e^{-\beta^2}}{\beta} d\beta$ is the exponential integral with tabulated values (Ingersoll et al. 1954) (-).

For $X \le 0.2$, the values of the exponential integral term can be obtained from the following approximation (Ingersoll et al. 1954):

$$I(X) = (2.303 \ \log_{10})/x + X^2/2 - X^4/8 - 0.2886 \tag{13.29}$$

For pipes with finite lengths, extending from z_1 to z_2, the infinite line source solution is given as (Ingersoll et al. 1954) (see Figure 13.4):

$$T(r,t) - T_{\infty} = \frac{\dot{q}_L}{4\pi \bar{k}_g} \int_{z_1}^{z_2} \frac{erfc\left(\sqrt{z^2 + r^2}/2\sqrt{\alpha_g \cdot t}\right)}{\sqrt{z^2 + r^2}} dz \qquad (13.30)$$

where

T_{∞} is the far-field ground/soil/rock temperature (°C).

$\bar{k}_g$ is the ground/soil/rock "effective" (average) thermal conductivity (W/m·K).

erfc is the error function (also called the Gauss error function), a complex function of a complex variable is defined as:

$$erf = 2/\sqrt{\pi} \int_0^m e^{-\beta^2} d\beta$$

β is the integration variable (-).

α_g is the ground/soil/rock thermal diffusivity (.....).

t is the time (s).

r is the radial coordinate (m).

z is the axial coordinate (m).

In cases where the radial heat transfer rate from or to the infinite line source is not constant but varies from month to month, the integral term in Equation 13.28 must be split into parts considering average heat transfer rates for a given time interval. The integration limits are then determined by integrand values corresponding to the beginning and the end of the particular time interval. The thermal interference between boreholes in close proximity can be determined by superposing the solution from each single borehole.

13.3.2.2 Infinite Cylindrical-Source Theory

The cylindrical-source theory (Carslaw and Jaeger 1959), similar to the line-source model, assumes that the vertical ground-coupled heat exchangers are infinite heat sources with finite radii (that, compared to depths, are much smaller), surrounded by infinite homogenous ground/soil/rock formations of constant thermos-physical properties along the depth (Figure 13.5).

By considering (or not) the borehole thermal mass, the cylindrical-source model, allows to predict the temperature distribution and/or the heat transfer rate around the vertically buried ground-coupled heat exchangers, as well as the brine temperature entering and leaving the geothermal heat pump(s) (Deerman and Kavanaugh 1991; Gu and O'Neal, 1995).

Assumptions as the followings are generally used with the infinite cylindrical-source models: (i) the two legs of U-tubes are approximated as a single pipe co-axial with the borehole having an equivalent diameter (for example, calculated as $D_{eq} = D\sqrt{2}$, where D is the diameter of each leg in perfect contact with the ground/soil/rocks formation); (ii) in other words, the geometry of boreholes is over-simplified assuming the borehole

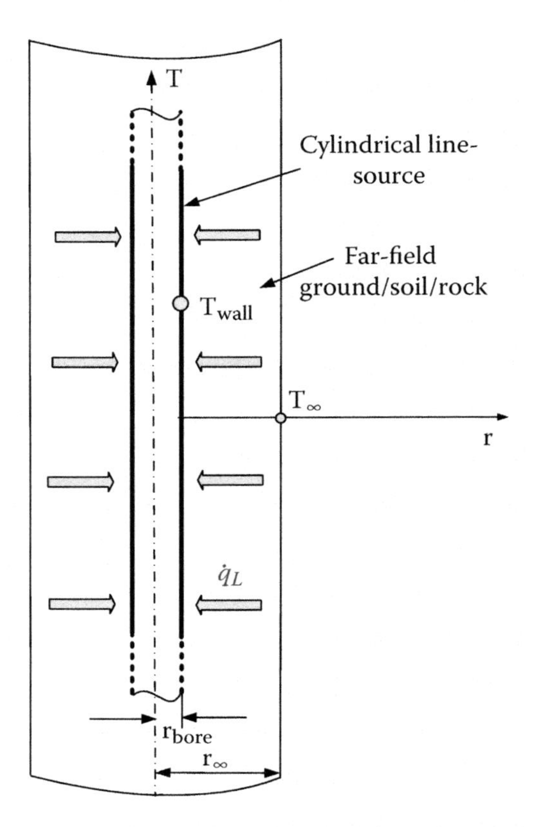

FIGURE 13.5 Schematic representation of an infinite-long cylinder embedded in a semi-infinite homogeneous ground/soil/rock medium with constant and isotropic properties; T_∞, ground/soil/rocks far-field undisturbed temperature; $\dot{q}_L$, density of heat flux; r, radii; r_{bore}, borehole radii; r_∞, ground/soil/rock far-field radii (Notes: when the diameter of the cylinder is zero, the cylindrical heat-source becomes a line heat-source).

ground-coupled heat exchangers as a single infinite isolated pipes with perfect contact with the borehole surrounded by an infinite solid ground/soil medium with constant properties; (iii) the cylinder transfers heat at constant rates from/to uniform surrounding infinite ground/soil/rock formations; (iv) the borehole is subjected to constant heat transfer by pure thermal conduction from (or to) the surrounding ground/soil/rocks for which the far-field temperature is the undisturbed temperature (T_∞); (v) groundwater movement in the earth, the short circuit heat transfer within the borehole introduced by the temperature difference between the two adjacent legs of U-tubes, and thermal interference between adjacent boreholes are neglected; (vi) the heat transfer in the axial direction is neglected; (vii) heat losses (or gains) from the borehole field to the ambient air and to the ground/soil/rocks region below the borehole field are not accounted for; (viii) the thermal mass and properties of the materials (brine, pipes, grout) inside the borehole and of the geothermal thermal carrier, are (or not) ignored; and (ix) heat transfer from the brine to the borehole wall is approximated as a steady-state process and, thus, the infinite cylindrical-source model has limited accuracy for dynamic simulation at short time steps.

Based on the governing equation of the transient heat conduction along with the given boundary and initial conditions, the temperature distribution surrounding the ground/soil can be given in the cylindrical coordinate as follows:

$$\alpha_g\left(\frac{\partial^2 T_g}{\partial r^2} + \frac{1}{r}\frac{\partial T_g}{\partial r}\right) = \frac{\partial T_g}{\partial t} \tag{13.31}$$

$-2\pi r_b \bar{k}_g \frac{\partial T_g}{\partial r} = \dot{q}_L$ for $r = r_b,\ t \geq 0$

$T - T_\infty = 0$ for $t = 0,\ r \geq r_b$

where

α_g is the ground thermal diffusivity (m^2/s).

T_g is the ground/soil/rocks temperature (°C).

r is the radial coordinate (m).

t is the time (s).

T_∞ is the (far-field) initial ground/soil/rock temperature (°C).

$\dot{q}_L$ is the (borehole) heat exchanger heat flux per unit length (W/m).

$\bar{k}_g$ is the ground/soil/rock effective (average) thermal conductivity (W/m·K).

Assuming a constant heat flux along the borehole, the governing equation of cylindrical-source solution that yields the temperature difference between the outer cylindrical surface and the undisturbed far-field ground/soil/rock temperature can be written in terms of Bessel's functions (Carslaw and Jaeger 1959; Ingersoll et al. 1954):

$$T(r,\ t) - T_\infty = \dot{q}_L/\bar{k}_g \cdot B(Fo_{rb},\ r/r_b) \tag{13.32}$$

where

$T(r,\ t)$ is the ground/soil temperature at any selected distance, r, from the infinite line source (Note: selecting a distance equal to the borehole diameter represents the borehole surface temperature) (°C).

T_∞ is the ground initial uniform temperature (or far-field undisturbed ground temperature, T_∞) (°C).

$\bar{k}_g$ is the ground/soil/rocks effective (average) thermal conductivity (W/m·K).

$\dot{q}_L$ is the linear heat flux transferred (W/m).

$\dot{q}_L$ is the constant linear heat transfer rate applied on unit length of borehole; when the ground/soil/rock acts as a heat source, $\dot{q}_L > 0$ (W/m).

$B(Fo_{rb},\ r/r_b)$ is the cylindrical-source Bessel's function depending on Fo_{rb} and dimensionless radius (r/r_b) given as tabulated values (-).

Fo_{rb} is Fourier number based on borehole heat exchanger radius r_b:

$$Fo_{rb} = \alpha_g \cdot t/r_b^2$$

α_g is the ground/soil/rock thermal diffusivity (m^2/s).
t is the time (s).
r_b is the borehole radius (m).

13.3.3 Numerical Models

Typical vertical ground-coupled heat exchanger fields consist of one or several boreholes of which time-variable thermal power evolution depends on factors as: (i) profiles of building thermal loads; (ii) borehole thermal resistances; (iii) instantaneous heating coefficients of performance and cooling energy efficiency ratios of geothermal heat pumps; (iv) brine entering and leaving temperatures; and (v) instantaneous values of the ground/soil/rocks temperatures resulting from previous heat flux pulses.

Numerical models (as finite element, finite difference, and finite volume) evaluate the borehole thermal response with higher levels of flexibility, details, and accuracy of a great number of ground-coupled (borehole) heat exchanger configurations and geometries over both long and short time periods. The relative complex structure of numerical models require significant computer consuming time and more expenses when hourly or sub-hourly time steps are considered, especially for multiple borehole fields.

The finite-element methods are used for solving partial differential equations of steady-state and transient heat transfer processes as those that occur inside and outside multiple vertical (and horizontal) ground-coupled (borehole) heat exchangers (Al-Khoury et al. 2005; Al-Khoury and Bonnier 2006; Reddy 2006; He 2012).

Two-dimensional (e.g., SUTRA) and three-dimensional (e.g., AQUA3D and THETA) finite-element commercial software are used to model of mass and heat transfer by considering the groundwater flow around vertical ground-coupled heat exchangers (Chiasson 1999).

The numerical model SUTRA (saturated-unsaturated transport), for example, based on both two-dimensional finite-element and finite-difference methods to approximate the governing equations (Voss 1984), simulates saturated or unsaturated groundwater movement and the transfer of thermal energy in the groundwater and solid matrix of aquifers.

The finite-difference method is used to determine: (i) temperature distribution over time in ground/soil/rock formations at given distances from boreholes submitted to variable (heat pulse) thermal loads and (ii) thermal interference between boreholes in underground heat storage applications.

The numerical finite-difference model THETA (Kangas 1996) accurately simulates the three-dimensional coupled transport of fluid and energy in porous media, as well as the effect of groundwater flow on the performance of the groundwater heat pump systems using vertical wells.

The numerical finite-volume method represents and evaluate partial differential equations in the form of algebraic equations referring to small volumes surrounding each node point on given complex meshed geometries such as U-tube ground-coupled heat exchangers and standing columns (Patankar 1980, 1991; Toro 1999; Randall 2000; Young 2004).

Two-dimensional finite-volume models provide significantly more accurate representations of the ground-coupled (borehole) heat exchangers by separately modeling the individual borehole components. The thermal effects of U-tube pipes, the backfill grout, ground/soil/rocks, and the heat transfer fluid can be accounted for as separately interacting entities. The finite-volume method has also been applied to develop dynamic three-dimensional (3-D) models (that generally offer explicit and accurate representations of heat transfer and fluid temperature variation with depth) to investigate the characteristics of heat transfer inside and surrounding single and multiple (arrays of) borehole heat exchangers at both short and long time scales.

13.3.3.1 Long-Time Step Temperature Response Factors

Depending on the balance between the heat extraction and heat rejection from and to the earth, the ground/soil/rocks temperature in the neighborhood of the ground-coupled heat exchangers may rise or fall over a number of years because of the thermal response to both short- and long-term fluctuations of thermal loads due to operation of geothermal ground-source heat pumps. In the case of heating- or cooling-dominated buildings, such variations may result in lower performances of the geothermal heat pumps as the brine temperature changes in the same direction over time, potentially exceeding the optimum operating temperature ranges. Therefore, a design goal should be to predict the temperature rises or drops within acceptable limits over the technical life of the ground-source heat pump systems.

Any heat extraction and heat rejection varying in time and space can be decomposed into sets of heat pulses. The response to each unit step heat pulse can be superimposed to calculate the overall thermal response of boreholes to any heat extraction/rejection time series.

The spatial superposition aims to determine the response of pre-defined configuration of boreholes (characterized by their ratio of horizontal spacing to depth) to the unit heat pulses. Using the thermal response from a single borehole, and neglecting the thermal conductivity of the borehole components, such as the pipe wall and the grout, the spatial superposition of a predefined configuration of multiple boreholes can be performed to determine the response of borehole configuration to the unit step heat pulse.

The principle of superposition of such independent elementary unit steps serves as the basis for analytical models applied to steady-state long-time heat transfer processes.

The temperature fields from single boreholes can be superimposed in space to obtain the thermal response from the whole borehole field and, then, converted to a series of non-dimensional temperature long- and short-time response factors (called g-functions) in response to step extraction or rejection heat pulses (Eskilson 1986, 1987).

The temperature response to heat pulses of boreholes can be converted to a series of non-dimensional temperature response factors (called g-functions) which represents the thermal behavior of a specific borehole system as a function of time and heat input, i.e., describe the relationship between the average borehole wall temperature and the step heat extractions/rejections, independently on the ground/soil/rocks "effective" thermal conductivity (Eskilson 1987).

The temperature response factors can be generated by using both numerical and analytical approaches (Eskilson 1987; Yavuzturk and Spitler 1999).

The non-dimensional temperature response functions, also known as g-functions, is an approach proposed by Eskilson (1987) to calculate the temperature variation at borehole wall in response to step heat inputs. The temperature response of vertical boreholes to a single step heat input from minutes to decades is represented by a set of non-dimensional factors. By transferring the heat input function into piecewise constant step heat inputs, and superimposing with the corresponding temperature response function value at each step, the temperature variation at the borehole wall due to any arbitrary heat input function can be calculated. The temperature response function values are calculated by interpolating between the near temperature response factors.

Once the thermal response of the borehole field to a single-step heat pulse is represented with a g-function, the thermal response to any arbitrary heat extraction/rejection function can be determined by devolving the heat rejection/extraction into a series of step functions, and superimposing the thermal response to each step-function.

In other words, these factors allow the calculation of the temperature change at the borehole wall in response to step heat extraction/injection pulses, which can be determined by summing the responses of the previous step functions for a defined configuration of borehole heat exchangers (with different spacing/depth ratios) over long time scales. Using the step g-functions, the temperature response at the end of a given period of time and at any point may be determined by superimposing the responses to each individual thermal load.

By neglecting the thermal conductivity of different materials inside a borehole (including those of brines, pipes and grouts), the temperature responses of the borehole fields were converted to non-dimensional sets of thermal response factors, called long-time step g-functions, valid to time scales greater than a few hours ($t \geq 5r_b^2/\alpha$) and calculated for many (over 200) geometrical configurations of multiple borehole heat exchangers arranged in different shapes and with different the ratios of borehole spacing to borehole length.

Over long periods of time (from months to up to 25 years), the g-functions describe the thermal response of borehole fields (Eskilson 1987).

A significant advantage of the g-function method is that, regardless of the configuration of a ground-coupled heat exchanger and the time scale, a set of non-dimensional temperature response factors are sufficient to describe the thermal behavior of a borehole ground-coupled heat exchanger.

When the borehole outer wall temperature versus time response is non-dimensionalized, the resulting dimensionless temperature versus dimension-less time curve is the g-function. Individual response factors, which give the temperature response to a specific length unit step function heat pulse, are determined by interpolating the g-functions. Once the response factors have been determined, the response of ground-coupled loop heat exchangers to any heat rejection/extraction versus time profile can be determined by decomposing the heat rejection/extraction versus time profile into a set of unit-step functions. Then, the response of ground-coupled heat exchangers to each unit-step function can be superimposed to determine the overall thermal response.

Once the response to a step function is known, the response to any heat extraction/injection step can be determined by decomposing the heat extraction/injection into a series of unit-step functions. Then, by using the long-term response factors (g-functions) to each unit step, functions can be superposed to determine the overall thermal response. This technique is intended to provide the response of the ground/soil to heat rejection/extraction over longer periods of time (up to 25 years), but as the numerical model that provides the g-functions does not account for the local borehole geometry, it cannot accurately provide the shorter-term response.

Given the thermal response for a single value of Q, the response to any heat extraction/rejection value of Q can be determined by dividing the heat extraction/rejection into a series of step functions, and superimposing the response to each step function.

For example, consider four months of heat rejection (Figure 13.6). The first month's heat pulse (Q_1) is applied for the entire period time scale; the second (Q_2), from the second month onward; and so forth, with each heat pulse equaling that month's value less the value from the previous month. In other words, the step heat inputs Q_2, Q_3, and Q_4 are superimposed in time on to the basic heat pulse, Q_1.

As can be seen in Figure 13.6, the basic heat pulse from zero to Q_1 is applied for the entire duration of, for example, the four months, and is effective as $Q_{1'} = Q_1$. The subsequent pulses are superimposed in time a series of step pulses: $Q_{2'} = Q_2 - Q_1$ (effective for three months, from t_1 to t_3), $Q_{3'} = Q_3 - Q_2$ (effective for two months, from t_2 to t_3), and, finally, $Q_{4'} = Q_4 - Q_3$ (effective for one month). Thus, the borehole wall temperature at time t_3 can be determined by adding the temperature responses of the three step functions.

The superposition principle determines the temperature of the borehole wall at the end of the nth month due to the heating and cooling loads by summing the responses of the previous step function:

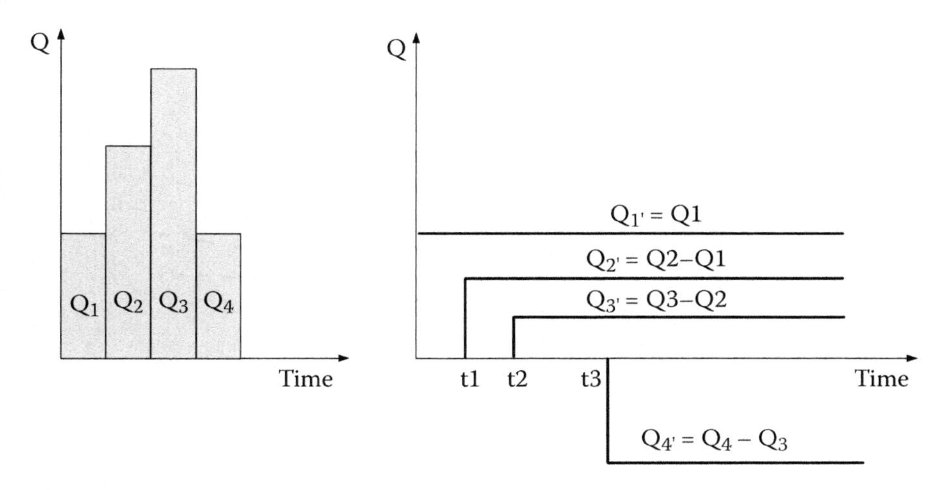

FIGURE 13.6 Superposition representation of four months of linear step heat inputs (pulses) in time.

$$\bar{T}_{borehole}^{wall} = T_\infty + \sum_{i=1}^{n} \frac{(q_{L,i} - q_{L,i-1})}{2\pi \bar{k}_g} \cdot g\left(\frac{t_n - t_{i-1}}{t_s} \cdot \frac{r_{borehole}}{L_{borehole}}\right) \quad (13.33)$$

where

$\bar{T}_{borehole}^{wall}$ is the average borehole wall temperature (°C).

T_∞ is the undisturbed (far-field) ground/soil/rock temperature (°C).

i is the index denoting the end of a time step (e.g., the end of the first hour or second month, etc.) (-).

$q_{L,i}$ is the individual heat extraction/injection pulse per unit length of borehole at step i (W/m).

$\bar{k}_g$ is the ground/soil/rock effective (average) thermal conductivity (W/mK).

t is the time (s).

$t_s = H_{borehole}^2/9\alpha$ is the time (characteristic) scale (s).

$r_{borehole}$ is the borehole radius (m).

$H_{borehole}$ is the active borehole depth (m).

α is the ground thermal conductivity (m^2/s).

For a fixed value of borehole spacing (B) (e.g., 5 m) and for a fixed number of boreholes and borehole field aspect ratio (B/H), the borehole wall temperature (T_b in Figure 13.7) is given by (Eskilson 1987):

$$T_b = T_g - \dot{q}/2\pi\bar{k}_g \cdot g(t/t_s, r_b/H) \quad (13.34)$$

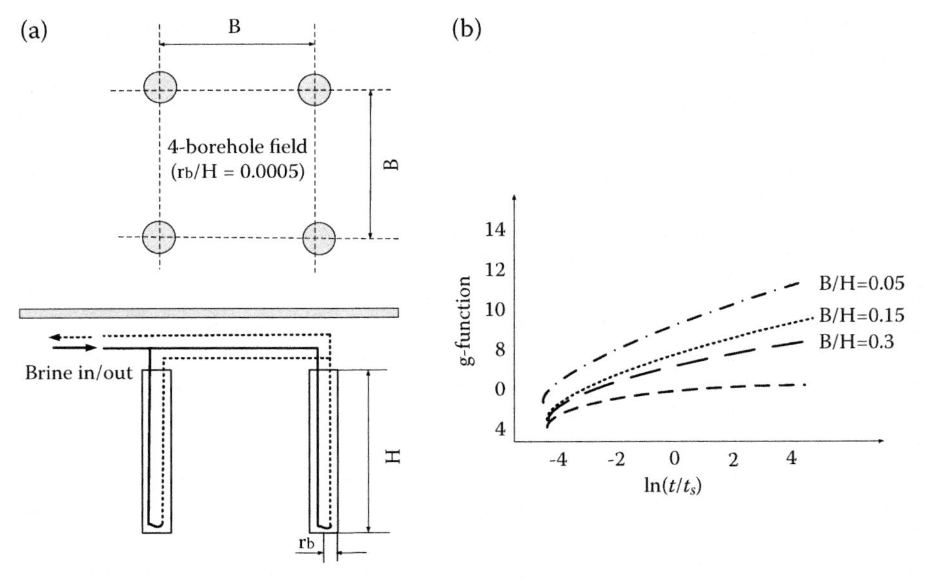

FIGURE 13.7 (a) Schematic of a four-borehole field; (b) typical set of temperature response g-functions.

where

T_b is the borehole wall temperature (°C).

T_g is the ground temperature (°C).

$\dot{q}$ is the linear heat flux (W/m).

$\bar{k}_g$ is the ground/soil/rock effective (average) thermal conductivity (W/m·K).

$g(t/t_s, r_b/H)$ is Eskilson's g-function (-).

t is the time (s).

$t_s = H^2/9\alpha_g$ is a characteristic time (s).

H is the borehole depth (length) (m).

The temperature response factor curves (g-functions) are graphically plotted as functions of non-dimensional time [$\ln(t/t_s)$] (where t is the operation time, in seconds; t_s is the time scale, in seconds) for various arrangements of multiple borehole fields and aspect ratios r_b/H (where r_b is the borehole diameter and H is the borehole depth), compared to the temperature response factor curve for a single borehole (Figure 13.8).

A typical set of temperature response factors is shown in Figure 13.7 for a four-borehole field arranged in a rectangle manner.

As can be seen in Figures 13.7 and 13.8, the non-dimensional response curves (g-functions): (i) provide the relation between the heat extracted from the ground/soil/rocks and the mean temperature at the borehole wall as a function of time and borehole heat exchanger field geometry; (ii) predict the thermal response and the performance of a borehole heat exchanger over the time scale from a single month to several years (i.e., over long times).

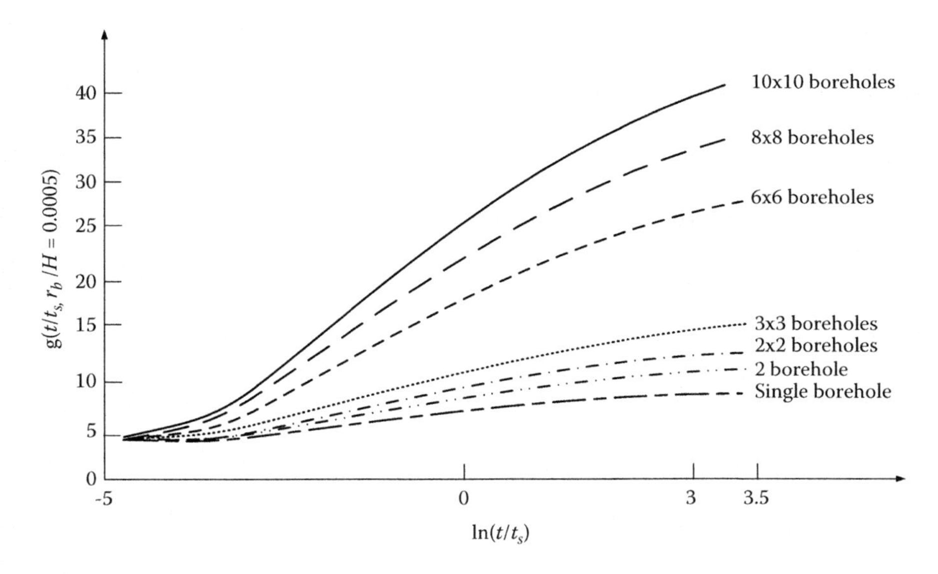

FIGURE 13.8 Temperature response factors (g-functions) for various multiple borehole configurations with a fixed ratio of B/H = 0.1 between the borehole spacing and the borehole depth, where B is the borehole spacing (m) and H is the borehole depth (m) compared to the temperature response curve for a single borehole.

13.3.3.2 Short-Time Step Temperature Response Factors

The long-time-step g-function model (Eskilson 1987) provides an efficient solution to simulate borehole fields with defined configurations over long time scales, ranging from one month to up to 25 years. However, due to the simplification of borehole geometry and the neglect of the thermal properties of all the components inside the borehole, this model is not suitable for accurate simulation of borehole heat exchangers over shorter time scales.

Some of the limitations of Eskilson's long-time step model can be summarized as follows (Yavuzturk and Spitler 1999, 2001): (i) the numerical model used in developing the long-time step g-functions approximates the borehole as a line source of finite length, so that the borehole end effects can be considered, and it is only valid for times greater than $(L_{borehole}^2/9\alpha)$; for a typical borehole, that might imply times from 3 to 6 hours; however, for a model that is suitable for energy simulation, it is highly desirable that the solution be accurate down to an hour and below; (ii) much of the provided data does not cover periods of less than one month; (iii) short-time thermal effects are not considered; for example, time steps less than two hours are not used; (iv) applied to limited number of borehole configurations and every change in borehole depth requires a change of the borehole field spacing; (v) cannot accurately and reliably account for temperature variations in boreholes during relatively short time steps (e.g., hourly, daily, and weekly) which limits the study of time-of-day utilities' demand-side energy management.

If the ground-coupled heat exchangers are submitted to short-term (peak) thermal loads (e.g., if brine temperatures rise rapidly, let say 5–10°C, in 1–2 hours) can dampen the short-term thermal response at the borehole wall where thermal conductivity of grout and thermal mass of the brine have significant impacts limiting the minimum or maximum entering brine temperature to the (borehole) ground-coupled heat exchanger.

The short-term behavior of ground-coupled heat exchangers is important for the design and the energy analysis of ground-source heat pump systems. Using short time-step response factors that account for thermal capacity and thermal resistance of individual boreholes, including the pipe walls, grout, and brine flow rates, the evaluation of system energy consumption and electrical demand on hour-by-hour basis (or shorter time) intervals becomes possible for commercial/institutional buildings that have time-of-day electricity rates.

The long time-step model has been extended to transient heat transfer in ground-coupled heat exchangers a two-dimensional (radial-angular) numerical finite-volume method that provides detailed representation of the borehole geometry and allow the thermal properties of pipes and grout to be considered precisely in order to develop g-functions suitable for shorter time periods. For typical ratios of borehole radius to borehole depth, the short time-step g-function data correspond to time steps between 2.5 minutes and 200 hours (Young 2004).

Short time-step methodology, accounting for tube, grout, and flow-related convective thermal resistances, allows the calculation of brine temperatures on minutely or hourly basis based on variable heat extraction and rejection rates, thermal characteristics of the ground/soil formation, and ground-loop heat exchanger geometry.

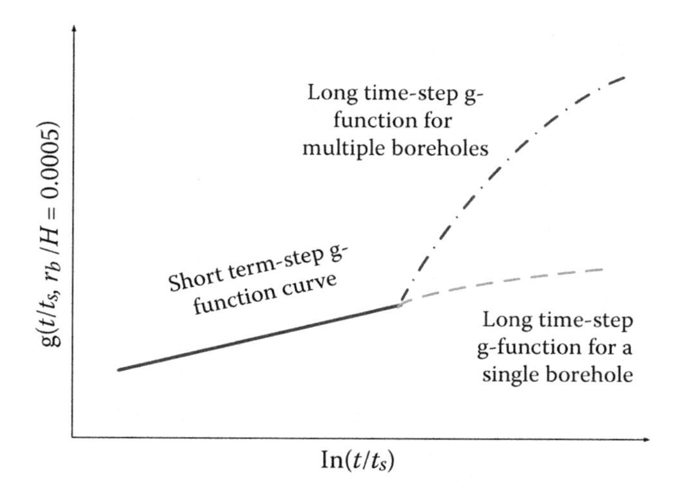

FIGURE 13.9 Short time-step g-function curve as an extension of the long time-step g-functions plotted for a single borehole and multiple (e.g., 8 × 8) borehole field; t, time; t_s, time scale.

The short time-step g-function curve blends into the long-time step g-functions developed by Eskilson, as shown in Figure 13.9.

As can be seen in Figure 13.9, the short-time-step g-function plotted side by side with the long time-step g-function values for a single borehole and a multiple (e.g., 8×8) borehole field line up well with Eskilson's long time-step g-function (Yavuzturk and Spitler 1999).

REFERENCES

Al-Khoury, R., P.G. Bonnier. 2006. Efficient finite element formulation for geothermal heating systems Part II: Transient. *International Journal for Numerical Methods in Engineering* 67:725–745.

Al-Khoury, R., P.G. Bonnier, R.B.J. Brinkgreve. 2005. Efficient finite element formulation for geothermal heating systems Part I: Steady state. *International Journal for Numerical Methods in Engineering* 63:988–1013.

Allen, J.R. 1920. Theory of heat loss from pipe buried in the ground. *Journal ASHVE (American Society of Heating and Ventilating Engineers)* 26:455–469, 588–596.

Bose, J., J. Parker, F. McQuiston. 1985. *Design/Data Manual for Closed-Loop Ground Coupled Heat Pump Systems*, Oklahoma State University, Oklahoma, OK.

Carslaw, H.S., J.C. Jaeger. 1959. *Conduction of Heat in Solids, Second Edition*, Clarendon Press, Oxford, UK.

Chiasson, A. 1999. Advances in Modeling of Ground-Source Heat Pump Systems. Master Thesis, Oklahoma State University, Stillwater, OK.

Deerman, J.D., S.P. Kavanaugh. 1991. Simulation of vertical U-tube ground-couple heat pump systems using the cylindrical heat source solution. *ASHRAE Transactions* 97:287–295.

Eskilson, P. 1986. *Superposition Borehole Model. Manual for Computer Code. Lund, Sweden*, Department of Mathematical Physics, Lund Institute of Technology, Lund, Sweden.

Eskilson, P. 1987. Thermal Analysis of Heat Extraction Boreholes. Doctoral Thesis, University of Lund, Department of Mathematical Physics, Lund, Sweden, June.

Fan, R., Z.L. Ma. 2006. Heat transfer analysis of geothermal heat exchanger under coupled thermal conduction and groundwater advection. *Journal of HV&AC* 36(2):6–10.

Fossa, M. 2011. The temperature penalty approach to the design of borehole heat exchangers for heat pump applications. *Energy and Buildings* 43(6):1473–1479.

Fourier, J. 1878. *The Analytical Theory of Heat*, Cambridge University Press, Cambridge. Translated with notes by Alexander Freeman, London, UK.

Gu, Y., D.L. O'Neal. 1995. Analytical solution to transient heat conduction in a composite region with a cylindrical heat source. *Journal of Solar Energy Engineering, Transactions of the ASME (American Society of Mechanical Engineers)* 117(3):242–248.

Gu, Y., D.L. O'Neal. 1998. Development of an equivalent diameter expression for vertical U-tubes used in ground-coupled heat pumps. *ASHRAE Transactions* 104(2):347–355.

Hart, D.P., R. Couvillion. 1986. *Earth Coupled Heat Transfer*, Publication of the National Water Well Association, Volume 92, pp. 22–25, Westerville, OH.

He, M. 2012. *Numerical Modelling of Geothermal Borehole Heat Exchanger Systems. Thesis submitted in partial fulfilment of the requirements of De Montfort University for the award of Doctor of Philosophy*. Institute of Energy and Sustainable Development De Montfort University, Leicester, UK, February.

Holman, J. P. 1997. *Heat Transfer*. 8th Edition, McGraw-Hill, New York.

Incropera, F.P., D.P. DeWitt, T.L. Bergman, A.S. Lavine. 2007. *Fundamentals of Heat and Mass Transfer*, Wiley, New York.

Ingersoll, L.R., H.J. Plass. 1948a. Theory of the ground pipe heat source for the heat pump. *Heating, Piping and Air Conditioning* 20:119–122.

Ingersoll, L.R., H.J. Plass. 1948b. Theory of the ground pipe source for the heat pump. *ASHVE (American Society of Heating and Ventilating Engineers) Transactions* 54:339–348.

Ingersoll, L.R., O.J. Zobel, A.C. Ingersoll. 1948. *Heat Conduction with Engineering and Geological Application*, McGraw-Hill, New York, 278 p.

Ingersoll, L.R., O.J. Zobel, A.C. Ingersoll. 1954. *Heat Conduction with Engineering, Geological, and Other Applications*, McGraw-Hill, New York, US.

Ingersoll, L.R., F.T. Adler, H.J. Plass, A.C. Ingersoll. 1954. Theory of earth heat exchangers for the heat pump. *ASHVE (American Society of Heating and Ventilating Engineers) Transactions* 56:167–188.

Kangas, M.T. 1996. Thermo-hydraulic analysis of ground as a heat source for heat pumps using vertical pipes. *Journal of Energy Resources Technology* 118:300–305.

Kelvin, T.W. 1882. *Mathematical and Physical Papers, Volume 4: Hydrodynamics and General Dynamics. 2012. William Thomson, Edited by Sir Joseph Larmor*, Cambridge University Press, Cambridge, UK.

Patankar, S.V. 1980. *Numerical Heat Transfer and Fluid Flow*, Hemisphere, New York.

Patankar, S.V. 1991. *Computation of Conduction and Duct Flow Heat Transfer*, Innovative Research, Inc., Maple Grove, MN.

Randall, L. 2000. *Finite Volume Methods for Hyperbolic Problems*, Cambridge University Press, Cambridge, UK.

Reddy, J.N. 2006. *An Introduction to the Finite Element Method, 3rd Edition*, McGraw-Hill, New York.

Rottmayer, S.P., W.A. Beckman, J.W. Mitchell. 1997. Simulation of a single vertical U-tube ground heat exchanger in an infinite medium. *ASHRAE Transactions* 103(2):651–659.

Toro, E.F. 1999. *Riemann Solvers and Numerical Methods for Fluid Dynamics*, Springer-Verlag, New York.

Voss, C.I. 1984. *SUTRA: A Finite-element Simulation Model for Saturated-Unsaturated Fluid-density Dependent Groundwater Flow with Energy Transport or Chemically Reactive Single Species Solute Transport*, U.S. Geological Survey, Reston, VA.

Young, R. 2004. Development, Verification, and Design Analysis of the Borehole Fluid Thermal Mass Model for Approximating Short Term Borehole Thermal Response. Master Thesis. Oklahoma State University. Stillwater, OK.

Yavuzturk, C., J.D. Spitler, S.J. Rees. 1999. A transient two-dimensional finite volume model for the simulation of vertical U-tube ground heat exchangers. *ASHRAE Transactions* 105(2):465–474.

Yavuzturk, C., J.D. Spitler. 1999. A short time step response factor model for vertical ground loop heat exchangers. *ASHRAE Transactions* 105(2):475–485.

Yavuzturk, C., J.D. Spitler. 2001. Field validation of a short time-step model for vertical ground loop heat exchangers. *ASHRAE Transactions* 107(1):617–625.

Zeng, H.Y., N.R. Diao, Z.H. Fang. 2003a. Heat transfer analysis of boreholes in vertical ground heat exchangers. *International Journal of Heat and Mass Transfer* 46(23):4467–4481.

Zeng, H.Y., N.R. Diao, Z.H. Fang. 2003b. A finite line-source model for boreholes in geothermal heat exchangers. *Heat Transfer Asian Research* 31(7):558–567.

14 Horizontal Closed-Loop (Indirect, Secondary Fluid) Ground-Source Heat Pump Systems

14.1 INTRODUCTION

Closed-loop (indirect, secondary fluid) ground-source heat pump systems with horizontal ground-coupled heat exchangers consist of networks of sealed high-density polyethylene pipes (at relatively low cost, without corrosion risks, and easy to install) buried in shallow trenches.

In cold and moderate climates, they use the relatively stable very low temperature (which, up to 2-m depth, varies accordingly to the geographical location, for example, from 20–25°C in south of the United States to 3–5°C in north of Canada) and energy storage capacity of undisturbed ground/soil being attractive alternatives for heating and cooling residential as well as commercial/institutional buildings. This type of ground-source heat pump systems relies upon the heat transfer from the ground/soil by means of secondary working fluids (generally, antifreeze mixtures, called brines) that are pumped through geothermal heat pumps to extract heat (in the heating mode) and reject heat (in the cooling mode) from/to buildings. After these heat transfer processes, the brines are pumped back to the ground-coupled horizontal heat exchangers to be reheated or re-cooled, respectively.

14.2 RESIDENTIAL AND SMALL-SCALE COMMERCIAL/ INSTITUTIONAL BUILDINGS

Since larger land areas are required for installation, horizontal closed-loop (indirect, secondary fluid) ground-coupled heat exchangers are generally applied for small residential and commercial/institutional buildings, particularly in rural and sub-urban regions.

Such systems consist of the following main components (see Figures 14.1a and 14.1b): (i) at least one horizontal ground-coupled heat exchanger; (ii) at least one reversible mechanical vapor compression brine-to-air or brine-to-water geothermal heat pump with or without a desuperheater (for domestic hot water preheating) linked to both an underground closed-loop heat exchanger and to the building heat distribution (by air or hot water) system; (iii) a brine circulating pump; (iv) a brine flow meter; (v) instrument ports for brine temperature and refrigerant pressure/ temperature measurements; and (vi) an internal building heating/cooling

DOI: 10.1201/9781003032540-14

distribution system that delivers space heating and cooling via, for example, air-handling units, underfloor heating pipes, or radiators.

14.2.1 Operating Modes

In cold and moderate climates, by circulating brines through horizontal closed loop (indirect, secondary fluid) ground-coupled heat exchangers, heat extraction (in the heating mode), and heat rejection (in the cooling mode) are accomplished via reversible geothermal heat pumps. The temperature of brines entering the geothermal heat pumps can vary over extended temperature ranges going from about –5°C (in the heating/winter mode) to up to 35–45°C (in the cooling/summer mode). Inside the reversible geothermal heat pumps, the brine circulates through brine-to-refrigerant heat exchangers that act as refrigerant evaporators (in the heating mode) (Figure 14.1a) and as refrigerant condensers (in the cooling mode) (Figure 14.1b).

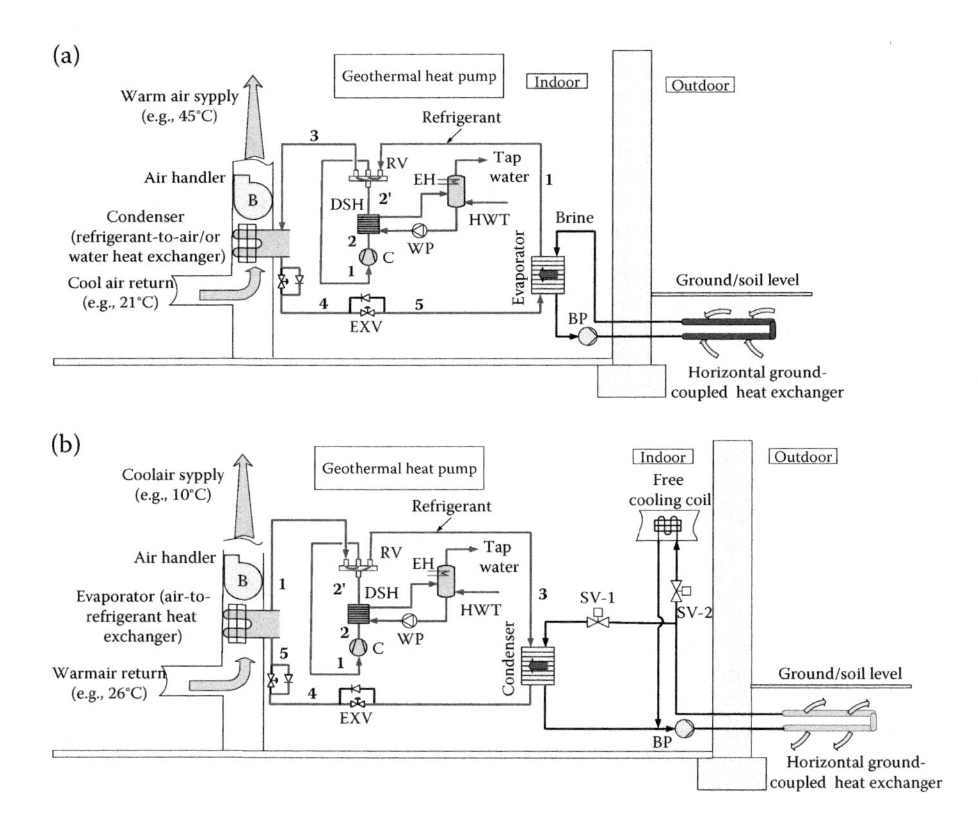

FIGURE 14.1 Horizontal closed-loop (indirect, secondary fluid) ground-source heat pump system for residential or small-scale commercial/institutional buildings operating in: (a) heating mode; (b) cooling mode. B, blower; BP, brine circulation pump; C, compressor; DSH, desuperheater; EH, electrical heater; EXV, expansion valve; HWT, hot water storage tank; P, brine pump; RV, reversing valve; SV, solenoid valve; WP, water circulation pump.

The thermodynamic process of such reversible geothermal heat pumps consists in the following thermodynamic processes (see Figures 14.1a and 14.1b): (i) the refrigerant leaving the evaporator as a slightly superheated vapor (state 1) is polytropicaly compressed (after passing through the four-way reversing valve RV) through the compressor C up to the further superheated vapor (state 2) depending on the setting of the expansion device that determines the degree of superheat at the evaporator outlet; the electrical energy input in the compressor is converted to shaft work to raise the pressure and temperature of the refrigerant; by increasing the vapor pressure, the condensing temperature is increased to a level higher than that of the heat source; (ii) in the desuperheater (DSH), the superheated refrigerant vapor is cooled (process 2–2') by transferring heat to cold water in order to preheat and store it (in a hot water storage tank [HWT], prior to being distributed to domestic hot water consumers; (iii) in the condenser (process 3–4), the superheated vapor is first cooled, then undergoes a two-phase condensation at constant temperature and pressure and, finally, the resulted saturated liquid is sub-cooled at constant pressure to a lower temperature in order to reduce the risks of flashing within the expansion valve; in other words, sub-cooling of the liquid refrigerant ensures that only liquid and no vapor bubbles enter the expansion valve; if vapor bubbles were to enter the expansion valve, there would be less liquid refrigerant available at the evaporator inlet, while the compressor still has to compress vapor that did not contribute to the evaporator thermal capacity representing a loss in the system overall efficiency; on the other hand, too much sub-cooling indicates that a considerable portion of the condenser volume is filled by a single-phase liquid, thus, the area available for heat rejection from the condensing refrigerant is smaller than it could be; as a consequence, the saturation temperature is increased along with the saturation pressure and, as a result, the compressor work increases; the degree of sub-cooling at the condenser outlet is primarily determined by the amount of refrigerant charged to the geothermal heat pump; (iv) after the condenser, the liquefied sub-cooled refrigerant (state 4) enters the expansion valve (EXV) where an expansion process at constant enthalpy (i.e., drop in pressure accompanied by a drop in temperature) (process 4–5) takes place in order to reduce the refrigerant pressure at a level corresponding to an evaporating temperature below to the heat source temperature; the expansion valve controls the refrigerant flow into the evaporator in order to ensure its complete evaporation, maintain an optimum superheat in order to avoid the liquid refrigerant to enter the compressor and, also, to avoid excessive superheat that may lead to overheating of the compressor; (v) the refrigerant then enters the evaporator in a two-phase state (state 5), absorbs (recover) heat from the heat source (brine), and undergoes change from a liquid-vapor mixture to saturated vapor at constant pressure and temperature (process 5–1); inside the evaporator, the saturated vapor is slightly superheated up to state 1 before entering the compressor; a reasonable vapor superheat is desirable to ensure that all the refrigerant flow rate is evaporated in order to achieve full evaporating capacity and, also, to supply to the compressor only refrigerant superheated vapor without any liquid entrained that could damage the compressor; however, excessive superheat may lead to overheating of the compressor; at this point, the geothermal heat pump thermodynamic cycle restarts.

Usually, on/off and/or programmable electric/electronic thermostats are used to efficiently control the operation of geothermal heat pump(s). The desuperheater, generally used in applications where the building cooling loads exist or are dominate, is a refrigerant superheated vapor-to-water heat recovery heat exchanger installed between the compressor and the four-way reversing valve. It operates only when the geothermal heat pump is running, having a small thermal power output (about 10–15% of the total geothermal heat pump thermal power output). This means that, in heating mode, the desuperheater leads to a small reduction in compressor thermal power output. When the space heating need is satisfied, the geothermal heat pump turns off, and no energy is available at the desuperheater for hot water production, thus an auxiliary electrical heater (EH) (Figures 14.1a and 14.1b) is required inside the hot water storage tank. With very low cold water flow rates, output temperatures as high as about 65–70°C can be achieved. However, in practice, even it contribute to improve the geothermal heat pumps' seasonal and year-round performance factors, the maximum output temperatures of cold water leaving the desuperheater is around 35–45°C, i.e., much lower than the required domestic hot water supply temperature (up to 60–65°C to reduce the risk of Legionella). Therefore, to boost the temperature of the domestic hot water before delivery, two solutions could be adopted: (i) insert an auxiliary electric heater (EH) into the hot water storage tank (HWT) linked to the geothermal heat pump's desuperheater (as shown in Figures 14.1a and 14.1b); the auxiliary immersion heater should not be able to operate at the same time as the desuperheater; or (ii) include a second (series) conventional electric hot water reservoir to re-heat (and store) the domestic hot water (not shown in Figures 14.1a and 14.1b); if possible, it should make use of electricity during daily reduced tariff periods.

In the heating mode (Figure 14.1a), the brine pump (BP) circulates the antifreeze mixture (brine) through the horizontal-coupled heat exchanger that extracts heat from the ground/soil/rock (acting as a heat source), and through the geothermal heat pump's evaporator where the refrigerant vaporizes by extracting heat from the heat carrier fluid, while the geothermal heat pump condenser, located inside the building, supplies heat rejects it into the building's indoor air (or water) distribution system acting as a heat sink medium. The brine-to-air or brine-to-water geothermal heat pump recovers heat from the brine at very low temperatures (ideally, between –5°C and 0°C, while the temperature of the far-field undisturbed ground/soil could rest at 5–10°C and even higher) and deliver heat at temperatures up to 35–45°C to the house indoor air (via warm air-handling units) or to hot water serving for space heating (e.g., through underfloor heating systems in the new, low-energy, highly insulated houses).

In the cooling mode, the cycle is reversed (Figure 14.1b). The building excess sensible and latent heat is dissipated (rejected) in the ground/soil/rock via the horizontal ground-coupled heat exchanger. In areas with suitable earth structure and favorable thermal conditions, it could be possible to partially store the rejected cooling energy in the ground/soil/rock to be used during the next heating season. In cold and moderate climates, during summer cooling-dominated periods, and depending on the size of buildings, it is possible to provide free (passive) cooling by bypassing the condenser of geothermal heat pump(s) and, thus, circulating the brine directly from the ground-coupled heat exchanger through, for example, one or several fan convectors (see Figure 14.1b). When the geothermal heat pumps operate in the cooling mode, there are two potential issues: (i) because the quantity of heat rejection is usually higher than that of heat extraction during heating operation, the length of ground-coupled heat exchanger required for cooling operation is generally larger than that required for heating operation; this means that, usually, in horizontal ground-coupled heat exchangers the ground-coupled heat exchanger should be sized for cooling operation; (ii) because of heat rejection, the ground/soil is heated up and its moisture migrates away from the ground-coupled heat exchanger; this process results in a drop of the thermal conductivity of the surrounding ground/soil structure and, partly, in an increase of the thermal resistance between the ground/soil and the horizontal ground-coupled heat exchanger; these conditions may remain even at the beginning of the next heating season, lowering both heat transfer rates and the geothermal heat pump(s) heating coefficients of performance; moisture migration could be avoided if the ground/soil heat exchangers are installed below the groundwater table or in regions with abundant rain and/or snow precipitations.

Using combined heating and cooling operation of horizontal (indirect, secondary fluid) ground-source heat pump systems with reversible geothermal heat pump units allows heat extracted during the heating season to be, in part, rejected and stored in ground/soil during the cooling operation in summertime. In this way, at the beginning of the next heating season (with high temperatures resulting from the heat rejection during summertime, and, also, and at the beginning of the next cooling season, with low temperatures resulting from the heat extraction during the heating season), the ground/soil average temperatures are more favorable to achieve better heating and cooling energy performances.

14.3 LARGE-SCALE COMMERCIAL/INSTITUTIONAL BUILDINGS

Horizontal closed-loop (indirect, secondary fluid) ground-source heat pump systems are used in large-scale commercial/institutional buildings with nominal cooling capacities above 70 kW when sufficient spaces (as large outdoor car parking areas, free open agricultural or athletic fields, etc.) are available to install below-grade ground-coupled heat exchangers. Such systems can be provided with separate (zonal) horizontal shallow ground-coupled heat exchanger fields supplying several building thermal zones (Figure 14.2a), or with unique direct connected (via reverse-return piping) ground-coupled heat exchangers (Figure 14.2b). As for residential and small-scale commercial/institutional buildings, the underground heat exchangers are typically constructed of thermally fused plastic tubing installed in horizontal trenches (filled with the removed ground/soil, sand, or other high conductive filling materials) in which brines are circulated. The pipes are connected together in parallel or series, and the individual run-outs returned to the building, where they are first connected to the buildings' internal closed loops, separated or not by intermediate heat exchangers and, then, to brine-to-air and/or brine-to-water geothermal heat pumps that might be distributed on building indoor fluid (brine or water) closed loops (Figure 14.3a) or installed in central mechanical room(s) (Figure 14.3b).

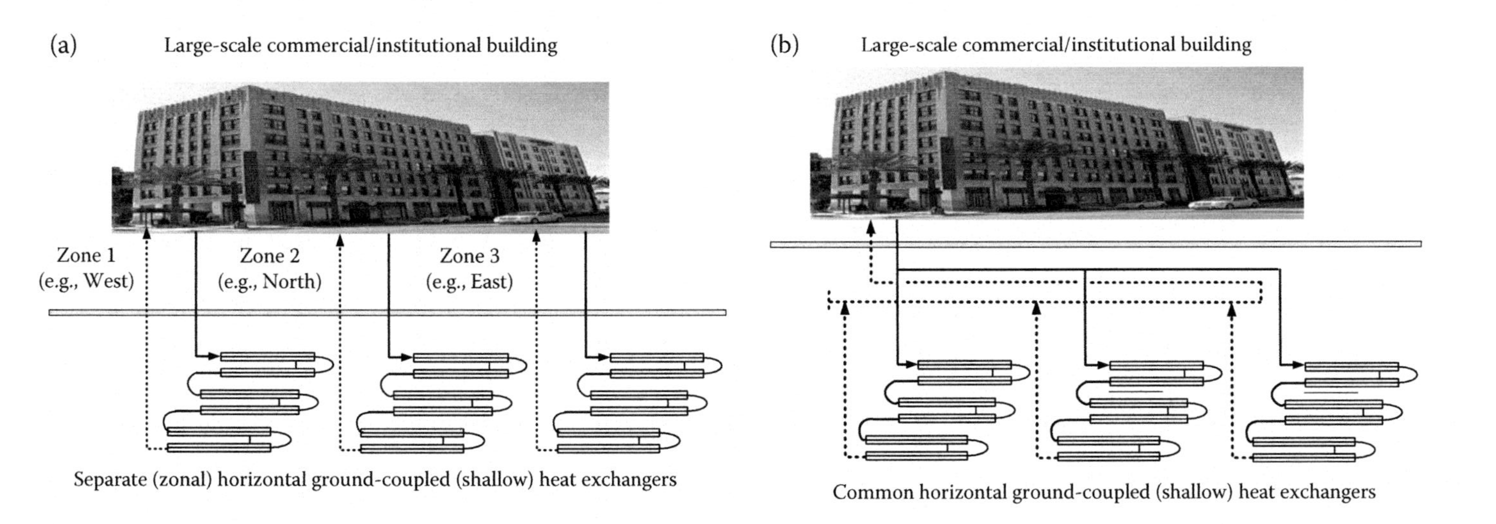

FIGURE 14.2 Schematics of horizontal ground-coupled heat pump systems for large-scale institutional/commercial building; (a) with separate (zonal) direct connected horizontal ground-coupled heat exchanger fields for different building thermal zones; (b) with common reverse-return connected horizontal ground-coupled heat exchangers (building view retrieved from: https://www.nps.gov/tps/how-to-preserve/briefs/14-exterior-additions.htm, accessed March 9, 2020) (Notes: drawings not to scale; shown horizontal ground-coupled heat exchanger sections connected in series).

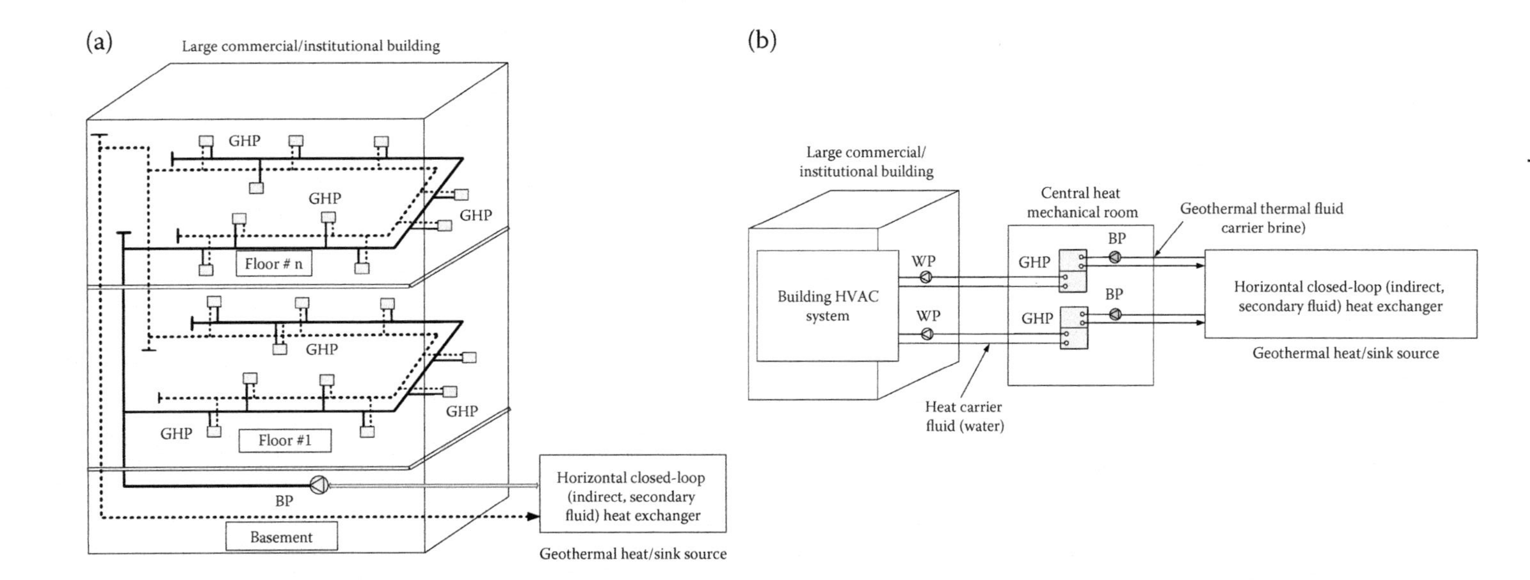

FIGURE 14.3 Schematics of geothermal heat pumps arrangements in large commercial/institutional buildings: (a) distributed and (b) central. BP, brine circulation pump; GHP, geothermal heat pump; HVAC, heating, ventilating, and air conditioning; WP, water circulation pump (Note: not all components shown).

Distributed geothermal heat pumps are optimum for large-scale constructions such as: (i) office buildings that, generally, have significant core and perimeter areas; cooling loads recovered from core areas can be transferred to perimeter zones where heating is simultaneously required; the diversity of thermal loads present in such buildings may result in significantly reduced sizes of geothermal heat pumps; (ii) hotels and motels, including restaurants, swimming areas, kitchens, laundry room, and meeting rooms, where heat recovered during the day cooling operation can be stored and used to warm guest rooms during the night; (iii) apartments and condominiums that can achieve individual metering of tenants' heat consumptions while offering heating and/or cooling from any geothermal heat pump unit at any time; (iv) schools and universities that have simultaneous periods of light and heavy thermal loads throughout classrooms; additionally, core-located laboratories, kitchens, and cafeterias can allow to recover heat and use it to heat perimeter-located classrooms and other spaces.

14.4 HEAT AND MASS TRANSFER

Shallow ground/soils are relatively complex structures consisting of layers that interact with variable atmospheric conditions which, depending on the thermal properties of the ground/soil (e.g., temperature, thermal conductivity, and moisture content), brine temperature, pipe thermal resistance, and the distance between trenches and the buried pipes, may cause temporal changes of their temperature profiles. In addition, the overall thermal efficiency of horizontal closed-loop (indirect, secondary fluid) ground-source heat pump systems is influenced by factors such as (Gan 2016; Larwa and Krzysztof 2020): (i) building thermal loads over short- (e.g., hourly and daily) and long-term (e.g., yearly or multiple years) operation; (ii) brine flow rate and circulation pump(s) performances; (iii) configuration, depth, and length of buried ground-coupled heat exchangers; (iv) pipe material, diameter, and wall thermal resistance; (v) temperature difference between the ground/soil and ambient air; (vi) undisturbed ground/soil temperature; (vii) properties of ground/soil and backfill material (e.g., thermal conductivity and diffusivity, and moisture content); and (viii) magnitude of groundwater natural movement (velocity).

The analysis of rather complex transient hear transfer processes at the ground/soil surface and, also, in trenches around the horizontal ground-coupled heat exchangers is usually performed by assuming simplifying assumptions such as the following (Bharadwaj and Bansal 1981; Neupauer et al. 2018): (i) even if the ground/soil is usually layered and inhomogeneous, it is treated as an infinite or semi-infinite homogeneous porous medium with constant thermal diffusivity and "equivalent" thermal conductivity; (ii) temperature at the ground/soil surface is variable periodically over time; (iii) deep geothermal gradient is zero, or negligible; (iv) thermally undisturbed ground/soils have uniform initial temperatures; (v) boundary conditions for the wall of buried pipes are considered either constant heat flux or constant temperatures; (vi) groundwater movement is generally assumed to be homogeneous and parallel to the ground/soil surface; advection resulting from the groundwater flow plays a beneficial role in reducing large changes in ground/

soil temperatures; (vii) ground/soil has large thermal mass, which causes relatively slow responses to changes in temperature variations at its surface; (viii) due to the large thermal mass of ground/soil, a lag exists between the ambient air and ground/soil surface temperatures; (ix) the large amplitude of ground/soil temperature fluctuations is dampened with depth, seeking constant temperatures; and (x) in cold and moderate climates, the amount of heat extracted from the ground/soil during the heating season prevails over the heat supplied to the ground/soil beyond this period; if, over annual periods, the amounts of heat extracted and supplied are comparable, the problems of long-term temperature changes of the ground/soil do not occur.

14.4.1 Ground/Soil Surface

The ground/soil can be treated as a geophysical complex structure consisting of subsurface layers, in which there are interactions related to changing weather conditions, and deeper layers, in which these impacts do not occur. Because the horizontal ground-coupled heat exchangers are placed in shallow trenches, their thermal performances are influenced by the varying conditions at the ground/soil surface. In other words, because the superposition of the ambient thermal conditions on the heating and cooling building thermal loads transferred via horizontal ground-coupled heat exchangers, the average temperature, as well as the heat storage capability of ground/soils are periodically disturbed, further decreasing (in heating-dominated operating modes) or increasing (in cooling-dominated operating modes) over short- and/or long-term periods of time. Knowledge of surface interaction with the ambient climate conditions that influence the daily, seasonal, and annual ground/soil temperature distributions along the depth is very important in order to prevent under- or over-sizing of horizontal ground-coupled heat exchangers.

The heat and mass transfer at the ground/soil surface include several thermal processes mainly due to: (i) variable ambient air temperature, (ii) variable solar radiation, (iii) variable wind speed, and (iv) irregular precipitations (rain, snow, etc.).

The exchange of energy between the overlying atmosphere and Earth's surface (and vice versa) involves, among many others, thermal processes such as: (i) absorption and emission of natural solar electromagnetic radiation, (ii) heat transfer towards or away from the surface within the atmosphere, (iii) thermal conduction of heat energy within the ground/soil, and (iv) evaporation of water stored in the ground/soil and/or condensation of atmospheric water vapor onto the surface.

Under natural conditions, heat and mass transfer at the ground/soil surface that can be treated as a plate with a finite thickness, is a cyclic thermal process where daily and seasonally variable thermal processes occur. Usually, in cold and moderate climates, the range of ground/soil temperature changes, resulting from daily thermal cycles at the ground/soil surface does not exceed maximum depths of 2 m (i.e., the burial that are usual for most of horizontal ground-coupled heat exchangers).

Inside the ground/soil, the temperature distribution presents characteristics such as the following (Khatry et al. 1978; Krarti et al. 1995; Popiel et al. 2001; Gwadera et al. 2017): (i) due to the thermal inertia, the amplitude of changes in the ground/soil temperature decreases with an increasing depth; (ii) the amplitude of changes in the ground/soil temperature caused by diurnal temperature variations on the surface decreases to zero at depths of about 2 m, while the amplitude caused by seasonal temperature changes decreases with increasing distance from the surface of the ground/soil to a significantly lesser degree; (iii) at depths up to 2 m, the ground/soil temperature is sensitive to short time changes of weather conditions; (iv) in shallow zones extending from 2 m to about 10 m for dry light ground/soils, or to about 20 m for moist heavy sandy ground/soils, the temperature is almost constant, close to the average annual air temperature, and depends mainly on the seasonal variations of weather conditions; (v) in deeper zones (below about 10 or 20 m), the ground/soil temperature is practically constant, rising slowly with depth according to the local deep geothermal gradient, of which effects on the operation of ground-coupled heat exchangers is usually ignored; effectively, the average value of the geothermal gradient on Earth is approximately 0.03 K/m, whereas the average value of the thermal conductivity of the ground/soil is of the order of 1.5 W/m·K; thus, the average deep geothermal flux density is about 0.045 W/m^2, a much smaller value than the other heat fluxes on the surface of the ground/soil; in other words, under natural conditions, the so-called undisturbed temperature of the ground/soil occurring below about 10–20 m from the surface is approximately constant and the natural geothermal gradients have negligible impacts; and (vi) solar radiation varies with time and location (for example, in cold climates as those prevailing in Scandinavian countries, the average annual net insolation of the order of 100 W/m^2).

On the surface of the ground/soil, where both heat and mass transfer simultaneously occur, the following daily, seasonally, and annually variable heat fluxes usually exist (Krarti et al. 1995; Mihalakakou et al. 1997; Gwadera et al. 2017) (Figure 14.4): (i) solar incident (direct) heat flux; (ii) solar incident (direct) heat flux absorbed by the ground/soil; (iii) convective heat flux; (iv) long-wave sky radiation heat flux; (v) reflected (scattered) solar incident (direct) heat flux; (vi) evaporative (moisture phase change) heat flux; and (vii) sensible heat flux from precipitations.

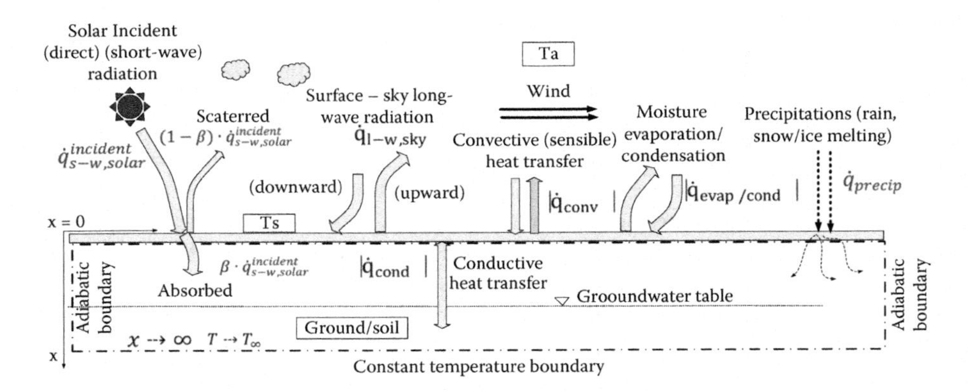

FIGURE 14.4 Main heat flux densities at the surface of the ground/soil; l-w, long-wave; s-w, short-wave; T_a, temperature of ambient air; T_s, temperature of ground/soil surface; T_∞, temperature of far-field undisturbed ground/soil; β, coefficient taking into account the degree of solar radiation absorption by the ground/soil.

In Figure 14.4, the left- and right-hand boundaries are considered adiabatic, the bottom boundary can be specified as a constant temperature, while the top boundary represents the ground/soil surface where heat fluxes due to varying weather conditions.

By neglecting the energy storage, which usually is small compared to the heat fluxes, the daily energy balance at the ground/soil surface (x = 0), can be written as follows (Krarti et al. 1995; Salah El-Din 1999; Mihalakakou et al. 1997; Larwa 2019) (Figure 14.4):

$$|(\dot{q}_{cond})_{x=0}| = -k_g (dT/dx)_{x=0} = \beta \cdot \dot{q}_{s-w,solar}^{incident} - \dot{q}_{s-w,solar}^{reflected} + |\dot{q}_{l-w,sky}| + |\dot{q}_{conv}| + |\dot{q}_{latent}| + \dot{q}_{precip} \tag{14.1}$$

where

$|(\dot{q}_{cond})_{x=0}|$ is the density of conductive heat flux resulting from the algebraic summation of the right side listed heat fluxes transferred into the ground/soil from the surface (or in the opposite direction, depending on its minus/plus sign) (W/m^2); the rate of heat transfer by conduction depends on the capacity for the ground/soil to conduct energy according to its thermal conductivity and the gradient of temperature with distance into the ground/soil; the thermal conductivity of ground/soil depends on the type and relative volume occupied by ground/soil constituents; air is a poor conductor of heat, so dry ground/soil with more air spaces have lower thermal conductivity; the thermal conductivity of dry ground/soil varies, but it is approximately 0.1, 0.25, and 0.3 W/m·K for organic, clay, and sandy ground/soil, respectively; if the ground/soil isnearly saturated with water, the thermal conductivity is approximately 0.5, 1.6, and 2.4 W/m·K, respectively, for the three mentioned ground/soil types, there is positive conduction into the ground/soil when the surface is warmer than the ground/soil below, and the conduction is negative when heat is transferred upward to the colder ground/soil surface; as the sun comes up, the surface is warmer than the ground/soil below, so heat is transferred downward and is stored in the ground/soil; as net radiation decreases in the afternoon, the ground/soil surface will cool relative to the earth below, and heat is conducted upwards towards the surface (i.e., the heat flux is negative); this negative heat flux continues during the night as heat is transferred upwards to replace lost energy at the colder ground/soil surface; on an hourly basis, the ground/soil heat flux density can change considerably but, on a daily basis, the amount of energy going into the ground/soil is generally about the same as the quantity leaving the ground/soil; over longer terms, there is a slight deficit each day during the autumn, so the ground/soil gradually loses energy and cools; the density of conductive heat flux at the ground/soil surface resulting from the algebraic summation of the described above fluxes (see Equation 14.1), is transported (as a conductive flux) into the ground/soil (or in the opposite direction), depending on the sign; it is direct proportional with the temperature gradient as defined by Fourier's law:

$$\dot{q}_{cond} = -k grad T = -k \left(\frac{dT}{dx} \right)_{x=0} \tag{14.2}$$

where

$\dot{q}_{cond}$ is density of conductive heat flux (W/m^2).

k is the ground/soil thermal conductivity (W/m·K).

T is the ground/soil temperature (K).

x is the direction of conductive heat transfer (m).

If $(dT/dx)_{x=0} \leq 0$, then $\dot{q}_{cond} \geq 0$ and the conductive heat flux is directed from the ground/soil surface to the interior ground/soil body.

$\dot{q}_{s-w,\ solar}^{incident}$ is the total daily average density of incident (direct) short-wave solar radiation that reaches the horizontal surface of the ground/soil (a part of this flux is absorbed by the ground/soil, and the other part is reflected) (W/m^2).

$\beta \cdot \dot{q}_{s-w,\ solar}^{incident}$ is the daily average density of incident (direct) short-wave solar radiation absorbed by the ground/soil (W/m^2).

β is a coefficient (also known as albedo value) that takes into account the degree of solar short-wave radiation absorption by the ground/soil; it depends on the ground/soil moisture content and cover type (e.g., 0.12 for asphalt, 0.20 for concrete, and 0.15 for bare soil) (–).

$\dot{q}_{s-w,\ solar}^{reflected} = (1 - \beta)\dot{q}_{s-w,\ solar}^{incident}$ is the daily average density of reflected (scattered) short-wave solar incident (direct) radiation (W/m^2).

$(1 - \beta)$ is the reflection coefficient that depends on the nature of the ground/soil surface, moisture content (wet/dry), and cover type.

$|\dot{q}_{l-w,sky}|$ is the daily average density of long-wave (surface-sky) radiation heat flux from the ground/soil upward (–) or towards (+) the sky (W/m^2).

$|\dot{q}_{conv}|$ is the daily average density of convective heat flux at the ground/soil surface (W/m^2).

$|\dot{q}_{latent}|$ is the daily average density due to water (moisture) phase change (evaporation or condensation) (latent) heat flux (W/m^2).

$\dot{q}_{precip}$ is the daily average density of sensible heat flux due to precipitations (rain, snow/ice melting) (W/m^2).

The direction of energy flux to or from the ground/soil surface provides positive and negative signs in Equation 14.1. Any thermal flux downward to the surface adds to the surface energy and, therefore, is considered positive (+). Any thermal flux away from the surface removes energy and it is considered negative (–). For example, downward short-wave radiation from the sun and sky radiation is positive, whereas short-wave radiation is reflected upward from the surface is negative.

During the day, the solar radiation dominates, warming the ground/soil and causing a downward energy flux into the ground/soil. On the other hand, during the night, no solar shortwave exists and a net loss of energy to space is achieved. Both latent and sensible heat fluxes are negative because the ground/soil is cooled and water vapor is condensing, leading to a positive gradient in temperature and specific humidity. The subsurface ground/soil is, thus, warmer than the surface, so there is an upward thermal flux from the subsurface to the surface. However, under natural conditions, the yearly average flux conducted from the surface of the ground/soil to its deeper layers is compensated by a heat flux in opposite direction during yearly cycles and amounts to:

$$- k_g \left(\frac{dT}{dx} \right)_{x=0} = 0 \qquad (14.3)$$

14.4.1.1 Solar Incident (Direct) Short-Wave Radiation

The solar radiation (i.e., electromagnetic energy transfer resulting from oscillation of electric and magnetic fields), transports huge amounts of energy from the sun to the Earth's surface through an empty (vacuum) space. The sun, with surface temperature as high as 6,000 K, emits short wavelengths of the electromagnetic energy within the range of 0.15 to 4.0 μm (1.0 μm = 10^{-6} m). Much of this high-energy short-wavelength radiation is absorbed or reflected as it passes through the atmosphere's clouds, etc. (Duffie and Beckman 2006).

At the top of the atmosphere, the solar incident (direct) solar radiation is often referred to as short-wave radiation is about 1,380 W/m^2 on a surface normal to the beam. However, the direct solar beam is attenuated as it passes through the atmosphere due to absorption and scattering, so the solar radiation at the Earth's surface (for the sun directly overhead in cloud-free conditions) is reduced to about 1,000 W/m^2. Some of the radiation scattered from the direct beam by molecules and aerosols reaches the surface as diffuse solar radiation. The solar incident radiation flux is strongly variable in time, having a zero value at night and significantly higher values during summer months than during winter months.

14.4.1.2 Sky Long-Wave Thermal Radiation

Although much cooler, the ground/soil surface radiates long-wave energy (in the range of 3 to 10 μm) according to the fourth power of its absolute temperature and fraction of cloud cover sky. If the ground/soil surface is assumed a horizontal grey body of which thermal emissivity does not depend on the temperature with non-selective radiative properties, the average net density of daily heat flux of long-wave radiation can be calculated as follows (Duffie and Beckman 2006):

$$|\dot{q}_{l-w,\ sky}| = \varepsilon \cdot \sigma \cdot f_{cloud} \cdot |T_g^4 - T_{sky}^4| \qquad (14.4)$$

where

$\dot{q}_{l-w,\ sky}$ is the average net density of daily heat flux of long-wave radiation (W/m^2).

ε is the thermal emissivity (or emittance) (defined as the ratio of the radiation emitted by the ground/soil surface to the radiation emitted by the sky approximated as a blackbody at the same temperature; or the fraction of maximum possible energy emitted at a particular temperature) of the ground/soil surface (–); most of ground/soil surfaces has emissivity values between 0.9 and 1.0.

$\sigma = 5.67 \cdot 10^{-8}$ is the Stefan-Boltzmann constant (W/m^{2}K^4).

f_{cloud} is the fraction of sky covered by clouds (–).

T_g is the absolute temperature of the ground/soil surface (K).

T_{sky} is the effective sky absolute temperature (K).

For approximate calculations, the average value of net long-wave thermal radiation from the ground/soil surface can be taken as about 63 W/m^2.

The sky temperature (variable during the day and night) depends on the ambient air temperature and relative humidity according with the following empirical formula (Duffie and Beckman 2006):

$$T_{sky} = T_a\,[0.711 + 0.0056 \cdot T_{dp} + 0.000073 \cdot T_{dp}^2 + 0.013 \cdot \cos(15t)]^{1/4} \quad (14.5)$$

where
T_{sky} is the sky temperature (°C).
T_a is the ambient air temperature (°C).
T_{dp} is the ambient air dew-point temperature (°C).
t is the time measured since midnight (hours).

The net radiation, the net amount of radiant energy that is retained by the surface (i.e., the sum of all gains and losses of radiation to and from the surface), includes solar incident short-wave radiation downward (that adds energy to the surface) and upward, as well as long-wave radiation downward and upward. The sum of these components can be positive, as happens during the daytime, or negative, as happens during the night.

14.4.1.3 Convective Heat Transfer

The convective heat transfer between the ground/soil surface and ambient air depends on the ground/soil surface and ambient air temperatures, wind speed, and vegetation height (Gan 2013).

The density of convective heat flux exchanged between the ambient air and the ground/soil surface (and vice versa) can be determined by using Newton's law (Staniek and Nowak 2016):

$$|\dot{q}_{conv}| = \bar{h}_{g,\,conv}|\bar{T}_a - \bar{T}_g| \quad (14.6)$$

where
$\dot{q}_{conv}$ is the density of convective heat flux (W/m^2).
$\bar{h}_{g,\,conv}$ is the average convective heat transfer coefficient at the ground/soil surface (W/m^2K).
$\bar{T}_a$ is the average temperature of the ambient air (°C).
$\bar{T}_g$ is the average temperature of the ground/soil surface (°C).

If $\bar{T}_a \geq \bar{T}_s$, then $\dot{q}_{conv} \geq 0$ and the convective heat flux is directed from the ambient air to the ground/soil surface. If $\bar{T}_a \leq \bar{T}_s$, then $\dot{q}_{conv} \leq 0$ and the convective heat flux is directed from the ground/soil surface to the ambient air.

The convective heat transfer coefficient between the ground/soil surface and the environment ($\bar{h}_{g,\,conv}$) depends on the wind velocity at the ground/soil surface according to the following empirical correlations (Ozgener et al. 2013):

$$h_g = 5.6 + 3.9 \cdot w_{wind} \quad \text{for } w_{wind} \le 5m/s$$
$$h_g = 7.2 \cdot w_{wind}^{0.72} \quad \text{for } w_{wind} \ge 5m/s \tag{14.7}$$

where

w_{wind} is the wind velocity (m/s).

14.4.1.4 Latent (Evaporation/Condensation) Heat Transfer

Heat is also transferred between the surface of the ground/soil and the environment as a result of the evaporation of water vapor from the ground/soil surface and the air moisture condensation to the ground/soil surface, two simultaneous heat and mass transfers caused by differences in the partial pressures of water vapor on the surface of the ground/soil and in the bulk of the ambient air.

The density of latent heat flux lost from the ground/soil surface by water evaporation can be expressed as follows:

$$|\dot{q}_{evap/cond}| = f_{evap/cond} \cdot h_{latent}^{water} \cdot G_{evap} \tag{14.8}$$

where

$\dot{q}_{evap/cond}$ is the density of evaporation (or condensation) heat flux (W/m^2).

$f_{evap/cond}$ is the coefficient of evaporation (or condensation) rate that takes into account the condition that the rate of evaporation (or condensation) of from/to the ground/soil surface is lower than the rate of evaporation (or condensation) from/to the Earth's surface; for example, for evaporation, this factor varies from 0.1 to 0.2 for dry ground/soils and up to 0.4–0.5 for humid soils (–).

h_{latent}^{water} is the enthalpy (latent heat) of water evaporation (J/kg) (defined as the amount of energy needed to vaporize a unit mass of water: $\simeq 2{,}450 kJ/kg$).

G_{evap} is the water evaporation velocity (kg/m^{2}s).

14.4.1.5 Precipitation (Sensible) Heat Transfer

The sensible heat transfer flux induced at the ground/soil surface by precipitation associated with the amount of rainfall and melting snow can be expressed as (Geiger 1971):

$$|\dot{q}_{precip}| = c_{precip} \cdot \rho_{precip} \cdot |T_{precip} - T_g| \cdot r_{precip} \tag{14.9}$$

where

$\dot{q}_{precip}$ is the heat flux density of precipitations (rain, melting snow, etc.) (W/m^2).

c_{precip} is the specific heat of rain or melting snow (J/kg·K).

ρ_{precip} is the mass density of the precipitation (kg/m^3).

T_{precip} is the precipitation-water temperature (C).

T_g is the ground/soil surface temperature (C).

r_{precip} is the rate of the precipitation (m/s).

A negative $\dot{q}_{precip}$ value signifies a loss of heat from the ground/soil surface to the rainwater.

Precipitation-induced sensible heat which is transferred between the land surface and rainwater/melting snow can be extremely large during precipitation events. Thus, the local surface temperature can be sharply altered on an hourly to daily timescale. However, precipitation-induced sensible heat is commonly neglected in current models because of its small magnitude on long timescales.

14.5 TEMPERATURE OF GROUND/SOIL

The temperature of ground/soil influences the heat transfer between the ground/soil and ground-coupled heat exchangers of ground-source heat pump systems as well as the operating temperatures of geothermal heat carrier fluids. The seasonal temperatures of the ground/soil slightly increases with increasing depth, but, practically, remain quasi-constant over the year at depths higher than 2 m, which represent ideal conditions for horizontal ground-coupled heat exchangers.

Some factors affecting the shallow ground/soil temperatures are the following: (i) climate conditions (as solar radiation, ambient air temperature, wind, rain, and snow); (ii) surface condition (terrain slope orientation, and cover vegetation); (iii) ground/soil structure and thermal properties (as thermal conductivity and moisture content); (iv) moisture freezing/thawing around the horizontal ground-coupled heat exchangers; and (v) snow cover, that, in cold and moderate climates, acts as natural insulation of the ground/soil, and reduce the variation of ground/soil temperature in winter; for stable and thick snow covers it is beneficial to bury the horizontal ground-coupled heat exchangers at depths of 0.30–0.50 m, while for unstable snow conditions and low ambient temperatures, deeper buried heat exchangers (at 0.50–0.90 m underground) are recommended.

The characteristics of ground/soil temperature are as follows (Kasuda and Archenbach 1965; Xing 2004; Al-Khoury 2012): (i) is influenced by heat fluxes occurring on the surface of the ground/soil; (ii) depends on the characteristics of the ground/soil surface and weather conditions, particularly the amount of solar radiation; in cold climates, the mean annual air temperature is lower than the mean annual ground/soil temperature, mainly due to the insulating effect of moisture evaporation and snow cover; in regions where there is a relatively deep and continuous winter snow cover, the mean annual ground/soil temperature may exceed the mean annual air temperature by as much as 5°C, thus, contributing to improve the annual energy performances especially of shallow closed-loop horizontal ground-coupled heat exchangers; (iii) the temperature of the ground/soil surface remains almost in phase with that of the ambient air; up to about 2 m below the ground/soil surface, changes in the ground/soil temperature are present mainly because the influence of seasonal atmospheric thermal conditions; (iv) at depths below 2 m, ground/soil temperatures are essentially constant and relatively stable throughout the year; due to the thermal inertia of the ground/soil, the amplitude of changes in the ground/soil temperature decreases with an increasing depth; below the surface, however, the maximum or minimum temperatures occur later than the corresponding values at the surface, the time lag increasing linearly with depth; for example, at a depth of 3–4 m, the maximum ground/soil temperature could occur about 2–3 months later than the average maximum temperature of the surface in summer; in addition to annual cycles, ground/soil temperature undergoes daily and monthly cycles and cycles associated with changes in the weather; these variations are confined to the near-surface region, daily cycles penetrating about 0.5 m and weather cycles about 1 m below the surface; diurnal temperature changes on the surface decreases to zero at a depth of about 1 m, while the amplitude caused by seasonal changes decreases with increasing distance from the surface of the ground/soil to a significantly lesser degree; the undisturbed ground/soil temperature will remain constant throughout the year below 10 m; above 10 m, the ground/soil temperature will change with the season; in cold and moderate climates, at depths of less than 2 m, where the ground/soil temperature is subject to solar radiation, seepage, and precipitations at the surface, the ground/soil average temperature generally follows the ambient air temperature seasonal variations with time delays of up to one month; (v) compared with ambient air, the temperature of undisturbed ground/soil is usually higher during the heating-dominated (winter) seasons and lower during the cooling-dominated (summer) periods; (vi) in the heating mode (winter), the heat extraction results in a decrease of ground/soil temperature around the ground-coupled heat exchangers; if the ground/soil temperature decreases under 0°C, antifreeze thermal fluid carriers are used; (vii) in the cooling mode (summer), during the heat rejection, the temperature of the surrounding ground/soil increases, which has a detrimental impact on the ground/soil thermal conductivity close to ground-coupled heat exchangers, leading to a reduction of heat transfer because of drier ground/soil; in certain cases, the contact between the tubes and the ground/soil isn't adequately achieved and the air will replace the water diffused far from the ground-coupled heat exchanger.

The ground/soil surface temperature cyclic variations can be linked to the ambient air temperature, both of parameters being dependent on annual average temperature fluctuation amplitude, and phase angles and may be described as follows (Williams and Gold 1976):

$$T_g = \bar{T} + A \cdot cos\,(2\pi t/t_0) \tag{14.10}$$

where

T_g is the ground/soil surface temperature at a given time (°C).

$\bar{T}$ is the average ground/soil surface temperature for the period, involving one or more complete cycles of variation (°C).

A is the difference between the maximum and minimum temperatures for the period (°C).

t is time (s).

t_0 is the time for one complete cycle (s).

In the Cartesian coordinates, Fourier's heat conduction partial differential equation is used to predict the temperature distribution of ground/soil considered as a semi- or infinite body with homogeneous and isotropic physical properties and without internal heat sources (Parton and Logan 1981):

$$dT/dt = \alpha_g (d^2T/dx^2) \tag{14.11}$$

where

T is the ground/soil temperature (°C).

t is the time (s).

$\alpha_g = k_g/c_v$ is the ground/soil thermal diffusivity (m^2/s).

k_g is the ground/soil thermal conductivity (W/m·K).

c_v is the volumetric ground/soil specific heat (J/m^3K).

x is the position coordinate (m).

In the case of a horizontal ground-coupled heat exchanger buried at a certain depth (considered as a heat source), the heat transfer in the ground/soil is described by the following equation:

$$dT/dt = \alpha_g \cdot (d^2T/dx^2) + \dot{q}_{vol}/\bar{c}_{vol} \tag{14.12}$$

where

T is the ground/soil temperature (°C).

t is the time (s).

α_g is the ground/soil thermal diffusivity (m^2/s).

x is the coordinate (m).

$\dot{q}_{vol}$ is the volumetric heat flux density, generated by the horizontal ground-coupled heat exchanger coupled with geothermal heat pump(s) that extract heat from the ground/soil during the heating season and supplies heat to the ground/soil during the summer (W/m^3).

$\bar{c}_{vol}$ is the ground/soil average volumetric specific heat (J/m^3K).

Based on the assumption that the ground/soil is homogeneous, isotropic, constant humidity, and constant thermal conductivity half-space limited by the ground/soil surface, in which temperature fluctuates harmonically according to variable meteorological conditions and with boundary conditions: (i) $T_s = f(t)$ at $x = 0$; and (ii) $T = T_\infty$ at $x \rightarrow \infty$ (i.e., at depths higher than 10–20 m), the solution of the heat transfer Equation 14.11, giving the daily average temperature of ground/soil as a function of position coordinate x and time t, is (Carslaw and Jaeger 1959, 1980; Zheng et al. 1993; Mihalakakou 2002; Popiel 2002; Mihalakakou 2002; Kotani and Sugita 2005):

$$T(x, t) = \bar{T}_g - A_s \exp(-x/L) \cdot \cos(\omega t - P_s - x/L) \tag{14.13}$$

where

$T(x, t)$ is the daily average temperature of ground/soil as a function of position coordinate x and time t (°C).

x is the position coordinate (m).

t is the time (days).

$\bar{T}_g$ is the annual average temperature of the ground/soil surface (°C).

A_s is the amplitude of daily average temperature of the ground/soil surface (°C).

(Note: The values of the annual average ground/soil surface temperature ($\bar{T}_g$) and the amplitude of the ground temperature [A_s] depend mainly on ambient climatic conditions.)

L is the damping depth, a constant characterizing the decrease in amplitude with an increase in distance from the ground/soil surface (m), defined as:

$$L = \sqrt{2a/\omega} \tag{14.14}$$

where

$a = k_g/c_v$ is the ground/soil thermal diffusivity (m^2/s).

k_g is the ground/soil thermal conductivity (W/m·K).

c_v is the volumetric ground/soil specific heat (J/m^{3}K).

$\omega = 2\pi/365$ is the frequency of ground/soil surface daily temperature fluctuations (1/day).

P_s is the phase angle related to time elapsed from the beginning of the calendar year until the daily temperature reaches the minimum value (T_{min}) (rad):

$$P = \omega T_{min} \tag{14.15}$$

For a ground/soil with constant thermal properties, the annual temperature variation has an amplitude decaying exponentially with the distance from the Earth's surface at a rate determined by the time necessary for one annual cycle can be expressed with the following equation (Williams and Gold 1976) (Figure 14.5):

$$T(x, t) = \bar{T}_g + A \cdot exp[-x\sqrt{\pi/\alpha_g t_{cycle}}] \cdot cos[2\pi t/t_{cycle} - x\sqrt{\pi/\alpha_g t_{cycle}}] \tag{14.16}$$

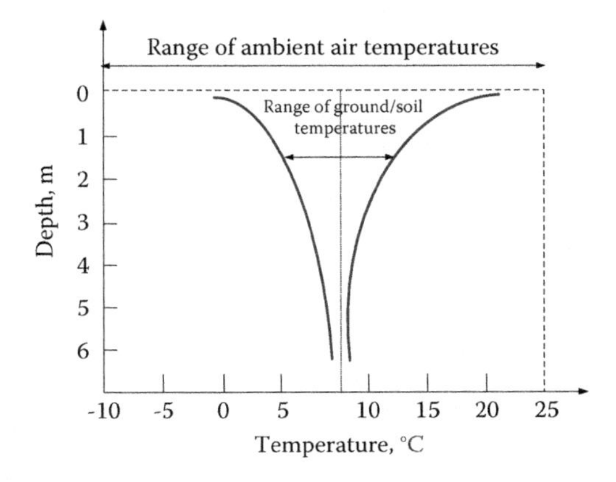

FIGURE 14.5 Typical depth dependence of the annual range of ground/soil temperatures.

where

$T(x, t)$ is the monthly average undisturbed ground/soil temperature at the depth x of and time t of the year (°C).

x is depth below the ground/soil surface (m).

$\bar{T}_g$ is the annual average temperature different depth and time (°C).

A is the difference between the maximum and minimum temperatures for the period (°C).

$\alpha_g = k_g/\rho_g \cdot c_{p,g}$ is ground/soil thermal diffusivity (m^2/s).

k_g is the ground/soil thermal conductivity (W/m·K).

ρ_g is the ground/soil mass density (kg/m^3).

$c_{p,g}$ is the ground/soil specific heat (J/kg·K).

t_{cycle} is the period of ground/soil temperature cycle (8,760 hours); the amplitude of a temperature variation at the surface is normally about equal to that of the corresponding one for air.

If ground/soil temperature observations are available for a site, Equation 14.15 can be used to determine the mean annual ground/soil temperature and thermal diffusivity for subsequent calculations.

Figure 14.6 shows a plot of yearly ground/soil temperatures (for 0.96-m and 1.83-m depths). Although this figure refers to Earth's temperatures and type of soil at a particular location, a similar plot could be made at any location.

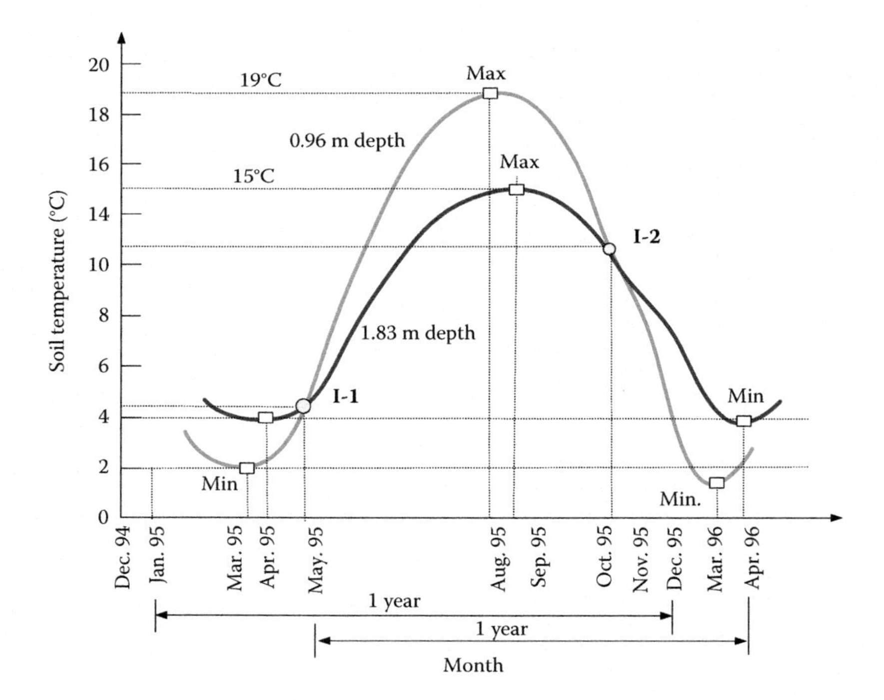

FIGURE 14.6 Typical annual variation of ground/soil undisturbed temperature at 0.96- and 1.83-m depth in a cold climate; I_1 and I_2, ground/soil temperature seasonal inversions (Note: x-axis not to scale).

14.6 HEAT TRANSFER AROUND HORIZONTAL BURIED PIPES

Most of the analytical models readily available for estimating heat transfer from and to underground horizontal pipes either consist of steady-state or transient solutions heat conduction solutions based upon the assumption that the earth surrounding the pipes is homogeneous, has constant thermal properties, and its temperature at reasonable distances is constant and unaffected by the existence of the buried pipes. However, these assumptions are unrealistic because thermal properties as well as Earth's temperature changes with respect to time and space due to seasonal change of the ground/soil surface temperature and, also, due to surface water seepage and the movement of groundwater in porous ground/soil structures. These processes strongly affect the combined heat transfer process by conduction and advection, which is significant for the long-term temperature response of horizontal ground-coupled heat exchangers. The more groundwater the ground/soil contains, the better the heat transfer. In order to significantly facilitate the prediction of long-term temperature changes with the distance from the ground/soil surface in which horizontal ground-coupled heat exchangers are embedded, it is beneficial to use simplified one-dimensional linear heat conduction models describing the heat transfer between the surface of the horizontal ground-coupled heat exchanger and the surrounding ground/soil. For that, it is necessary to know the thermal properties of the ground/soil, the heat flux extracted from or supplied to the ground/soil, the boundary conditions on the ground/soil surface, and the depth of pipes' installation (Kupiec et al. 2015).

14.6.1 Single Horizontal Pipe

In the case of one-pipe horizontal ground-coupled heat exchanger, the interaction of adjacent pipes is not accounted for and, consequently, boundary conditions of the first type assume that the temperature of pipe surface is time-constant.

Generally, the buried pipes gain or lose heat from/to the surface of the ground/soil when the pipes' temperatures were different from the ambient air temperatures. In addition, there is a substantial heat transfer between the buried pipe(s) and the ground/soil below. Thus, the assumption of isothermal ground/soil surface temperatures near ambient and the ignoring of heat transfer downward is likely to lead to oversized horizontal ground-coupled heat exchangers. The isothermal ground/soil surface case is approximated for real buried pipes by using the line integral method and the method of images as for the adiabatic case. In the isothermal case, however, the image (or mirror) is assumed to be a sink, with equal strength but of opposite sign to the original line source. approximates the effect of nearby influences, as those induced by ground/soil surface and/or nearby horizontally buried pipes. The important difference between the adiabatic and the isothermal surface cases is that the assumption that the ground/soil surface is adiabatic gives decreasing ground/soil thermal resistance values as the pipe is buried more deeply. On the other hand, with the isothermal assumption, the ground/soil thermal resistance values get larger with increasing depth of burial. For design purposes, which considers the extreme conditions, the ground/soil surface could be considered adiabatic (Figure 14.8b). This means that the ground/soil surface effects could be obtained by using the mirror image pipe of the same strength and the same sign as the buried pipe. Thus, the effect of the ground/soil surface on a buried pipe is to increase the ground/soil thermal resistance above that of a pipe buried infinitely deep. This ground/soil thermal resistance will decrease as the pipe is buried more deeply.

The solution for steady-state heat conduction from an underground pipe installed horizontally at a finite depth in homogenous ground/soil of constant property is based upon the potential flow theory and is obtained by the use of the "mirror-image" technique. Consider a long pipe with exterior radius, r_e, and constant temperature, T_0, on the exterior surface buried at a depth, h, in a homogeneous and isotropic ground/soil medium with thermal conductivity, k_g, and constant temperature, T_∞ (Figure 14.7).

According to this method, the buried pipe is considered a punctual heat source, P_1 (approximating the buried pipe), generating a heat flux ($+\dot{Q}$) and placed at the depth y_0 from the ground/soil surface considered as an adiabatic surface (i.e., a surface across which there is no heat transfer). If it is admited the existance of a symetrical ponctual fictive heat source P_2 (Figure 14.7), located at an equal distance y_0 on the other side of the ground/soil adiabatic surface, which has identical characteristics as the real buried pipe, but acting as an absorbent heat source ($-\dot{Q}$); thus, in all equidistant points from the two pipes, the temperature is constant. In this way, the system has as isothermal surface the plan $y = 0$ of temperature $\theta = 0$, if $\theta = T - T_s$ is the difference between the temperature T in any point M (x, y) and the ground/soil surface temperature (T_s) taken as a reference. At the real buried pipe surface, $\theta_0 = T_0 - T_s$. Because the two systems together consist of two parallel

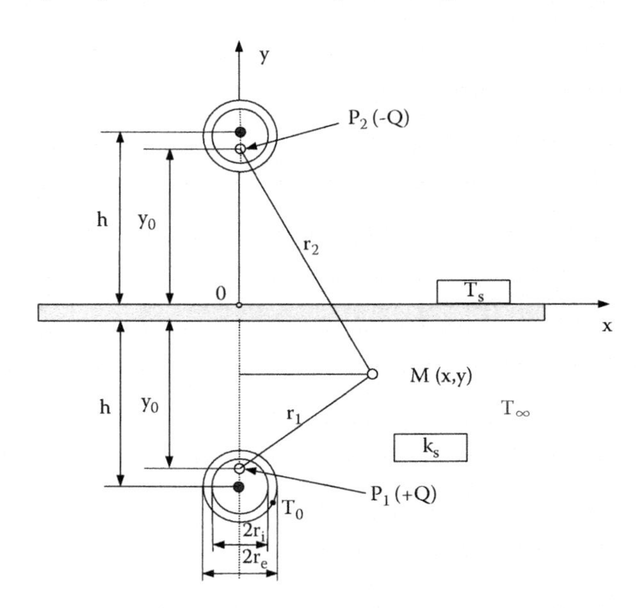

FIGURE 14.7 Mirror image method for a single buried pipe.

line sources of equal strength the plane which is equal distance from each must be adiabatic because of symmetry. The temperature distribution at any time can be obtained by solving for the temperature around each heat source and ignoring the other heat source. The two solutions are then added to give the temperature distribution for the single pipe near the adiabatic surface. The above method could be used to determine the temperature distribution and the ground/soil thermal resistance around a buried horizontal pipe if it is assumed that the surface of the ground/soil is adiabatic.

The heat flux transferred from the real buried pipe to the point M (x, y) is:

$$\dot{Q} = \frac{\pi(\theta_0 - \theta_1)L}{(1/2k_g)\ln(r_i/r_e)} \tag{14.17}$$

With the excess temperature of:

$$\theta_1 = \theta_0 - (\dot{Q}/2k_g L)\cdot \ln(r_i/r_e) \tag{14.18}$$

Considering now the symetric heat source (i.e., the fictive pipe), which absorbs a heat flux equal to that generated by the real buried pipe and has the excess temperature at the ground/soil surface of $-\theta_0$, the thermal influence on the point M (x, y) (i.e., the contribution of heat source, P_2) can be expressed as:

$$-\dot{Q} = \frac{\pi(-\theta_0 - \theta_2)L}{(1/2k_g)\ln(r_2/r_e)} \tag{14.19}$$

With the temperature excess:

$$\theta_2 = \theta_0 - (\dot{Q}/2k_g L)\cdot \ln(r_2/r_e) \tag{14.20}$$

Because the heat transfer equation is linear, and based on the superposition (of effects) principle, the simultaneous existence of heat sources P_1 and P_2 will produce in the point M(x, y) the excess of temperature of:

$$\theta = \theta_1 + \theta_2 = (\dot{Q}/2k_g L)\cdot \ln(r_2/r_1) \tag{14.21}$$

By expressing the radii r_1 and r_2 as functions of x, y, and y_0 (see Figure 14.7):

$$r_1 = \sqrt{x^2 + (y_0 + y)^2} \tag{14.22a}$$

$$r_2 = \sqrt{x^2 + (y_0 - y)^2} \tag{14.22b}$$

Equation 14.21 becomes:

$$\theta = (\dot{Q}/2k_gL)\cdot ln\sqrt{\frac{x^2 + (y_0 + y)^2}{x^2 + (y_0 - y)^2}} \tag{14.23}$$

Equation 14.21 can be written as:

$$e^{2\pi k_g L\theta/\dot{Q}} = \sqrt{\frac{x^2 + (y_0 + y)^2}{x^2 + (y_0 - y)^2}} = \sqrt{\rho} \tag{14.24}$$

where, for the isotherms $\theta = constant$, the left-side term has a constant value, $\sqrt{\rho}$.

The expression $\frac{x^2 + (y_0 + y)^2}{x^2 + (y_0 - y)^2} = \rho$ transformed in the following forms:

$$\begin{aligned} &x^2 + y_0^2 + 2y_0y + y^2 = \rho(x^2 + y_0^2 - 2y_0y + y^2) \\ &(\rho - 1)x^2 + (\rho - 1)y^2 - 2y_0(\rho + 1)y + (\rho - 1)y_0^2 = 0 \\ &x^2 + y^2 - 2\frac{\rho + 1}{\rho - 1}y_0y + y_0^2 = 0 \end{aligned} \tag{14.25}$$

represents a family of circles with centers having coordinates ($a = 0$; $b = (\rho + 1/\rho - 1)y_0$) and as radii:

$$R = \sqrt{a^2 + b^2 - c^2} = \sqrt{(\rho + 1/\rho - 1)^2y_0^2 - y_0^2} = [2y_0/(\rho - 1)]\sqrt{\rho} \tag{14.26}$$

The isotherm $\theta = 0$ corresponds to the circle of infinite radius for which $\rho = 1$, with the center located at $-\infty$ on the y-axis, being the line $y = 0$. The heat flow lines are normal to the lines of constant temperature and they do not cross the adiabatic surface (Figure 14.8a). It can be also seen that the isotherm $\theta = \theta_0$ corresponds to the exterior surface of the considered pipe, $r = r_e$, and the circle with a radius equal to zero corresponds to heat source P_1 located above the pipe center, because $(\rho + 1)(\rho - 1) \geq 0$ and $\rho \geq 1$. Thus, the centers of isotherm surfaces move down when the radius increases (see Figure 14.8b).

To determine the heat flux ($\dot{Q}$) transferred by the pipe to the sourrounding ground/soil medium, it is necessary to find the expression of $\sqrt{\rho}$ because, from Equation 14.26, the following results:

$$\dot{Q} = \frac{\pi\theta_L}{(1/2k_g)\cdot ln\sqrt{\rho}} \tag{14.27}$$

The ratio b/R generates a second-degree equation with an unknown parameter:

$$b/R = \frac{[(\rho + 1)/(\rho - 1)]y_0}{2y_0\sqrt{\rho}/(\rho - 1)} = (\rho + 1)/2\sqrt{\rho} \tag{14.28}$$

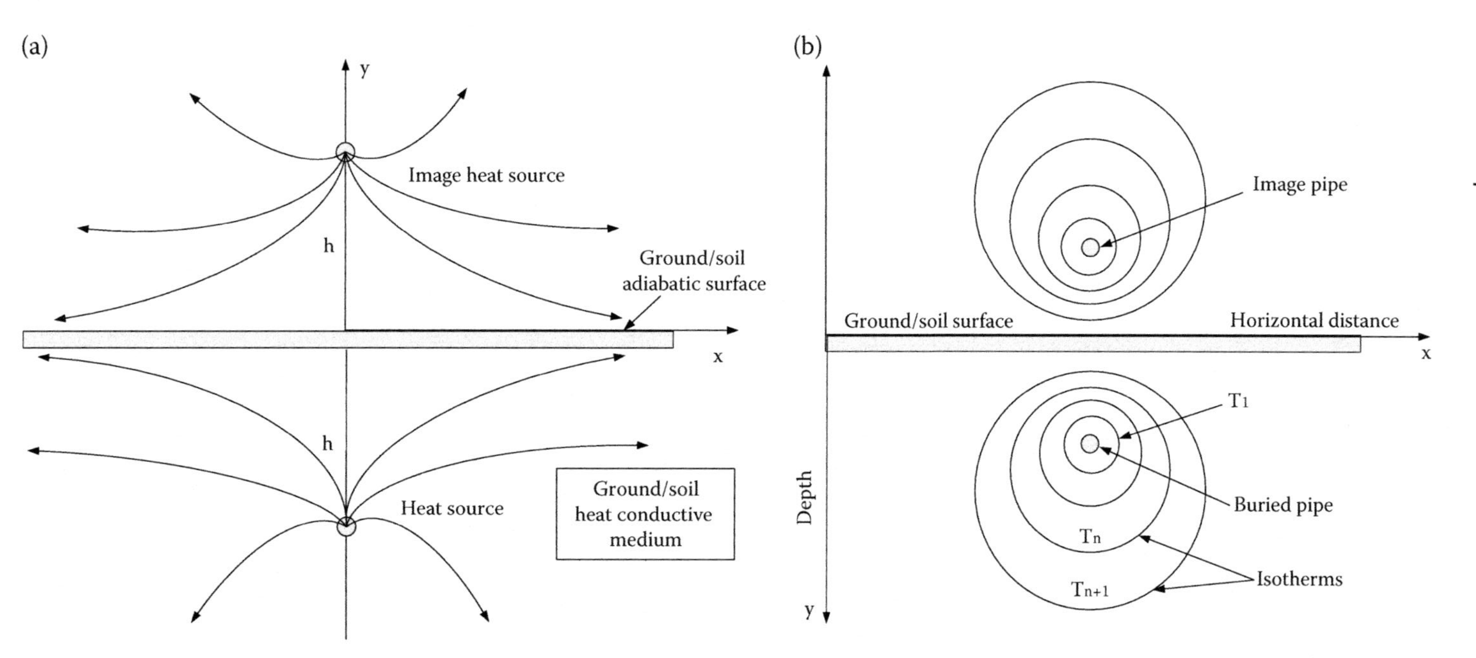

FIGURE 14.8 (a) Line of heat fluxes for two sources of equal strength in the case where the ground/soil surface is adiabatic with all heat transfer into the ground/soil; (b) temperature field predicted by image method (Note: ground/soil temperature is given in °C).

From Equation 14.28, the results are:

$$\sqrt{\rho} = b/R \mp \sqrt{(b/R)^2 - 1} \tag{14.29}$$

The sign minus doesn't correspond to the physical conditions of the proble. Thus, if $b = h$, $R = r_e$, and h/r_e is very large, utilizing negative sign results in $\rho = 0$ and, consequently, $\theta \le 0$, which contradicts the initial hypothesis that the ponctual source generates heat.

Therefore:

$$\sqrt{\rho} = h/r_e + \sqrt{(h/r_e)^2 - 1} \tag{14.30}$$

or

$$\sqrt{\rho} = 2h/d_e + \sqrt{(2h/d_e)^2 - 1} \tag{14.31}$$

Because on the outside surface of the pipe ($r = r_e$), $\theta = \theta_0 = T_0 - T_g$, Equation 14.27, considering the relation 14.31, becomes:

$$\dot{Q} = \frac{\pi (T_{fluid,\ in} - T_g)L}{(1/2k_g)\cdot ln\left[2h/d_e + \sqrt{(2h/d_e)^2 - 1}\right]} = \pi\theta_0 L/R_{cs} \tag{14.32}$$

where

$\dot{Q}$ is the heat flux transferred by the pipe to the sourrounding ground/soil medium (W).

$T_{fluid,\ in}$ is the average brine temperature inside the pipe (C).

T_g is the average temperature of surrounding ground/soil (C).

L is the length of buried pipe (m).

k_g is the average thermal conductivity of earth surrounding the pipe (W/m·K).

h is the depth of the pipe measured from the ground surface to the centerline of the pipe (m).

d_e is the pipe outer diameter (m).

R_{cs} is the combined thermal resistance of pipe – ground/sol:

$$R_{cs} = (1/2k_g)\cdot ln\left[2h/d_e + \sqrt{(2h/d_e)^2 - 1}\right] \quad (\text{m·K/W}) \tag{14.33}$$

If $\left(\frac{2h}{d_e}\right)^2 \gg 1$, the following simplified relation can be used:

$$\dot{Q} = \frac{\pi (T_0 - T_g)L}{(1/2k_g)\cdot ln\frac{4h}{d_e}} \tag{14.34}$$

If the temperature of the pipe interior wall (T_i) is known, from the equation of heat flux, the results are:

$$\dot{Q} = \frac{\pi\left(T_i - T_g\right)L}{(1/2k_g)\cdot \ln(d_e/d_i) + (1/2k_g)\cdot \ln(4h/d_e)} \tag{14.35}$$

where

d_i is the pipe inner diameter (m).

If the temperature of the fluid flowing inside the pipe ($T_{fluid,\ in}$) and the wall convective heat transfer coefficient (h_{in}) are known, the results are:

$$\dot{Q} = \frac{\pi\left(T_{fluid,\ in} - T_g\right)L}{1/h_{in}d_i + (1/2k_g)\cdot ln(d_e/d_i) + (1/2k_g)\cdot \ln(4h/d_e)} \tag{14.36}$$

14.6.2 Multiple Horizontal Pipes

Because in practice, several pipes are installed side-by-side in the same vicinity, the heat transfer around each pipe is affected by the presence of its neighbor. The effects of seasonal variation of Earth's temperature are usually treated as a quasi-steady-state process that includes the ground/soil thermal properties and pipes' depth of burial, sizes, and spacing. Depending on spacing, this thermal inference (or short-circuiting) phenomena also occurs between multiple trenches which may reduce the thermal performance of horizontal closed-loop (indirect, secondary fluid) ground-source heat pumps systems.

In the case of several horizontal pipes, that can be seen as multiple discrete heat sources with mutual potential thermal interferences, the heat flux exchanged between the pipes with diameter d buried at a depth h and spaced at p (Figure 14.9) can be also determined by applying the superposition principle. Based on Equation 14.21, for the ith pipe it can be written as followed:

$$\theta_i = \dot{Q}/2\pi k_g L\cdot \ln(r_{2i}/r_{1i}) \tag{14.37}$$

The summation of the effects gives:

$$\theta = \sum_{i=1}^{n} \theta_i = \dot{Q}/2\pi k_g L\cdot[\ln(r_{21}/r_{11}) + \ldots \ln(r_{2i}/r_{1i})\ldots \ln(r_{2n}/r_{1n})] \tag{14.38}$$

or

$$\theta = \dot{Q}/2\pi k_g L\cdot ln \prod_{i=1}^{n} (r_{2i}/r_{1i}) \tag{14.39}$$

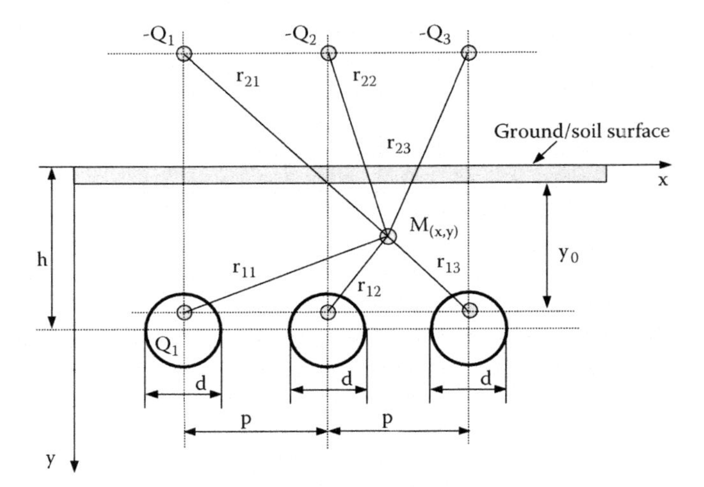

FIGURE 14.9 Mirror image method for multiple buried horizontal pipes.

The multiple pipe system considered in this section is shown schematically in Figure 14.9. The undisturbed earth temperature is designated by T_g, whereas the earth temperature at any point (x, −y) in the region of pipe heat transfer is designated by T_q. The difference in temperature, $T - T_q$, due to M number of heat sources (or sinks) can be obtained by the superposition of the mirror image technique employed for the single pipe problem in consistent units. The considerations refer to the ground-coupled heat exchanger with horizontal pipes positioned at one level.

14.7 FLOW INSIDE HORIZONTAL PIPES

14.7.1 Friction Losses

When the brine flows in a horizontal pipe, energy is dissipated by friction. The amount of energy lost depends on a number of factors such as the fluid's speed and viscosity. If the flow is turbulent, it can even depend on the roughness of the pipe walls. Losses due to friction will cause the pressure to decrease along the length of the pipe, therefore increasing the amount of power that a pump must deliver to maintain the flow. These losses can become significant in systems where long stretches of piping are employed (e.g., in heat exchangers, etc.).

The friction loss in uniform, straight sections of pipe, known as "major loss," is caused by the effects of viscosity, the movement of fluid molecules against each other or against the (possibly rough) wall of the pipe. It is greatly affected by whether the flow is laminar (Re < 2,000) or turbulent (Re > 4,000). In laminar flow, friction losses are proportional to fluid velocity that varies smoothly between the bulk of the fluid and the pipe surface, where it is zero. The roughness of the pipe surface influences neither the fluid flow nor the friction loss. In turbulent flow, friction losses are proportional to the square of the fluid velocity. A layer of chaotic eddies and vortices

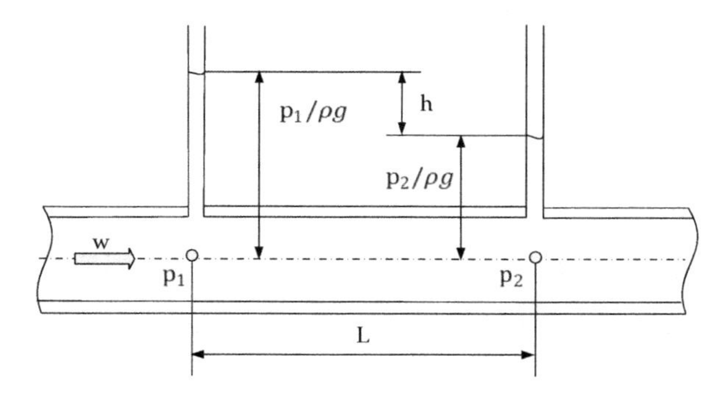

FIGURE 14.10 Pipe friction loss.

near the pipe surface, called the viscous sub-layer, forms the transition to the bulk flow. In this domain, the effects of the roughness of the pipe surface must be considered. It is useful to characterize that roughness as the ratio of the roughness height to the pipe diameter. In the smooth pipes, friction losses are relatively insensitive to roughness. In the rough pipes, friction losses are dominated by the relative roughness and is insensitive to Reynolds number. In the transition domain, friction loss is sensitive to both. For Reynolds numbers 2,000 < Re < 4,000, the flow is unstable, varying with time as vortices within the flow form and vanish randomly.

When the viscosity of the brine is taken into account, one-dimensional Bernoulli equation for viscous flow between sections 1 and 2 can be written as follows (Figure 14.10):

$$w_1^2/2g + p_1/\rho g + z_1 = w_2^2/2g + p_2/\rho g + z_2 + \Delta h_{loss} \quad (14.40)$$

where

w_1 is the brine velocity at the pipe inlet (m/s).
w_2 is the brine velocity at the pipe outlet (m/s).
p_1 is the brine pressure at the pipe inlet (mH_2O).
p_2 is the brine pressure at the pipe outlet (mH_2O).
z_1 is the pipe inlet elevation (m).
z_2 is the pipe outlet elevation (m).
ρ is the brine mass density (kg/m^3).
g is the gravitational acceleration (m/s^2).

Δh_{loss} is the hydraulic loss between two different cross sections along the pipe is equal to the difference of total energy ($w^2/2g + p/\rho g + z$) for this cross section. In horizontal pipes, $z_1 = z_2$ and, if the diameter of the pipe is constant, $w_1 = w_2$, the hydraulic loss is equal to the head of pressure drop (Figure 14.10):

$$\Delta h_{loss} = (p_1 - p_2)/\rho g \quad (14.41)$$

Inside horizontal straight tubes, the total pressure drop in single-phase flow is the result of the action of three factors, i.e., (i) wall friction (due to the fluid viscosity);

(ii) gravity (due to elevation) forces; and (iii) fluid momentum (due to the fluid acceleration) change (McQuiston and Parker 1994):

$$\Delta p_{loss} = \Delta p_{frictional} + \Delta p_{momentum} + \Delta p_{gravitational} \quad (14.42)$$

where

Δp_{loss} is the total pressure drop, equal to final pressure less initial pressure (Pa).

$\Delta p_{frictional}$ is the frictional pressure drop, mainly depending on the liquid viscosity and mass flux (Pa).

$\Delta p_{momentum}$ is the momentum (acceleration) pressure drop (Pa).

$\Delta p_{gravitational}$ is the gravitational pressure drop (Pa).

For horizontal tubes with adiabatic flow conditions, the gravitational pressure drop can be neglected ($\Delta p_g = 0$) as well as the acceleration pressure drop that is generally much smaller than the frictional pressure drop. Under turbulent flow, the frictional pressure losses of the brine flowing in horizontal ground-coupled heat exchangers are roughly proportional to the square of the flow velocity and inversely proportional to the pipe diameter, according to the phenomenological Darcy–Weisbach equation:

$$\Delta h_{frictional} = f \cdot w^2/2g \cdot l/d \quad (14.43)$$

where

Δh_{loss} is the head loss (m).

f is Fanning friction factor (–) that depends on Reynolds number and relative roughness [$f = \varphi(Re, \varepsilon/d)$] that can be accurately found in the Moody chart.

w is the mean fluid velocity (m/s).

g is the gravitational acceleration (m/s^2).

l is the pipe length (m).

d is the pipe inner diameter (m).

Re is the Reynolds number ($Re = \bar{u}_* d/\nu$).

$\bar{u}$ is the fluid average velocity (m/s).

$\nu = \frac{\mu}{\rho}$ is the kinematic viscosity (m^2/s).

μ is the fluid absolute (dynamic) viscosity ($Pa.\ s$).

ρ is the fluid density (kg/m^3).

For turbulent in-tube flow of sub-cooled liquid and superheated vapor, the Blasius friction factor varies in the range $0.006 < f < 0.06$ and can be calculated as follows:

$$f = 0.0791/Re^{1/4} \text{ for } 2.1 * 10^3 < \text{Re} < 10^5 \quad (14.44)$$

The friction factor (f) can also be determined from the Moody diagram by assuming that the pure liquid is flowing in the tube at the mixture mass velocity. At qualities below 70%, it is generally low, but higher at qualities above 70%.

In circular tubes:

f is the Darcy friction factor, which is a function of the Reynolds number and the tube roughness (–).

L is the tube length (m).

ρ is the density of the fluid (kg/m^3).

$\bar{w}$ is the fluid average velocity (m/s).

d is the tube internal diameter (m).

For laminar flow inside circular tubes, the friction (or Moody) factor is only dependent on Reynolds, in the form:

$$f = 64/Re \tag{14.45}$$

The Reynolds number is expressed by:

$$Re = w \cdot d/\nu \tag{14.46}$$

where

$\nu = \mu/\rho$ is the kinematic viscosity.

μ is the viscosity of the fluid (kg·m/s).

According to the Moody chart, for laminar flow (Re < 2,100), f = 64/Re, which is independent of relative roughness. For turbulent flow ($4 \cdot 10^3 \leq Re \leq 10^6$), f is independent of the Reynolds number ($f = \varphi(\varepsilon/d)$, where ε/d is the relative roughness (–) and ε, the pipe surface roughness (–).

14.8 HEAT TRANSFER INSIDE HORIZONTAL PIPES

Because of heat extraction and heat rejection due to the geothermal heat pump(s) operation, the brine temperature down the length of horizontal buried pipes is not uniform. It increases in the heating mode (Figure 14.11a) and decreases in the cooling mode (Figure 14.11b).

The outlet temperature can be determined as follows:

$$T_{out} = T_{in} + \frac{\dot{Q}_{conv}}{\dot{m} \cdot \bar{c}_p} \tag{14.47}$$

where

T_{out} is the temperature of the brine leaving the pipe (°C).

T_{in} is the temperature of the brine entering the run of pipe (°C).

$\dot{Q}_s$ is the forced convective heat transfer rate (W).

$\dot{m}$ is the brine mass flow rate (kg/s).

$\bar{c}_p$ is the brine average specific heat (J/kg·K).

The rate of forced convection heat transfer can be expressed by Newton's law:

$$\dot{Q}_{conv} = \bar{h}_{conv} \cdot A \cdot |T_{in} - T_{out}| \tag{14.48}$$

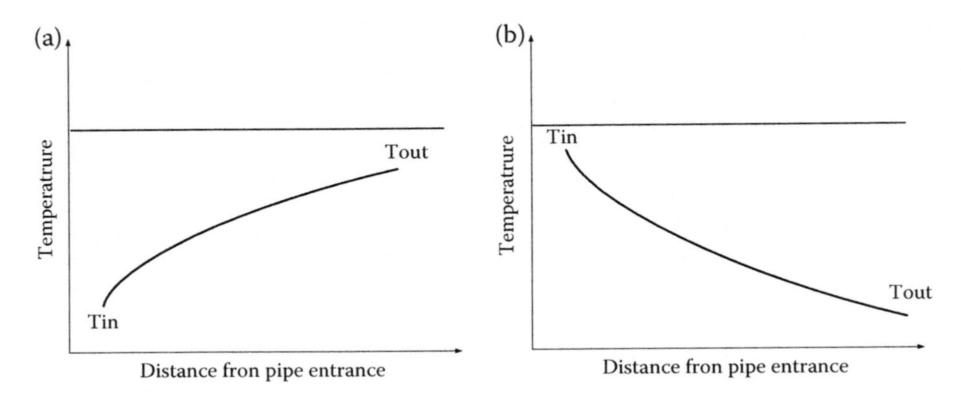

FIGURE 14.11 Temperature profile along a long, straight pipe: (a) heating mode and (b) cooling mode.

where

$\dot{Q}_{conv}$ is the forced convection heat rate (kW).

$\bar{h}_{conv}$ is the forced convection average heat transfer coefficient that varies along the flow direction; the mean value is determined by properly averaging the local heat transfer coefficients over the entire tube (W/m^{2}K).

A is the heat transfer area (m^2).

T_{in} is the brine inlet temperature (°C).

T_{out} is the brine outlet temperature (°C).

The non-dimensional convective heat transfer coefficient is determined as follows:

$$\bar{h}_{conv} = Nu \cdot k/d \tag{14.49}$$

where

Nu is the Nusselt number (-).

k is the brine thermal conductivity (W/m·K).

d is the pipe inner diameter (m).

For fully developed turbulent flow in smooth pipes (Re > 2,300), the Dittus-Boelter equation can be used (for stabilized flows and $Pr \geq 0.7$):

$$Nu = 0.023 \cdot Re_d^{0.8} \cdot Pr^n \tag{14.50}$$

where

$n = 0.3$ for cooled brine (-).

$n = 0.4$ for heated brine (-).

The physical parameters are chosen at the brine inlet and outlet arithmetic mean value.

REFERENCES

Al-Khoury, R. 2012. *Computational Modeling of Shallow Geothermal Systems*. CRC Press, Boca Raton, FL, US.

Bharadwaj, S.S., N.K. Bansal. 1981. Temperature distribution inside ground for various surface conditions. *Building and Environment* 16:183–192.

Carslaw, H.S., J.C. Jaeger. 1959. *Conduction of Heat in Solids*. Oxford University Press, Oxford, UK.

Carslaw, H.S., J.C. Jaeger. 1980. *Conduction of Heat in Solids*. 2nd Edition. Oxford University Press, Oxford, UK.

Duffie, J.A., W.A. Beckman. 2006. *Solar Engineering of Thermal Processes*. Wiley Inc., New York.

Gan, G. 2013. Dynamic thermal modelling of horizontal ground source heat pumps. *International Journal of Low Carbon Technologies* 8(2):95–105.

Gan, G. 2016. Dynamic thermal simulation of horizontal ground heat exchangers for renewable heating and ventilation of buildings. *Renewable Energy* 103 (10.1016/j.renene.2016.11.052. Accessed December 12, 2020).

Geiger, R. 1971. *The Climate Near the Ground*. Harvard University Press, Cambridge, UK.

Gwadera, M., B. Larwa, K. Kupiec. 2017. Undisturbed ground temperature – Different methods of determination. *Sustainability*, 1–14.

Kasuda, T., P.R. Archenbach. 1965. Earth temperature and thermal diffusivity at selected stations in the United States. *ASHRAE Transactions* 71(1):61–74.

Khatry, A.K., M.S. Sodha, M.A.S. Malik. 1978. Periodic variation of ground temperature with depth. *Solar Energy* 20(5):425–427.

Kotani, A., M. Sugita. 2005. Seasonal variation of surface fluxes and scalar roughness of urban land covers. *Agricultural and Forest Meteorology* 135(2005):1–21.

Krarti, M., C. Lopez-Alonzo, D.E. Clardidge, J.F. Kreider. 1995. Analytical model to predict annual soil surface temperature variation. *Journal of Solar Energy Engineering* 117:91–99.

Kupiec, K., B. Larwa, M. Gwadera. 2015. Heat transfer in horizontal ground heat exchangers. *Applied Thermal Engineering* 75:270–276.

Kusuda, T. 1973. Underground heat and chilled-water distribution systems. In *Proceedings of the Symposium on the Underground Heat and Chilled-Water Distribution Systems*. Washington, DC, November 26–27.

Larwa, B. 2019. Heat transfer model to predict temperature distribution in the ground. *Energies* 12:1–17.

Larwa, B., K. Krzysztof. 2020. Heat transfer in the ground with a horizontal heat exchanger installed – long-term thermal effects. *Applied Thermal Engineering* 164:1–11.

McQuiston, F.C., J.P. Parker. 1994. *Heating, Ventilating and Air Conditioning – Analysis and Design*, John Wiley & Sons, New York.

Mihalakakou, G. 2002. On estimating soil surface temperature profiles. *Energy and Buildings* 34:251–259.

Mihalakakou, G., M. Santamouris, J.O. Lewis, D.N. Asimakopoulos. 1997. On the application of the energy balance equation to predict ground temperature profiles. *Solar Energy* 60:181–190.

Neher, J.H. 1949. The temperature rise of buried cables and pipes. *AIEE (American Institute for Electrical Engineers) Transactions* 68(1):9–21.

Neupauer, K., S. Pater, K. Kupiec. 2018. Study of ground heat exchangers in the form of parallel horizontal pipes embedded in the ground. *Energies* 11:491–507.

Ozgener, O., L. Ozgener, J.W. Tester. 2013. A practical approach to predict soil temperature variations for geothermal ground heat exchangers applications. *International Journal of Heat and Mass Transfer* 62:473–480.

Parton, W.J., J.A. Logan. 1981. A model for diurnal variation in soil and air temperature. *Agricultural Meteorology* 23:205–216.

Popiel, C.O. 2002. Effect of surface cover on ground temperature season's fluctuations. *Foundations of Civil and Environmental Engineering* 2:151–164.

Popiel, C.O., J. Wojtkowiak, B. Biernacka. 2001. Measurements of temperature distribution in ground. *Experimental Thermal and Fluid Science* 25:301–309.

Salah El-Din, M.M. 1999. On the heat flow into the ground. *Renewable Energy* 18:473–490.

Staniek, M., Nowak, H. 2016. The application of energy balance at the bare soil surface to predict annual soil temperature distribution. *Energy and Building* 127:56–65.

Williams, G.P., L.W. Gold. 1976. *Ground Temperatures.* Canadian Building Digest Division of Building Research, National Research Council Canada, Ottawa (Ontario), Canada.

Xing, L. 2004. *Estimations of Undisturbed Ground Temperatures Using Numerical and Analytical Modeling*. Submitted to the Faculty of the Graduate College of the Oklahoma State University in partial fulfillment of the requirements for the Degree of Doctor of Philosophy.

Zheng, D., E.R. Hunt Jr., S.W. Running. 1993. A daily soil temperature model based on air temperature and precipitation for continental applications. *Climate Research* 2:183–191.

15 Closed-Loop Direct Expansion (Mono-Fluid) Ground-Source Heat Pump Systems

15.1 INTRODUCTION

The costs of residential closed-loop (indirect, secondary fluid) ground-source heat pumps, including those of borehole drilling, steel sleeves and grouting, vertical (or horizontal) ground-coupled heat exchangers, and thermal carrier fluids are relatively high. Even if partially compensated by higher evaporating temperatures and the elimination of outdoor fans, the additional electrical energy consumption required for secondary fluid pumping affects the seasonal performance factors of such systems compared to other, more conventional HVAC (heating, ventilating, and air conditioning) systems, as air-source heat pumps.

Closed-loop, direct expansion (mono-fluid) ground-source heat pump systems (also known as direct evaporation or direct exchange systems), generally installed in residential and small commercial/institutional buildings requiring thermal power outputs less than 20 kW, use artificial and natural refrigerants as thermal fluid carriers between the ground-coupled heat exchangers and refrigerant-to-air, refrigerant-to-water, or refrigerant-to-refrigerant geothermal heat pumps.

In other words, in such systems, the geothermal heat pumps' own working fluids (refrigerants) are circulated directly through horizontal (see Figure 15.2a), vertical (see Figure 15.2b), or diagonal (inclined) (see Figure 15.2c) heat exchangers (usually made of copper piping networks) placed underground and acting as evaporators (in the heating mode) and condensers (in the cooling mode) by directly exchanging heat with the earth (soil/ground/rock) without using antifreeze liquid (brine) circulation pumps and intermediary heat exchangers (Mei and Baxter, 1990; Halozan and Svec 1993; Halozan 1997; Nagano 2012).

For reversible direct expansion (mono-fluid) ground-source heat pump systems with horizontal or vertical/inclined ground-coupled heat exchangers, the energy sources are both solar and geothermal. The ground/soil/rock mass acts as a storage media. In the heating mode, the conversion of recovered heat from the ground/soil/rocks to useable thermal potential is provided by geothermal heat pump(s), and the heat distribution inside the building is achieved by refrigerant-to-air (air handlers) and/or radiant slabs via direct condensing (heating) coils. In the cooling mode, the cycle is reversed (Figure 15.1).

DOI: 10.1201/9781003032540-15

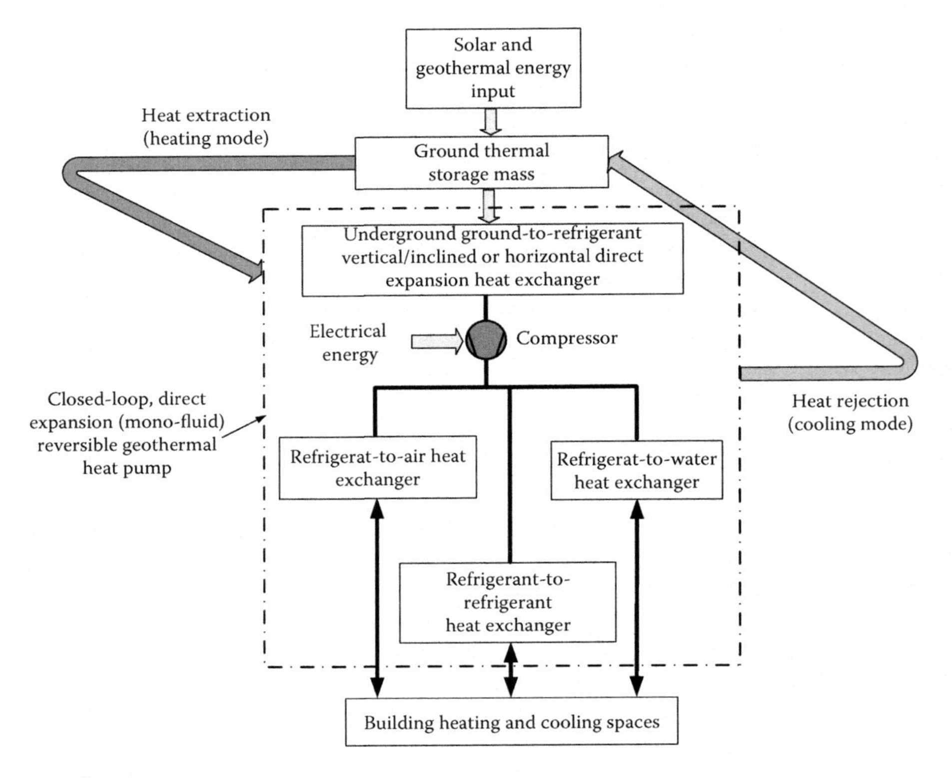

FIGURE 15.1 Schematic of energy fluxes of a reversible, closed-loop direct expansion (mono-fluid) ground-source heat pump system.

Direct-expansion ground-source heat pumps may achieve heating seasonal performance factors of at least 2.5 if the design of the system is made carefully. The choice of the refrigerants for horizontally and vertically installed direct expansion heat exchangers (i.e., R-410A or propane) is motivated by efficiency, reliability, environmental considerations, safety, and national regulations.

15.2 BASIC CONCEPTS AND OPERATING PRINCIPLE

According to the configuration of ground-coupled heat exchanger, there are at least three types of closed-loop direct expansion (mono-fluid) ground-source heat pump systems applied especially to residential and small-scale commercial/institutional buildings: (i) horizontal (Figure 15.2a), (ii) vertical (Figure 15.2b), and (iii) inclined (Figure 15.2c).

The most used material for the underground direct expansion heat exchangers is the copper alloy, a material that is strong, ductile, durable, and easy to be brazed and bent, owning excellent heat transfer properties (as a very high thermal conductivity). The factory- or field-assembled 12.7–28.5 mm ID tubes buried pipes are joined with soldered or brazed, wrought, or cast copper capillary socket-end fittings.

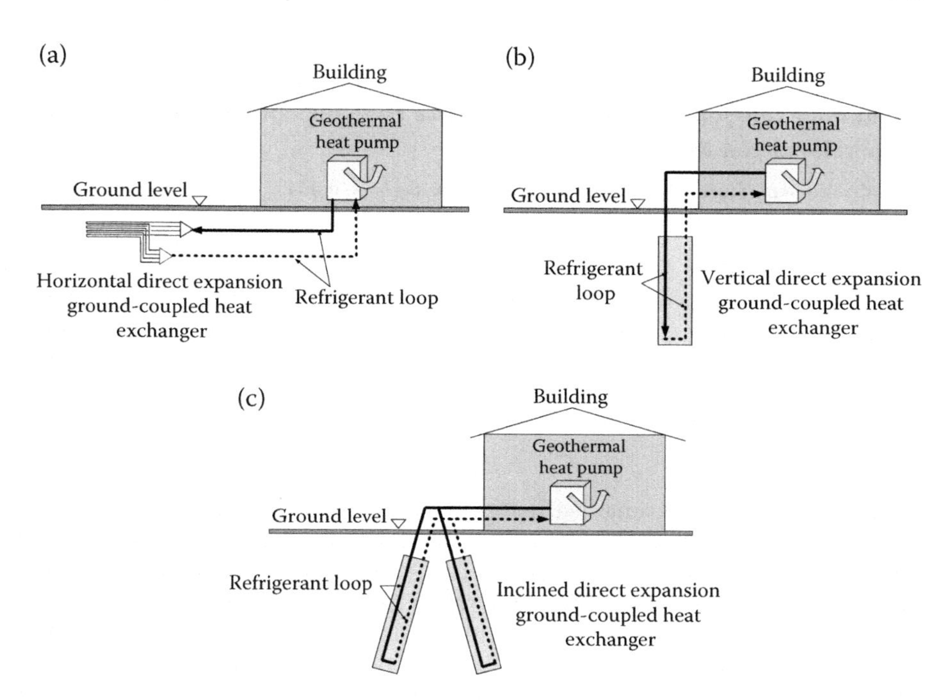

FIGURE 15.2 Basic concepts of ground-coupled heat exchangers used in closed-loop, direct expansion (mono-fluid) ground-source heat pump systems: (a) horizontal, (b) vertical, and (c) inclined.

In particularly corrosive grounds/soils or groundwater containing highly oxidizing chemical compounds, copper tubing might undergo electrochemical deterioration processes that result from a loss of electrons as they react with water and/or oxygen. To prevent corrosion, polyethylene-coated (with thin plastics forming protective films on their outer surfaces) copper pipes are generally utilized. In addition to the protective plastic films, the continuous copper tubes (without any connection) are provided with cathodic protections aiming to further protect the tube surface from corrosion by making it the cathode of an electrochemical cell. In that process, the copper tubes are connected to a sacrificial metal which will corrode in their place.

Thermodynamic cycles of real closed-loop direct expansion geothermal heat pumps are similar to those of conventional brine (water)-to-air or brine (water)-to-water mechanical subcritical vapor compression geothermal heat pumps that contain at least a compressor, an expansion device, a reversing valve, and brine(water)-to-air or brine(water)-to-water heat exchangers (see Chapter 8).

In the heating mode (see Figure 8.6a in Chapter 8), the liquid refrigerant coming through the liquid line from the geothermal heat pump's indoor coil (serving as the condenser) is expanded and enters the ground-coupled heat exchanger at relatively low pressure and temperature. Heat is then transferred from the relatively warmer ground/soil to the ground-coupled copper tubing that serves as the geothermal heat pump's evaporator. The absorbed heat changes the refrigerant from saturated liquid/vapor

TABLE 15.1
Evaporating Temperatures and Capacities as Functions of Ground/Soil Temperatures and Refrigerant Flow Rates

Ground/soil temperature (°C)	Refrigerant flow rate (kg/s)	0.006	0.009	0.012	0.015
10	Evaporating temperature (°C)	6.73	5.11	3.39	1.56
Evaporating capacity (W)	1,018	1,522	2,021	2,515	
5	Evaporating temperature (°C)	1.73	0.03	−1.79	−3.76
	Evaporating capacity (W)	1,007	1,506	2,000	2,487
0	Evaporating temperature (°C)	−3.30	−5.09	−7.04	−9.24
	Evaporating capacity (W)	996	1,489	1,974	2,457

mixture state to slightly superheated vapor state by a forced vaporization process. The superheated refrigerant vapor leaving the ground-coupled heat exchanger returns to the compressor where it is compressed to a higher pressure and temperature. The hot, superheated refrigerant vapor is then delivered to the geothermal heat pump's indoor coil (condenser) where it gives off heat to the cooler building's delivery fluid (air or water) with the assistance of an indoor fan or water circulating pump. The superheated refrigerant vapor gradually cools and then condenses back to the liquid state, and the cycle begins again.

In closed-loop direct expansion geothermal heat pump systems operating in the heating mode, evaporating temperatures, refrigerant flow rates, and thermal (heating) capacities vary with the ground/soil/rocks temperatures at which the heat is recovered. Table 15.1 shows, as an example, the measured evaporating temperatures (°C) and thermal (heating) capacities (W) of a 12-kW residential direct expansion heating-only geothermal heat pump as functions of ground/soil temperatures (°C) and refrigerant flow (kg/s) (Halozan and Svec 1993).

In the cooling mode (see Figure 8.6b in Chapter 8), the thermodynamic cycle is reversed. The superheated vapor refrigerant from the geothermal heat pump's compressor flows into the ground-coupled heat exchanger that serves as a condenser at relatively high pressure and temperature. Heat is transferred to the relatively cooler ground/soil/rock formation, the superheated vapor changes to sub-cooled liquid by a condensation process. The refrigerant liquid exiting the underground buried heat exchanger is returned to the geothermal heat pump indoor coil before being re-expanded to a lower pressure and temperature through an expansion valve, an orifice, or a capillary tube. This process enables to refrigerant to absorb heat from the building space with the assistance of the indoor (air or water) circulator (fan or pump). It can also dehumidify the indoor air if the vaporization temperature is lower than the room temperature dew point. As the refrigerant absorbs heat while passing through the indoor coil (evaporator), it vaporizes and exits as a saturated or superheated vapor towards the heat pump's compressor, and the cycle begins again.

Performances of closed-loop direct expansion ground-source heat pumps, usually expressed by the heating coefficient of performance ($\boldsymbol{COP_{heating}}$) (a dimension-less ratio obtained by dividing the condenser heat delivered by the electrical energy input, excluding any supplementary heat), cooling energy efficiency ratio (***EER****cooling*) (a ratio calculated by dividing the evaporator cooling latent and sensible capacities expressed in British thermal units per hour (Btu/h) (1 Btu/h = 0.29 W) by the electrical power input expressed in watts at a given set of rating conditions), and seasonal heating and cooling performance factors (SPF) (defined as the ratio of heat output or input over the heating and cooling season, respectively, to electricity used; both expressed in kWh) increase as the average temperature difference between the temperatures of condensation and evaporation processes decreases.

At low-temperature operation tests, the standard rating temperatures for heating and cooling shall be as follows (ANSI/AHRI Standard 870. 2005): (i) for heating: air temperature entering indoor heat exchanger of geothermal heat pump: 21.1°C dry-bulb, 15.6°C wet-bulb; air temperature surrounding the geothermal heat pump: 21.1°C dry-bulb; (ii) for cooling (not required for heating-only geothermal heat pump models), air temperature entering indoor heat exchanger of geothermal heat pump: 26.7°C dry-bulb, 19.4°C wet-bulb.

At high-temperature operation, the standard rating temperatures for heating and cooling shall be as follows (ANSI/AHRI Standard 870. 2005): (i) for heating: air temperature entering indoor heat exchanger of geothermal heat pump: 26.7°C dry-bulb, 19.4°C wet-bulb maximum; (ii) for cooling (not required for heating-only models): air temperature entering indoor heat exchanger of geothermal heat pump: 35°C dry-bulb, 21.7°C wet-bulb.

In Canada, for example, closed-loop direct expansion geothermal heat pumps shall have the following minimum net performances at standard rating conditions (including electrical input power to compressor, fan, and controls, but not supplementary heat) (CSA C748-94): (i) standard ***COP****heating* of 3.1; (ii) standard ***EER****cooling* of 13.0.

15.3 ADVANTAGES

Some of the advantages of closed-loop direct expansion ground-coupled heat pump systems over the conventional closed-loop (indirect, secondary fluid) horizontal and vertical ground-source heat pump systems are as follows (Mei and Baxter 1990; Halozan and Rieberer 2006; Minea 2005) (Note: not listed in importance order): (i) for residential and light commercial/institutional buildings, simple and more flexible installation not requiring anti-freezing heat carrier fluids, system flushing (because only sealed refrigerant circuits are need), and fluid circulating pumps (because the drive energy for the circulation of the refrigerant in the underground evaporators/condensers comes from the geothermal heat pumps' compressors); (ii) by using refrigerants as thermal energy fluid carriers, the freezing issue of secondary heat carrier fluids inside the ground-coupled heat exchangers is completely eliminated; (iii) potentially, reduce the required lengths of buried heat exchangers, borehole depth, and footprint (which may help offset the higher piping cost), as well as construction costs; (iv) do not require production (supply) and injection (return) wells as in the case of groundwater heat pump systems; (v) because the underground evaporator/condenser is directly installed in the ground/soil (horizontally or vertically) and the heat transfer from the earth to the geothermal heat pumps' refrigerant takes place directly, theoretically, the heat transfer losses could be minimized, the refrigerant evaporating/condensing temperatures being closer to the surrounding ground/soil temperatures; (vi) potentially, higher overall thermal efficiencies due to elimination of secondary fluid thermal carrier, circulation pump(s) (that are significant electrical energy consumers in indirect, secondary fluid and groundwater geothermal systems), and of intermediate heat exchanger(s), as well as due to the higher thermal conductivity of copper tubes compared to those of conventional polyethylene pipes; (vii) in the case of appropriate design and particular conditions (e.g., installation of direct expansion heat exchangers in moist sandy ground/soils and the high thermal conductivity of coated copper tubes), higher seasonal energy efficiency could be achieved in the heating mode; (viii) in certain favorable weather conditions, adequate ground/soil thermal properties, and proper installation and control strategies, reversible direct expansion geothermal heat pumps could also provide air cooling and dehumidification in the summer.

15.4 LIMITATIONS

While in some favorable installation conditions the direct expansion, closed-loop (mono-fluid) ground-source heat pump systems could achieve relatively high heating efficiencies, they also present significant design, technical uncertainties, economic and operational limitations that, in the majority, have not yet been addressed to make the systems viable for a large, widespread market.

Among these limitations, could be mentioned (Mei and Baxter 1990; Minea 2000): (i) without providing adequate velocity of the refrigerant inside ground-coupled heat exchangers, primarily in the heating mode, inadequate oil return to the geothermal heat pumps' compressors may occur; for adequate oil return, the refrigerant velocity has to be kept as high as possible, but as small as possible to achieve small pressure drops (which means also small drop in the vaporization temperatures); in horizontally installed direct expansion heat exchangers, the required velocity at the evaporator outlet has to be about 5 m/s, while in vertical direct expansion heat exchangers it has to be about 7 m/s; (ii) the maximum depth of vertical direct expansion heat exchangers is limited to about 30–40 m because with deeper wells, it could be more difficult to achieve low-pressure drops and refrigerant liquid return during the cooling mode when it condensates at relatively low temperatures; (iii) applications in large-scale commercial/institutional buildings are strongly limited because of the large quantities of refrigerant charges needed (these can be 10 times greater compared to those required for indirect, secondary fluid ground-source heat pump systems for similar nominal thermal capacities) and environmental restrictions related to eventual accidental leakages of refrigerant/oil mixtures; (iv) in addition, approximately three times as much refrigerant is required in the cooling mode as is needed in the heating mode, otherwise the system may provide insufficient heating capacity during the coldest (winter) seasons; (v) in the case of reversible direct expansion geothermal heat pumps, the migration of the refrigerant from the evaporators to the condensers, and vice-versa (which means passages from compact indoor condensers to large ground-coupled condensers and from large ground-coupled evaporators to compact indoor evaporators, changes that result in large refrigerant migrations within the geothermal heat pump circuit) can be managed by using relatively large (high- and low-pressure) refrigerant storage receivers; (vi) refrigerant liquid building-up during the cooling mode operation in both horizontal and vertical underground heat exchangers is a huge operational problem especially in cold climates where refrigerant condensation takes place at relatively low ground/soil temperatures and, thus, the refrigerant could be trapped inside during the geothermal heat pumps' standby periods; in such a situation, the geothermal heat pumps' compressors may not be able to move the refrigerant through the system properly because they could be unable to create enough discharge pressure to overcome the pressure drop in the liquid line when have to start in heating mode after cooling cycles or after standby periods; in these cases, special circuits and control strategies must be implemented to help remove the sub-cooled refrigerant from the horizontal and/or vertical direct expansion ground-coupled heat exchangers because, if the refrigerant liquid enters the compressor suction port, shut it down by the low-pressure cut-off and, even, damage the reciprocating compressors; therefore, the refrigerant management during the system start-ups and switching modes must be of maximum efficiency; (vii) due to the relatively large refrigerant charges and the risk of leakage, direct expansion ground-source heat pump systems with natural working fluids (as propane and isobutane) could be environmentally acceptable alternatives; (viii) also in the case of reversible direct expansion ground-source heat pump systems, the refrigerant liquid resulting inside the ground-coupled heat exchangers during the condensation process in the cooling mode of operation, or trapped inside during the geothermal heat pumps' standby periods, could not be removed by the aid of the too low compressor discharge pressure.

15.5 HORIZONTAL DIRECT EXPANSION GROUND-SOURCE HEAT PUMP SYSTEMS

The first closed-loop direct expansion (also known as direct evaporation) ground-source heat pump system with buried copper tubing was built in the late 1940s. Before the 1970s, such systems had been designed and in-field tested almost exclusively for heating-only operating mode. In 2002, in some countries as, for example, Austria, more than 60% of the heating-only heat pumps installed in residential market were direct expansion ground-source systems with horizontal ground-coupled coils due to their simplicity, relatively lower costs compared to closed-loop (indirect, secondary fluid) systems, relatively compact footprint and, potentially, to higher overall energy efficiency due to elimination of brine circulation pumps and improved heat transfer in the underground evaporators (Halozan and Rieberer 2002; Halozan et al. 2003; Halozan and Rieberer 2006).

15.5.1 Basic Concepts

Subcritical direct expansion mechanical vapor compression ground-to-air and ground-to-water direct expansion geothermal heat pumps generally used for heating residential and/or small commercial/institutional buildings, contain at least a compressor, an expansion device, an evaporator, a condenser, and, sometimes, a suction-line (liquid sub-cooling) internal heat exchanger. Such systems may operate in heating-only purposes (Figures 15.3a) and heating and cooling modes (Figures 15.3b).

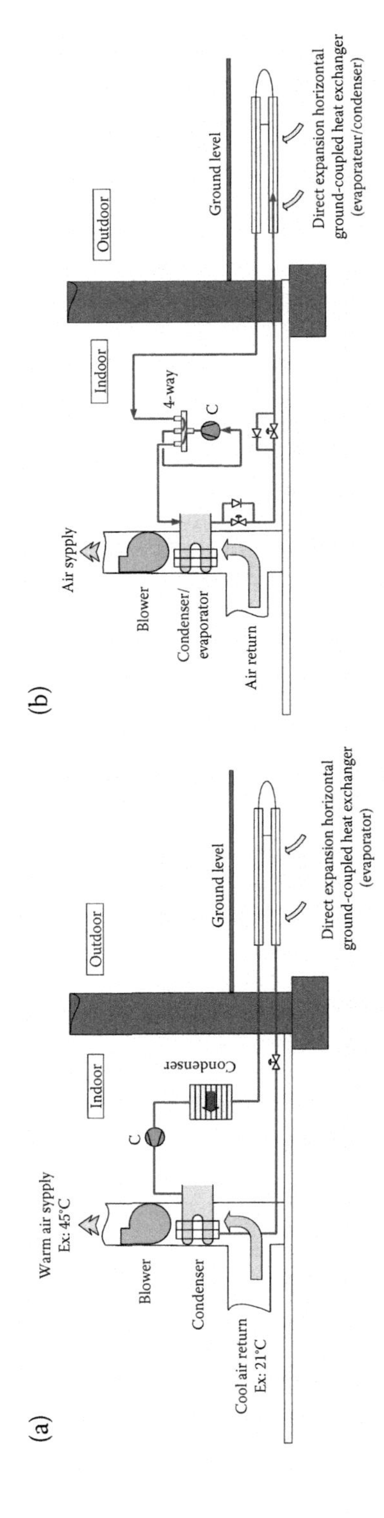

FIGURE 15.3 Schematic of closed-loop horizontal direct-expansion ground-source heat pump systems: (a) non-reversible, operating in heating mode only; (b) reversible, operating in both heating and cooling modes; C, compressor.

Depending on geographic location, direct expansion underground heat exchangers may consist of several parallel coated copper tubes containing one refrigerant supply tube and one refrigerant return tube (CSA C748.1994 – R-2009), each of maximum length of 55 m and interior diameters of 12–18 mm. The spacing between the horizontal underground loops (e.g., 0.6–0.8 m) must be kept large enough do not thermally influence one another. The tubes are usually buried in horizontal trenches at depths varying between 0.8 and 1.5 m and, in most cases, at about 0.3 m below the local freezing depth (Figure 15.4).

At such depths, the ground/soil temperature changes during the year. At the beginning of the heating season it is usually higher than the undisturbed ground/soil temperature (e.g., 15–17°C instead of 10–12°C). During the heating season, the ground/soil temperature generally drops below 0°C because of continuous or intermittent heat extraction, but moisture migration and frost formation around the coil increase the thermal conductivity and help to stabilize the ground/soil temperature. At the end of the heating season, natural thermal recharging of ground/soil starts, and heat is transferred from the ground/soil surface to the underground buried heat exchangers.

In cold and moderate climates, residential and/or small commercial/institutional direct expansion ground-source heat pump systems require 12.7 mm ID horizontal copper pipes (to be able to maintain sufficient refrigerant velocity in the heating mode to insure adequate oil return to the geothermal heat pump compressor) with lengths of about 30 m per kW of geothermal heat pump(s) nominal cooling capacity, as opposed to 40–45 m per kW for horizontal plastic ground-coupled closed-loop (indirect, secondary fluid) heat exchangers.

15.5.2 In-Tube Refrigerant Vaporization

As working fluids for horizontal closed-loop direct expansion ground-source heat pump systems, halocarbon refrigerants as HFC-134a (and its low-GWP replacements, as HFO-1234yf and HFO-1234ze(E)), R-404A, and hydrocarbons (as propane and isobutane) (Halozan 1995) are currently used. The content of this section applies to both ground-coupled direct expansion and to air (or water)-to-refrigerant evaporators of geothermal heat pumps.

15.5.2.1 Two-Phase Flow Patterns

The flow pattern of the liquid and vapor phases in internally non-enhanced (plain, smooth) horizontal round tubes during the two-phase flow of refrigerants depends on tube diameters, fluid mass velocities and heat flux densities, vaporization saturated temperatures, vapor titles, and thermos-physical properties such as density, viscosity, and surface tension (Collier 1981) (Figure 15.5). Multiphase flow regime transitions are rather unpredictable since they may depend on the wall's roughness or the initial (inlet) conditions. Hence, the flow pattern boundaries are not distinctive lines (or curves) but more or less poorly defined transition zones in almost all flow pattern maps.

If a single-phase, subcooled refrigerant liquid (title $x = 0$) enters a horizontal tube, a length of 30–50 diameters is necessary to establish fully developed turbulent

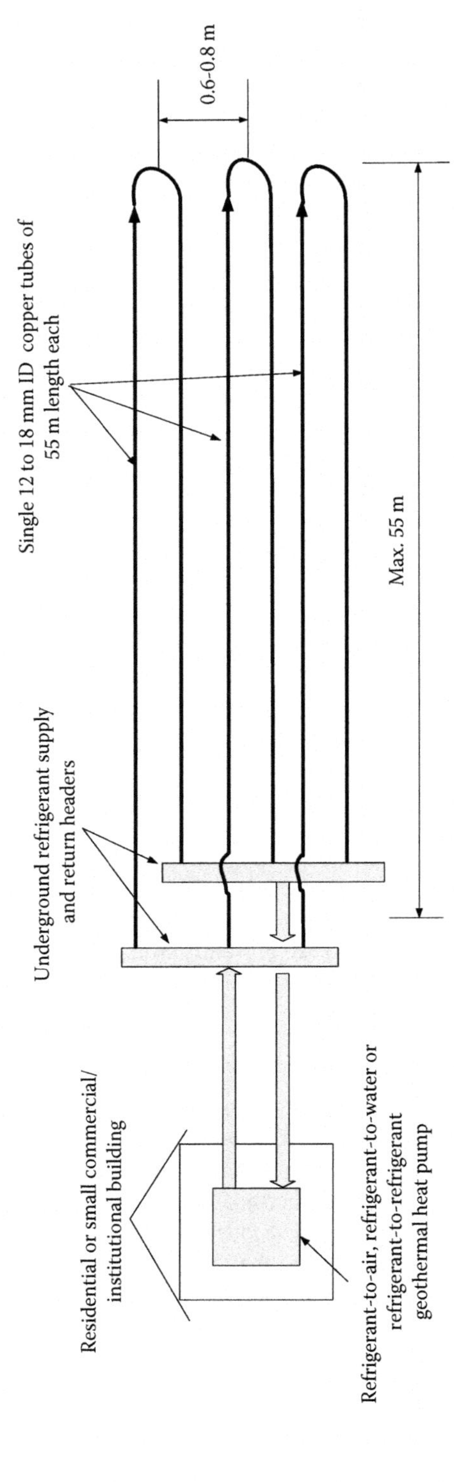

FIGURE 15.4 Typical construction of horizontal ground-coupled direct expansion heat exchangers (Note: plan view; schematic not to scale).

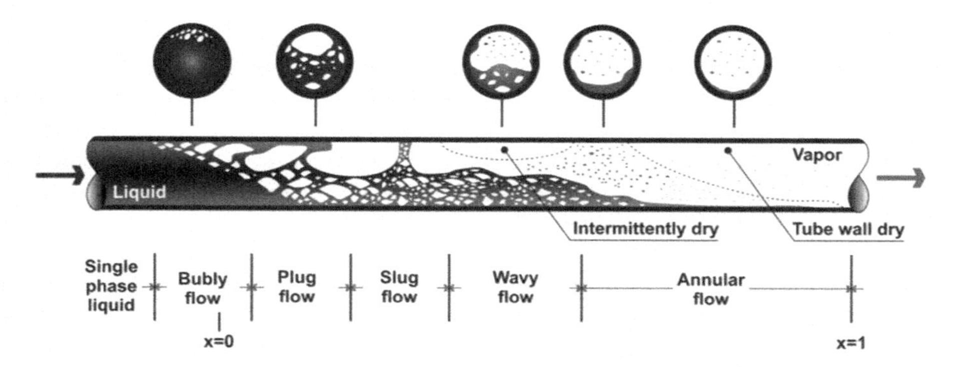

FIGURE 15.5 Vaporization flow regime map for an internally smooth, non-enhanced horizontal tube.

flow. In this region, the heat transfer occurs by forced convection. Once nucleate boiling is initiated, vapor bubbles that appear at the tube surface grow, are carried into the mainstream of the liquid, and are strongly influenced by the flow mass velocity.

In the bubbly flow zone, the vapor bubbles are dispersed in the liquid with a high concentration of bubbles in the upper half of the tube due to their buoyancy. When shear forces are dominant, the bubbles tend to disperse uniformly in the tube. Later, these large vapor bubbles coalesce and form a continuous core of vapor with the liquid flowing in an annular film along the wall. At high values of the mass flux, the annular flow pattern occurs over most of the tube length, whereas at low values the flow may become stratified with liquid moving along the bottom of the tube and vapor along the top. As the flow progresses, the liquid film on the walls becomes thinner and eventually disappears as all of the liquid evaporates.

As the vapor title (volume fraction) increases, individual bubbles coalesce to form plugs and slugs of vapor. In the plug flow regime, the liquid phase is continuous along the bottom of the tube, below the elongated vapor bubbles, smaller than the tube diameter. Slug flow regime occurs at higher vapor velocities, the diameters of elongated bubbles becoming similar in size to the tube diameter. The plug and slug flow regimes are followed by stratified and annular mist (not explicitly shown in Figure 15.5) regimes in which the liquid forms a film that moves along the inner surface, while vapor moves at a larger velocity through the core of the tube. In the stratified flow, at low liquid and vapor velocities, complete separation of the two phases occurs. The vapor goes to the top and the liquid to the bottom of the tube, separated by an undisturbed horizontal interface. By increasing the vapor velocity in the stratified flow, waves are formed on the interface and travel in the direction of flow. The amplitude of the waves is notable and depends on the relative velocity of the two phases. However, their crests do not reach the top of the tube. The waves climb up the sides of the tube, leaving thin films of liquid on the wall after the passage of the wave. This is a stratified-wavy flow. At high-vapor fractions (titles), the top of the tube with its thinner film becomes dry first, so that the annular film covers only part of the tube perimeter and thus this is then classified as stratified-wavy flow. Further increasing the gas velocity, these interfacial waves become large enough to wash the top of the tube. This is an intermittent flow regime characterized by large-amplitude waves intermittently washing the top of the tube with smaller amplitude waves in between. Large-amplitude waves often contain entrained bubbles. The top wall is nearly continuously wetted by the large-amplitude waves and the thin liquid films left behind. Intermittent flow is also a composite of the plug and slug flow regimes. At larger vapor flow rates, the liquid forms a continuous annular flow (film) around the perimeter of the tube, the liquid film being thicker at the bottom than the top. The interface between the liquid annulus and the vapor core is disturbed by small-amplitude waves and droplets may be dispersed in the gas core. Finally, the wall dry-out occurs at the top of the horizontal tube, first where the annular liquid film is thinner, and then the dry-out proceeds around the perimeter of the tube along its length until reaching the bottom where the liquid film disappears. Thus, dry-out in the horizontal tube takes place over a range of vapor qualities, beginning as an annular flow and ending when the fully developed mist flow regime is reached. With increasing vapor quality, the heat transfer coefficient decreases.

15.5.2.2 Pressure Drops

Pressure drops in a two-phase flow inside non-adiabatic smooth tubes where the velocities of the two phases differ, and the acceleration pressure drop may be very important, are due to the combination of interactions between the refrigerant and the walls of the tube, as well as between the liquid and vapor phases. Two-phase (vapor-liquid) pressure drops are usually much higher than would occur for either phase (liquid and vapor) flowing along at the same mass rate. They can be determined according to both homogeneous (where the two-phase mixture is treated as a single-phase flow and it is assumed that the liquid and vapor phases are flowing at the same velocity with averaged properties for density and viscosity) and separated flow models (where the two phases are distinct).

The continuity equation for each phase determines a relation between the liquid (w_f, m/s) and vapor (w_g, m/s) phase velocities, the void fraction (α), and the flow quality (x). The liquid- and vapor-phase velocities, respectively, are defined as follows (Griffith 1998):

$$w_f = \dot{V}_f / A(1 - \alpha) \tag{15.1}$$

$$w_g = \dot{V}_g / A \cdot \alpha \tag{15.2}$$

where

$\dot{V}_f$ is the liquid volume flow rate (m^3/s).
$\dot{V}_g$ is the vapor (gas) phase volume flow rate (m^3/s).
A is the tube cross-sectional area (m^2).
α is the void fraction (dimensionless).

The relationship between the void fraction, velocity ratio, and the vapor quality is (Griffith 1998):

$$(1 - \alpha)/\alpha = (\dot{V}_g/\dot{V}_f)(\rho_g/\rho_f)[(1 - x)/x] \quad (15.3)$$

where

α is the void fraction (dimensionless).

$\dot{V}_f$ is the liquid volume flow rate (m^3/s).

$\dot{V}_g$ is the vapor (gas) phase volume flow rate (m^3/s).

ρ_g is the vapor density (kg/m^3).

ρ_f is the liquid density (kg/m^3).

x is the vapor quality (dimensionless).

To calculate the pressure drop during convective vaporization in horizontal tubes, the Martinelli-Nelson correlation (1948) can be used:

$$\Delta p_{MN} = (2 \cdot f_{fo} \cdot G^2 \cdot L/d \cdot \rho_f)\left[\left(\frac{1}{x}\right)\int_0^x \phi_{fo}^2 dx\right] + (G^2/\rho_f)[(x^2/\alpha)(\rho_f/\rho_g) + (1 - x)^2/(1 - \alpha) - 1] \quad (15.4)$$

where

Δp_{MN} is the Martinelli-Nelson total two-phase pressure drop (Pa).

f_{fo} is the frictional coefficient considering the two-phase flow as a liquid flow (–).

G is the fluid mass flux ($kg/m^2.s$).

x is the vapor quality (–).

L is the tube length (m).

d is the tube inner diameter (m).

ρ_f is the liquid density (kg/m^3).

ρ_g is the vapor density (kg/m^3).

ϕ_{fo}^2 is the local two-phase frictional multiplier (–).

α is the void fraction.

As can be seen in Equation 15.4, to obtain total pressure drop during the two-phase vaporization process of fluid as water, it is required that the evaluation of the local two-phase frictional multiplier (ϕ_{f0}^2) and vapor void fraction (α) can be calculated with the following relations:

$$\phi_{fo}^2 = \phi_f^2 (1 - x)^{1.75} \quad (15.5)$$

$$\alpha = (1 + x^{0.8})^{-0.378} \quad (15.6)$$

where

x is the vapor quality (–).

$$\phi_f^2 = \left(1 + \frac{1}{x^{1/2}}\right)^2 \tag{15.7}$$

15.5.2.3 Heat Transfer

Inside horizontal tubes, the heat transfer depends on refrigerant properties, flow patterns and regimes, heat flux density, mass velocity, vapor quality, tube geometry, and one or more empirical constants. During the vaporization process of refrigerants consisting of heat addition that converts a liquid into its vapor, there are two interacting mechanisms: (i) nucleate boiling at the wall surface that dominates when the heat transfer rate is high or the refrigerant flow rate is low; this process occurs when vapor bubbles are formed at a heated solid surface; and (ii) convective (forced) vaporization that dominates when the refrigerant flow rates are relatively high; this process occurs when the refrigerant is superheated at the surface, and vaporizes at liquid-vapor interfaces within a flowing fluid. If nucleate boiling and convective (forced) evaporation occur simultaneously in a flowing fluid, the coexisting heat transfer mechanisms are simply described as vaporization where the flow is due to a bulk motion of the fluid, as well as to buoyancy effects characterized by rapid changes from liquid to vapor in the flow direction.

As heat is added to the refrigerant in the horizontal tubes of geothermal direct expansion heat pump evaporators, it progressively vaporizes, and the velocity increases until the refrigerant leaves the evaporator at saturated or superheated states. The variations in the heat transfer coefficient along the evaporator tubes are associated with different patterns of flow as the fraction of vapor and the velocity change along the tube (Figure 15.6).

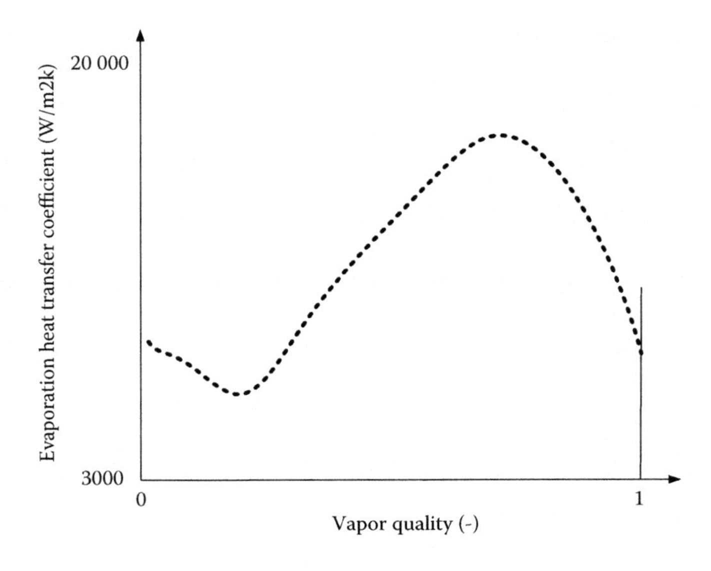

FIGURE 15.6 Relative variation of two-phase convective evaporation heat transfer inside a horizontal tube.

Correlations for the refrigerant-side heat transfer coefficients are classified into correlations for single-phase flows (i.e., subcooled liquid and superheated vapor) and for two-phase vaporization flows.

For single-phase flows, the following correlation is generally used (Dittus and Boelter 1930) is:

$$Nu = 0.023Re^{0.8}Pr^{n} \tag{15.8}$$

where

Re is Reynolds number (–).

Pr is Prandtl number (–).

n is 0.4 for the refrigerant being heated (e.g., inside a geothermal heat pump evaporator) and 0.3 for the refrigerant being cooled (e.g., inside a geothermal heat pump condenser).

Most of the available empirical correlations for heat transfer during in-tube two-phase vaporization of pure refrigerants combine nucleate boiling and convective evaporation contributions, most of them being restricted to one specified fluid, but not limited to one flow geometry.

Generally, two-phase vaporization models are classified as follows: (i) superposition models where the total heat transfer coefficient is the sum of nucleate boiling and bulk convective contributions (e.g., Chen 1966; Gungor-Winterton 1986, 1987); (ii) enhancement models, where the single-phase heat transfer coefficient of the flowing liquid is enhanced by a two-phase enhancement factor (e.g., Shah 1982, Kandlikar 1987); and (iii) asymptotic models (e.g., Liu and Winterton 1999).

Using the Foster and Zuber (1955) equation for pool boiling to predict the nucleate (pool) boiling heat transfer coefficient ($h_{nb} = h_{pool}$),Chen (1966) divided the heat transfer during vaporization in vertical tubes into (i) nucleate (pool) boiling contribution based on Foster and Zuber equation (1955); (ii) forced convection contribution based on the single-phase (liquid only) Dittus-Boelter (1930) equation.

The relatively simple superposition (additive) model assumes that the total heat flux density ($\dot{q}$) is the sum of nucleate boiling ($\dot{q}_{nb}$) and convective evaporation ($\dot{q}_{cv}$) components:

$$\dot{q} = \dot{q}_{nb} + \dot{q}_{cv} \tag{15.9}$$

The terms $\dot{q}_{nb}$ and $\dot{q}_{cv}$ are calculated at the pipe wall superheat value:

$$\Delta T_{wall} = T_{wall} - T_{sat} \tag{15.10}$$

where

T_{wall} is the pipe wall temperature (°C).

T_s is the refrigerant saturated temperature (°C).

By dividing Equation 15.9 by the wall superheat, the superposition model may be written in terms of the respective heat transfer coefficients:

$$\bar{h}_{tp} = h_{nb} + h_{cv} \quad (15.11)$$

where

$\bar{h}_{tp}$ is the superposed total heat transfer coefficient (W/m^2K).

h_{nb} is the nucleate boiling heat transfer coefficient (W/m^2K).

h_{cv} is the convective evaporation heat transfer coefficient (W/m^2K).

According to Chen (1966), the nucleate boiling and convective vaporization contributions can also be superposed to give the following overall two-phase heat transfer coefficient:

$$\bar{h}_{tp} = F * h_f + S * h_{nb} \quad (15.12)$$

where

$\bar{h}_{tp}$ is the overall two-phase heat transfer coefficient (W/m^2K).

$F < 1$ is the Chen's empirical two-phase enhancement factor (that reflects the influence of much higher vapor velocities and qualities, hence, forced convection heat transfer in the two-phase flow compared to the single-phase, liquid-only flow) correlated against the Lockhart-Martinelli parameter as follows:

$$F = 1 + 1.8 * X_{tt}^{-0.79} \quad (15.13)$$

h_f is the liquid heat transfer coefficient (W/m^2K).

h_{nb} is the nucleate boiling (or pool) heat transfer coefficient (W/m^2K).

$S < 1$ is the Chen's in-tube flow suppression factor that accounts for the thinner boundary layer at higher vapor qualities and showing the lower effective superheat existing in forced convective vaporization convection as opposed to nucleate boiling; it assumes that the liquid velocity suppresses nucleate boiling, empirically correlated as a function of the two-phase Reynolds number (Re_{tp}):

$$S = 1/(1 + 2.53 * 10^{-6} Re_{tp}^{1.17}) \quad (15.14)$$

where

Re_{tp} is the two-phase Reynolds number (–).

The vaporization heat transfer coefficients can be also calculated by using the boiling (Bo) and convection (Co) numbers defined as follows (Shah 1982):

$$Bo = \dot{q}/h_{fg} \cdot G \quad (15.15)$$

and

$$Co = [(1 - x)/x]^{0.8} \ (\rho_g/\rho_f)^{0.5} \quad (15.16)$$

where

$\dot{q}$ is the heat flux density (W/m^2).

h_{fg} is the refrigerant enthalpy (latent heat) of vaporization (kJ/kg).

G is the mass flux velocity (kg/m^2s).

x is the vapor quality (–).

ρ_g is the vapor density (m^3/kg).

ρ_f is the liquid density (m^3/kg).

For horizontal tubes, the Shah's enhancement model defines the two-phase heat transfer coefficient as:

$$h_{tp} = E * h_f \tag{15.17}$$

where

h_f is the single-phase heat transfer coefficient for the liquid phase flowing alone (W/m^2K).

E is the enhancement factor of the single-phase heat transfer coefficient which is a function of the boiling number (Bo), the convection number (Co), the Froude number accounting for partial wall wetting occurring in horizontal tubes:

$$E = f(Bo, \quad Co, \quad Fr_f) \tag{15.18}$$

where the Froude number is defined as follows:

$$Fr_f = G^2/g \cdot d \cdot \rho_f^2 \tag{15.19}$$

where

G is the mass flux velocity (kg/m^2s).

g is the gravitational acceleration (m/s^2).

d is the tube inner diameter (m).

ρ_f is the liquid density (m^3/kg).

15.5.3 In-Tube Refrigerant Condensation

In-tube condensation of refrigerants in reversible direct expansion geothermal heat pumps consists of sensible and enthalpy (latent heat) transfer from the desuperheated and, then, partially or totally condensed refrigerant vapor to the surrounding ground/soil (in the heating mode) or to the building indoor air through air-cooled fin-and-tube refrigerant condensers (in the cooling mode).

Underground refrigerant condensers are circular smooth pipes installed in horizontal trenches.

15.5.3.1 Flow Pattern

The flow pattern of refrigerant superheated vapor, vapor-liquid mixture, and subcooled liquid in horizontal tubes depends on parameters such as: (i) flow rate and quality, (ii) physical properties of the two phases, and (iii) tube geometry. The variety of the flow patterns reflects the different ways that the vapor and liquid phases are distributed in the tube, which causes pressure drops and heat transfer mechanisms to be different in the different flow patterns. Flow patterns of pure HFC refrigerants during two-phase condensation inside horizontal plain tubes are a complex phenomenon affected by simultaneous mass, momentum, and heat transfer rates (Rayleigh 1880).

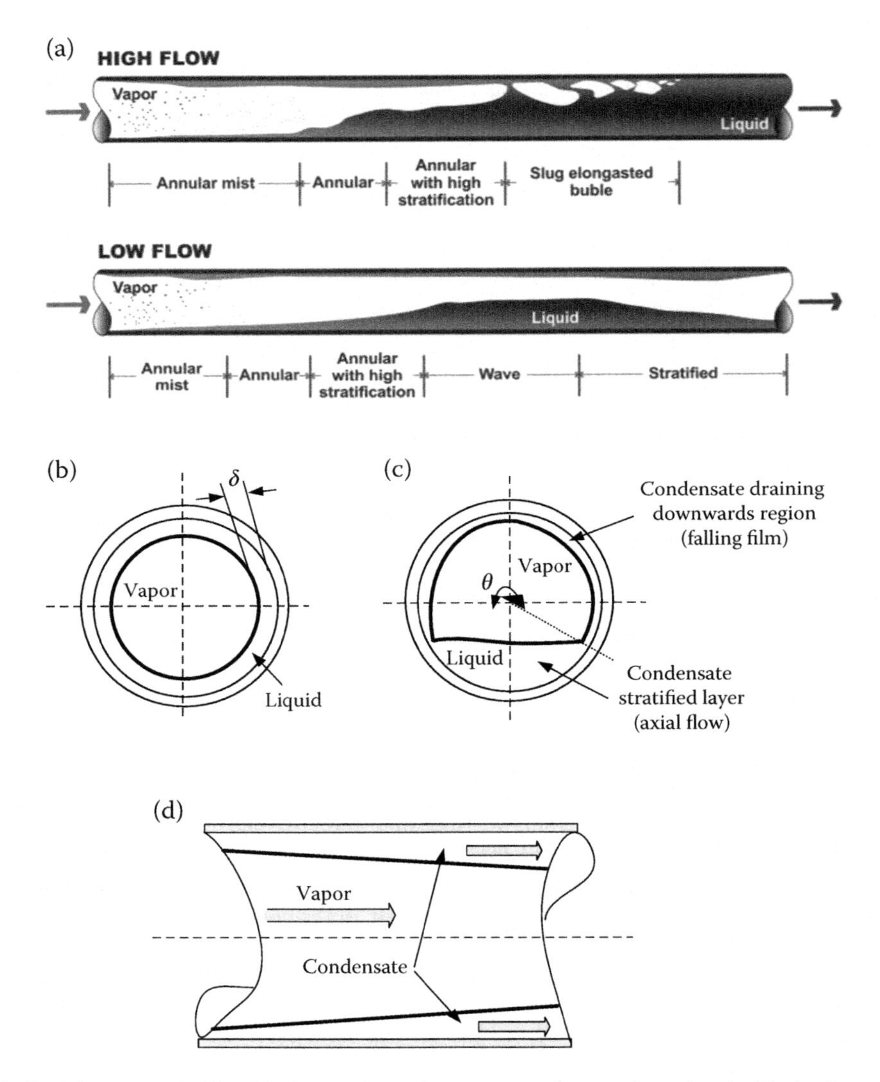

FIGURE 15.7 (a) Simplified two-phase flow patterns for condensation inside horizontal tubes; (b) cross-section of idealized annular flow regime; (c) longitudinal section of condensate flow for large vapor velocities; (d) cross-section of idealized stratified flow regime for low vapor velocities.

Figure 15.7a illustrates simplified flow patterns that typically occur during tube-side condensation of pure HFC (or similar) refrigerants. It can be seen that, during condensation of pure refrigerants inside horizontal tubes, there are several flow regimes similar to those for two-phase evaporation (Thome et al. 2003): (i) annular, (ii) stratified-wavy, and (iii) stratified.

During condensation inside horizontal tubes, the annular flow pattern is dominated by vapor shear, while stratified, wavy, and slug flows are controlled by gravity force.

Contrary to in-tube vaporization (see Section 15.2.2), in-tube condensation of refrigerants presents differences such as (Rohsenow 1973; Palen et al. 1979; Cavallini et al. 2003): (i) depending on the geothermal heat pump thermodynamic cycle, the refrigerant inlet vapor may be superheated, saturated ($x = 1$) or wet ($x < 1$); (ii) the top of the tube is wetted (coated) by a thin condensate (liquid) film in both annular and stratified regimes and, thus, no dry-out occurs.

At the entrance region annular flow is formed but this quickly transforms condensate film to intermittent flow with its characteristic large amplitude waves washing the top of the tube or to stratified-wavy flow with smaller amplitude waves. If liquid does not span the cross-section of the tube, vapor may reach the end of the tube without condensing. The vapor enters the condenser tube and forms a thin liquid film around the perimeter of the tube as an annular flow, a liquid layer in the bottom of the tube and a gravity-controlled condensing film around the upper perimeter as a stratified or stratified-wavy flow. The predominant regime in geothermal heat pump condensers is the annular flow. In fully developed annular flow, the condensate flows almost symmetrically distributed as a film on the inside tube wall. The flow converts to annular then stratified with the liquid flowing along the bottom of the tube. For annular flow, a uniform liquid film thickness of δ is assumed and the effects due to gravity ignored.

For annular flow (Figure 15.7a), for the sake of simplicity a uniform liquid film thickness of δ is assumed and the actual larger thickness of the film at the bottom than the top due to gravity is ignored.

At relatively low-mass velocities, the condensate forms on the upper portion of the tube wall and flows toward the bottom of the tube, while the vapor flows in the tube central core (Figure 15.7b). The annular-mist flow regime, has the appearance of an annular film with a mixture of vapor and mist in the core flow.

For simplicity's sake, the annular flow structure is also assumed to apply to the intermittent flow regime, which has a very complex flow structure, and, also, tentatively to the mist flow regime (assuming the impinging droplets create an equivalent unsteady liquid film).

At higher vapor velocities, the two-phase flow regime becomes annular. The vapor occupies the core of the annulus, diminishing in diameter as the thickness of the outer condensate layer increases in the flow direction.

At high mass flow rates and qualities, the flow regime is annular, where the liquid film is on the perimeter of the wall, the vapor is in the central core and some liquid is entrained in the vapor from the tips of waves on the interface of the film. As condensation proceeds along the tube, the vapor velocity decreases and thus there is a corresponding decrease in vapor shear on the interface and the liquid film becomes thicker at the bottom of the tube than at the top. New condensate formed increases to the thickness of the liquid film. As the quantity of liquid increases along the tube, slug flow is encountered and still further along all the vapor is finally converted to liquid.

For high flows, the first condensate produced forms an annular film. Some of the condensation is entrained as droplets in the high-velocity vapor flow. Farther along the tube, the vapor velocity falls due to condensation, which gives a corresponding fall in the vapor shear. This has two effects (i) the entrainment decreases, since there is less shear to break off the droplets; (ii) gravity forces become relatively more important, causing marked asymmetry in the film; with increasing condensate flow, slug flow is eventually achieved, which gives way to plug or elongated bubble flow; the tube runs full of liquid beyond the point at which all the liquid is condensed.

The wavy flow exists when the portion of the tube wall near the interface between the liquid pool and the vapor is affected by waves. There are more subcategories of these flow regimes that appear to be a transition between two of those mentioned above.

Slug flow appears when interfacial waves grow sufficiently in amplitude to block the entire cross-section at some transversal sections. When slug flow is present, there are large waves in the flow bridging the full cross section of the tube, and as condensation continues, the slugs coalesce into a predominantly liquid flow with large bubbles within. This is referred to as the plug flow regime. Slug and plug flows occur at the end of the condensation process.

Flow conditions within horizontal tubes during refrigerant condensation are complicated and depend strongly on the velocity of the vapor flowing through the tube.

If this velocity is small, condensation occurs in the manner depicted by Figure 15.7c for a horizontal tube. The condensate flows from the upper portion of the tube to the bottom, then flows in a longitudinal direction with the vapor.

For low-mass flow rates, the flow patterns is initially the same but the annular flow pattern gives way to wave flow followed by stratified flow. In the stratified region the liquid flows to the end of the tube under the action mainly of a hydraulic gradient.

At low flow rates, the flow is stratified. There is a film of condensate formed by film-wise condensation that drains from the top of the tube towards the bottom under the force of gravity. The film flow is laminar and primarily downwards when the vapor core velocity is low. If the vapor shear is sufficient and the onset to turbulence has been surpassed, then a turbulent film is formed whose predominant flow direction is axial.

The fully annular stratifying flow regime (where fresh condensate is continually formed at the top of the tube and, then, stratifies under the influence of gravity) with all the liquid normally in the lower portion of the tube, will have a thin layer of condensate around the upper perimeter. Stratified-wavy flow (not explicitly shown in Figure 15.7 is characterized by waves on the interface of the stratified liquid flowing along the bottom of the tube with film condensation on the top perimeter. In stratified flow, where no interfacial waves on the stratified liquid flowing along the bottom of the tube with film condensation on the top perimeter that drains into the stratified liquid. In stratified-wavy flow, the interfacial waves are small and do not reach the top of the tube and, hence, the upper perimeter would remain dry if not for the condensate that forms, again assuming that the stratified liquid creates an annular truncated ring. Thus, the angle θ varies between its maximum value of θ_{strat} at the threshold to fully stratified flow and its minimum value of zero at the threshold to annular flow. Importantly, these three simple geometries yield a smooth geometrical transition from one flow structure to another. For fully stratified flow, the stratified geometry is converted to an equivalent geometry with the same angle of stratification and cross-sectional area occupied by the liquid, but with the liquid distributed as a truncated annular ring of uniform thickness, δ. In stratified-wavy flow, the interfacial waves are small in amplitude and do not reach the top of the tube. Hence, the top perimeter of the tube is not wetted by the stratified liquid but only by the condensate that forms on this part of the exposed tube perimeter.

The most part of pressure drop occurs in gas-liquid flow during condensation (being much higher than would occur for either phase flowing along at the same mass rate) consist of three major contributions: (i) frictional, (ii) accelerational, and (iii) gravitational pressure drop. For horizontal tubes, the gravitational pressure drop is eliminated and the accelerational component is much smaller than the frictional pressure drop. Therefore, only frictional pressure drops are taken in consideration.

15.5.3.2 Heat Transfer

In direct expansion condensers of geothermal heat pumps, the mechanism of condensation is complex since the flow regimes change as the refrigerant passes from the superheated vapor state at the inlet up to the sub-cooled liquid state at the outlet (Figure 15.8). In the desuperheating and sub-cooling processes, the heat transfer coefficients are predicted based on known correlations of single-phase superheated vapor and sub-cooled liquids.

During two-phase condensation processes, the heat transfer coefficients are affected by various factors and parameters, such as (Cavallini et al. 2003): (i) flow

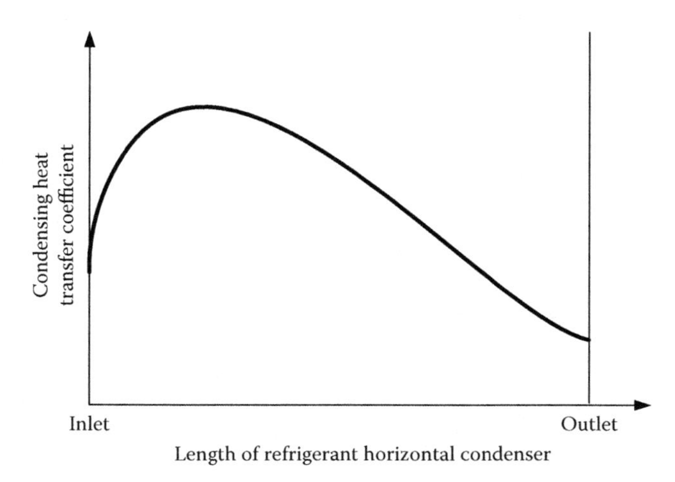

FIGURE 15.8 Variation of condensing heat transfer coefficient inside a horizontal tube.

regime (where the inertial and gravitational forces are of relative importance), (ii) refrigerant properties, (iii) tube diameter, (iv) saturation temperature, (v) heat flux density, (vi) mass velocity, and (vii) vapor quality.

As can be seen in Figure 15.8, at the entrance to the tube with its superheated vapor content, the coefficient is low, which is typical of convection heat transfer with a vapor. The coefficient increases once surface condensation begins and is usually at its highest value during annular flow. As more and more condensed liquid flows with the vapor, the surface available for condensation decreases. Near the end of the condenser tube (when all or most of the vapor has condensed) the coefficient drops quite low, because the process has approached that of convection heat transfer to a liquid.

For heat transfer during condensation inside horizontal plain tubes, the available semi-empirical condensation correlations are divided into three categories: (i) gravity-driven, (ii) annular flow-driven, and (iii) stratified and annular flow-driven.

The gravity-driven flow regime includes stratified, wavy, and slug flow regions, which are lumped together primarily because the dominant heat transfer mechanism is conduction across the film at the top of the tube. Consequently, this type of condensation is commonly referred to as film condensation.

The annular flow regime represents the situation where the interfacial shear stress dominates the gravitational force and results in a nearly symmetric annular film with a high-velocity vapor core. Some condensing heat transfer coefficients have been determined as functions of the interfacial vapor-liquid shear stress and of the liquid film thickness requiring prediction of the frictional pressure gradient. At high-mass velocity and low quality, where stratified or wavy flow could prevail with substantial convective heat transfer in the bottom part of the tube heat transfer in the liquid pool is not negligible. In fully developed sheet-dominated annular flow, there is a thin condensate film on the entire tube wall, while the vapor phase flows in the central core, and heat transfer is governed by vapor shear and turbulence.

The condensation heat transfer coefficients increase with the mass flux and vapor quality. At high-mass fluxes, the condensation heat transfer coefficients rise significantly with the vapor quality, but it is less sensitive to mass flux in low-vapor qualities; for high-mass fluxes and vapor qualities, there are general annular flows. In annular flow regime, forced convection which is mainly drove by velocity plays a leading role to condensation heat transfer; the liquid and vapor velocities increase with the mass flux convective effects. In addition, as the liquid film thickness becomes thinner when the vapor quality increases, the thermal resistance decreases. Thus, the condensation heat transfer coefficient is large in high-mass fluxes and vapor qualities; for low-mass fluxes and vapor qualities, the flow patterns are the gravity-driven flow; and there is a thick liquid layer in the bottom of the tube, which is the biggest part of the thermal resistance; the factors affecting the heat transfer for this regime are the liquid film thickness; therefore, the condensation heat transfer coefficient is insensitive to mass flux and vapor quality in low-mass fluxes and vapor qualities.

In the case of stratified flow dominated by gravity forces, a thick condensate layer flows at the bottom of the tube, while a thin liquid film forms on the wall in the upper portion of the tube. Heat transfer through the thin film is treated by the classical Nusselt (1916)-type analysis, while at very low-mass velocities, heat transfer through the thick condensate layer can either be neglected or treated as a convective process.

The convective condensation heat transfer coefficient (h_{cond}) is applied to the perimeter wetted by the axial flow of liquid film, which refers to the entire perimeter in annular, intermittent, and mist flows but only the lower part of the perimeter in stratified-wavy and fully stratified flows. The axial film flow is assumed to be turbulent.

Nusselt's (1916) falling film theory is used to obtain the film condensation heat transfer coefficient for laminar flow of a falling film (h_{film}) (which is essentially always the case for the tube diameters used in direct expansion ground-coupled heat exchangers). By assuming that, and ignoring the effect of vapor shear on the falling film, the mean film condensation heat transfer coefficient (h_{film}) is applied to the upper perimeter that would otherwise be dry for stratified-wavy and fully stratified flows (see Figure 15.7d):

$$h_{film} = 0.728\left[\frac{\rho_f(\rho_f - \rho_g)g \cdot h_{fg} \cdot k_f^3}{\mu_f \cdot d \cdot (T_{sat} - T_{wall})}\right]^{1/4} \tag{15.20}$$

Heat transfer coefficients for stratified flows are functions of the wall temperature difference while those for annular flow are not.

Since heat exchanger design codes are typically implemented assuming a heat flux in each incremental zone along the exchanger, it is more convenient to convert Equation 15.20 to heat flux using Newton's law of cooling, such that the heat flux version of Nusselt's equation where the local heat flux density ($\dot{q}$) is given by the expression:

$$h_{film} = 0.655\left[\frac{\rho_f(\rho_f - \rho_g) \cdot g \cdot h_{fg} \cdot k_f^3}{\mu_f \cdot d \cdot \dot{q}}\right]^{1/3} \tag{15.21}$$

where the constant 0.655 comes from $0.728^{4/3}$.

15.5.4 Heat Transfer Around the Horizontal Tubes

15.5.4.1 Single Tube

In the case of only one pipe of the horizontal direct expansion ground-coupled heat exchanger the interaction of pipes isn't take into account and first type of boundary conditions, i.e., that the temperature of pipe surface is time-constant, can be assumed. In order to determine the ground/soil's isothermal temperature flield and the steady-state heat flux transferred trough the pipe wall and dissipated into the surrounding ground/soil, the method of conform transformation associated with mirror image is usually used.

The conform transformation, which is permitted to transform a part of the plan $(x_a - x_b)$ into a circular contour, is of the following form (Leonachescu 1981):

$$lnW = aZ \tag{15.22}$$

where

$$Z = x + iy$$

a is a constant (–).

$$W = u + iv$$

$$W = R \cdot e^{i\Phi}$$

If one note, thus:

$$ln\ R \cdot e^{i\Phi} = a(x + iy) \tag{15.23}$$

$$lnR = ax \tag{15.23a}$$

$$\Phi = ay \tag{15.23b}$$

According to Equations 15.23a and 15.23b, the isotherms $x = constant$ in the complex plan Z become cercles $R = contant$ in the plan W. The lines of thermal flux $y = constant$ are replaced in the plan W by angles $\Phi = constant$, which are lines passing by the axes' origin O. This conform tranformation allows to determine the heat flux, for example, between the flux lines y_1 and y_2 over the depth L (Figure 15.9). In the planes Z and W, the total heat flux will be:

$$\dot{Q}_{12} = \dot{q}(y_2 - y_1) \cdot L = \dot{q}/a \cdot (\Phi_2 - \Phi_1) \cdot L \tag{15.24}$$

where

$\dot{q}$ is the heat flux density (W/m^2).

For $\Phi_2 - \Phi_1 = 2\pi$, the total heat flux becomes:

$$\dot{Q}_{12} = 2\pi \cdot L/a \cdot \dot{q} \tag{15.25}$$

where

$\dot{Q}_{12}$ is the total heat flux (W)

L is the ground/soil depth (m).

a is a constant (–).

$\dot{q}$ is the heat flux density (W/m^2).

Since the density of heat flux $\dot{q}$ is constant, on the basis of Fourier's law, results by integration in plan Z (see Figure 15.9):

$$T = T_0 - \dot{q}/k \cdot (x - x_0) \tag{15.26}$$

where

k is the ground/soil thermal conductivity (W/mK).

Based on Equations 15.23a and 14.23b, this results in plan W (Figure 15.9):

$$T = T_0 - \dot{q}/ak \cdot lnR/R_0 \tag{15.27}$$

By using Equation 15.25, results:

$$T = T_0 - \dot{Q}_{12}/2\pi kL \cdot lnR/R_0 \tag{15.28}$$

In the case of a ground/soil horizontally buried pipe, its outer surface is considered as an isotherm with temperature T_s.

With $\Theta = T - T_s$, Equation 15.28 becomes:

$$\Theta = \Theta_0 - \dot{Q}/2\pi kL \cdot lnR/R_0 \tag{15.29}$$

The heat flux generated by a horizontal buried pipe as well as the ground/soil temperature in a given point P (x, y) could be determined by using Kelvin's imagine method (Leonachescu 1981; Chiriac et al. 1988).

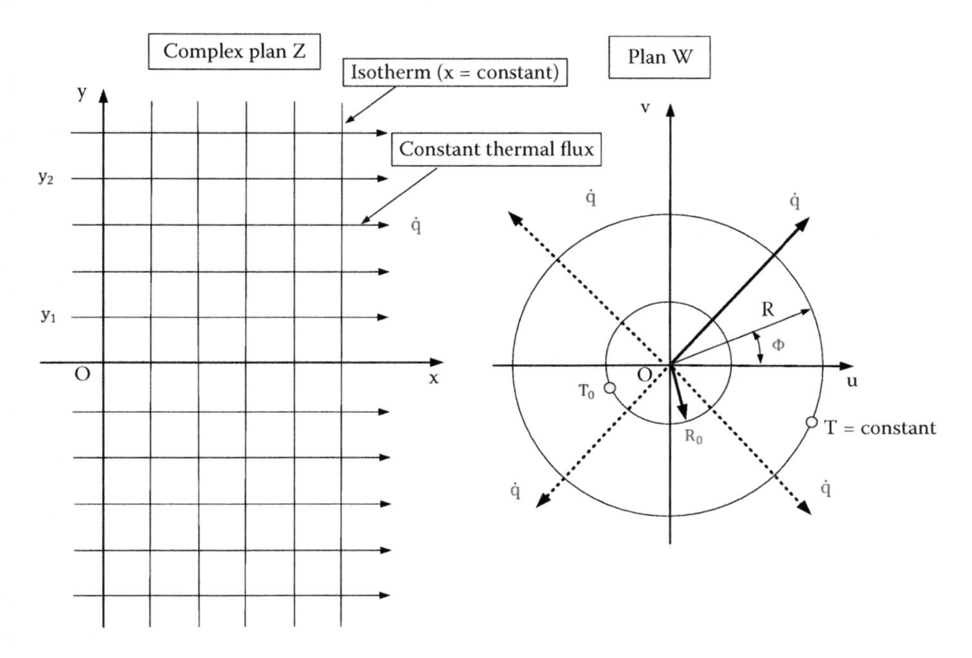

FIGURE 15.9 Principle of the conform transformation.

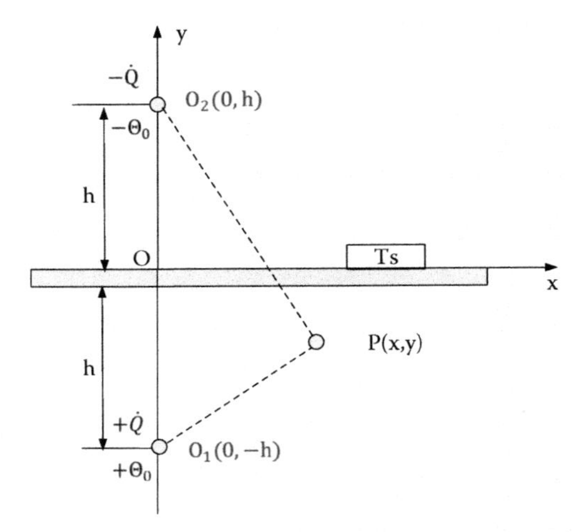

FIGURE 15.10 Kelvin's mirror method in the case of a punctual heat source.

The study of heat transfer through ground/soil in the case of a punctual heat source is performed by the aid of Kelvin mirror method.

Consider a punctual heat source placed in the ground/soil at the depth h in the point O_1 and its image O_2 qui emits the heat flux $-\dot{Q}$ at the temperature $-\Theta_0$. According to Equation 15.29, the effect of the heat source and of its image in the point P(x, y) are (Figure 15.10):

$$\Theta_1 = \Theta_0 - \dot{Q}/2\pi kL \cdot lnR_1/R_0 \tag{15.30a}$$

$$\Theta_2 = -\Theta_0 + \dot{Q}/2\pi kL \cdot lnR_2/R_0 \tag{15.30b}$$

The temperature at point P(x, y) results from the sum of the effects:

$$\Theta = \dot{Q}/2\pi kL \cdot lnR_2/R_1 \tag{15.31}$$

where the vector radii are:

$$R_1^2 = x^2 + (y + h)^2 \tag{15.32a}$$

$$R_2^2 = x^2 + (y - h)^2 \tag{15.32b}$$

From Equation 15.31, the results are:

$$\frac{x^2 + (y - h)^2}{x^2 + (y + h)^2} = \exp\left(\frac{4\pi kL}{\dot{Q}} \cdot \Theta\right) = c^2 \tag{15.33}$$

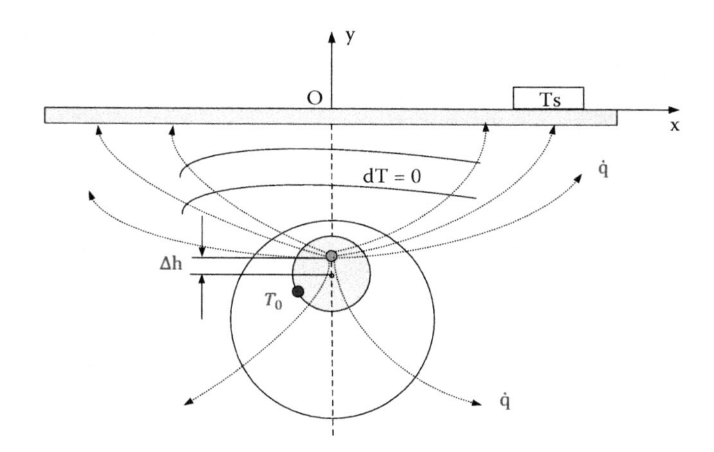

FIGURE 15.11 Familly of isotherms with centers on axis y.

For Θ = *constant*, the following family of isotherms is obtained:

$$x^2 + \left(y - h\frac{1 + c^2}{1 - c^2}\right)^2 = \frac{4c^2h^2}{(1 - c^2)^2} \tag{15.34}$$

This represents a family of circles having a center on the azis Oy (Figure 15.11):

$$O\left(0,\ h\frac{1 + c^2}{1 - c^2}\right) \tag{15.35}$$

And as variable radius:

$$r_c = 2c \cdot h/(1 - c^2) \tag{15.36}$$

For $r_c \dashrightarrow \infty$, it is obtained that the plane isotherm T_s corresponds to the ground/soilsurface. Another particular isotherm is the isotherm T_0 coincides with the outer surface of the buried pipe. The heat flux lines are normal on the isotherms and form a family of circles with centers with variable radius located on axis Ox.

Currently, the distance Δh is negligible and, thus, the heat flux can be calculated as the ratio of the ordinate of the isotherm family's center and the isotherm's radius, based on Equations 15.35 and 15.36: $y_c/r_c = (1 + c^2)/2c$, from which the constant c can be calculated as:

$$c = y_c/r_c \mp \sqrt{(y_c/r_c)^2 - 1} \tag{15.37}$$

By replacing Equation 15.33 and by considering $y_c \backsimeq h$ and $r_c = d/2$ (which correspond to isotherm T_0,) and taking the positive sign for physical reasons, one obtains:

$$\dot{Q} = \frac{2\pi kL\,(T_0 - T_s)}{ln\,[2h/d + \sqrt{(2h/d)^2 - 1}]} \tag{15.38}$$

$$L \gg d.$$

where

The thermal resistance of the pipe – ground/soil assembly is thus:

$$R_{thermal} = (1/2\pi kL)\,ln\,[2h/d + \sqrt{(2h/d)^2 - 1}] \tag{15.39}$$

For $h \gg d$, the heat flux could be approximated as:

$$\dot{Q} \backsimeq \frac{2\pi kL\,(T_0 - T_s)}{\ln\,(4h/d)} \tag{15.40}$$

where

d is the pipe diameter (m).

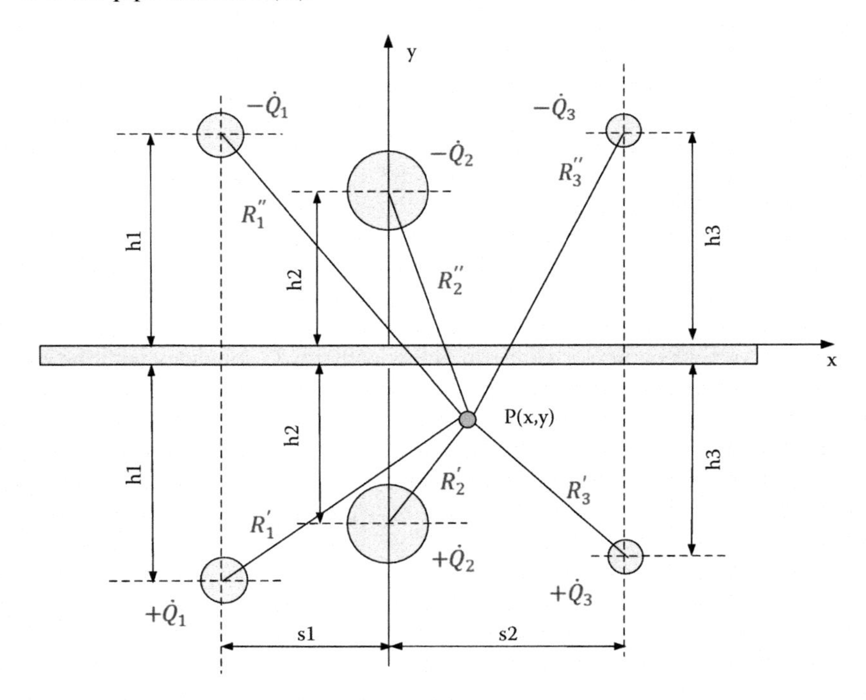

FIGURE 15.12 Schematic of multiple buried pipes.

15.5.4.2 Multiple Tubes

In the case of several pipes buried in the ground/soil (Figure 15.12), one can write:

$$\Theta = \sum_1^n \Theta_k = \sum_1^n \dot{Q}_k / 2\pi kL \cdot ln\,(R_k''/R_k') \tag{15.41}$$

Some terms of Equation 15.41 could be negative or even have nil value. For the general case represented in Figure 15.12, the solution of Equation 15.41 is unknown. However, for particular cases, there are solutions more or less exact. For identical pipes having the same diameter, buried depth, spacing, and temperature, Equation 15.41 becomes:

$$\Theta = \dot{Q}/2\pi kL \cdot ln\,(\Pi \cdot R_k''/R_k') \tag{15.42}$$

which can be also written as (Leonachescu 1981):

$$\dot{Q} = 2\pi kL\,(T_0 - T_s)/ln\,[2s/\pi d \cdot sh\,(2\pi h/s)] \tag{15.43}$$

15.6 VERTICAL DIRECT EXPANSION GROUND-SOURCE HEAT PUMP SYSTEMS

In the case of limited land area (e.g., garden, parking) around new or existing single-family houses or small commercial/institutional buildings, the concept of vertical and/or near vertical (inclined or diagonal) direct expansion ground-source heat pump systems have been developed, tested, and even commercialized as alternatives to horizontal direct expansion systems (Minea and Kaye 2008; Minea 2008).

Such systems use the geothermal heat pumps' working fluids (refrigerants) that circulate directly in vertical and/or inclined ground-coupled heat exchangers in order to either extract (in the heating mode, by vaporization) or reject (in the cooling mode, by condensation) heat from/to the ground/soil, respectively.

15.6.1 Basic Concepts

Direct expansion ground-source heat pump systems with vertical (or inclined) ground-coupled heat exchangers, that recover solar radiation collected by the ground/soil, consist of the following main components: (i) one (or more) refrigerant-to-air, refrigerant-to-water, or refrigerant-to-refrigerant geothermal heat pump(s) with or without desuperheater (dedicated to domestic hot water preheating); (ii) U-tubes with inner diameters varying between 19 mm and 38 mm inserted into up to 45 m deep vertical (Figure 15.13a) or inclined (Figure 15.13b) boreholes acting as direct expansion ground-coupled heat exchangers; inclined configurations involve drilling several non-vertical boreholes at angles from a common entry point, or nearby entry points at sites that may have very limited access at the ground/soil surface for installing multiple vertical boreholes; the total lengths of U-shaped tubes depend on the required heat input/output of the

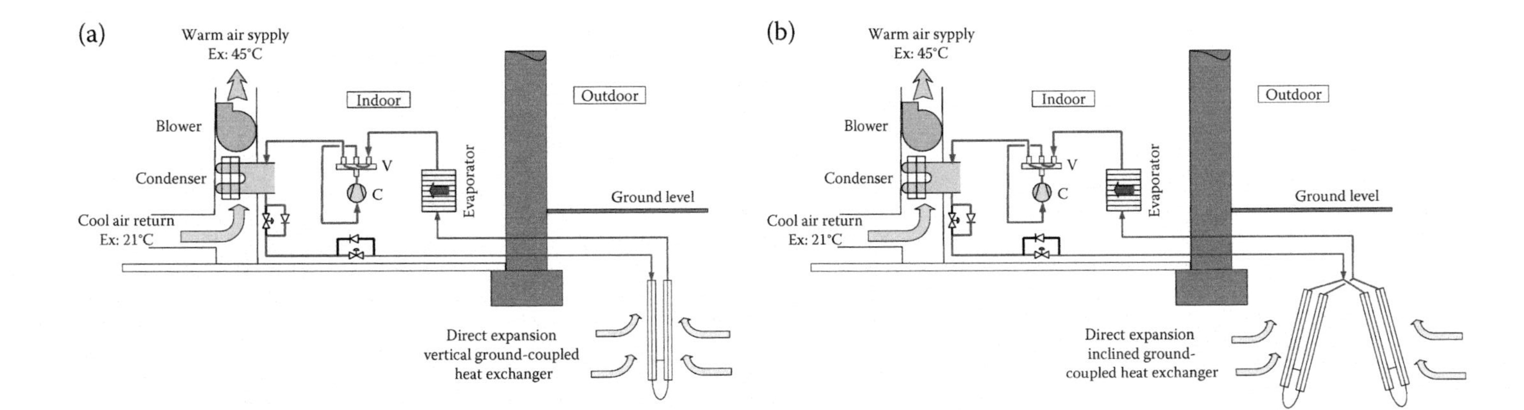

FIGURE 15.13 Schematic of a reversible ground-to-air direct expansion ground-source heat pump system with: (a) vertical; (b) inclined (diagonal) ground-coupled heat exchangers; C, compressor (Note: schematic operating in heating mode).

geothermal system; and (iii) an internal building heating/cooling distribution system that delivers space heating and cooling via, for example, air-handling units, underfloor heating pipes, or radiators.

The appropriate choice of the working fluids (refrigerants), motivated by efficiency, reliability, safety, regulations, and environmental considerations is one of the most important challenge for vertical/inclined direct expansion ground-source heat pump systems (Halozan 1995).

As working fluids, these building and cooling systems generally use refrigerants as HFC-134a (and its low-GWP replacements, e.g., HFO-1234yf) or R-404A and hydrocarbons (as propane and isobutane) (see Section 8.3 for information about the thermophysical properties of these fluids).

15.6.2 Kaye's Improved Concept

Kaye's improved concept is an improved vertical direct expansion ground-source heat pump system called "staged cooling" that can efficiently work in both heating (Figure 15.14) and cooling (Figure 15.15) modes by using at least three parallel vertical and/or inclined ground-coupled U-tubes which can be individually operated (Kaye 1995).

As can be seen in Figures 15.14 and 15.15, the "staged cooling" direct expansion geothermal heat pump concept contains: (i) at least three boreholes where, inside each of them, one copper U-tube (designed to provide 3.5 kW of geothermal heat pump nominal cooling capacity) is inserted; (ii) a compressor; (iii) a suction liquid accumulator; (iv) a four-way valve allowing for reversing the direction of refrigerant flow; (v) a refrigerant vapor header; (vi) a scavenging line connected between the upstream line of the suction liquid accumulator and the refrigerant vapor header; (vii) a refrigerant liquid header; (viii) an indoor coil that can act as a refrigerant condenser in the heating mode (see Figure 15.14) and as a refrigerant evaporator in the cooling mode (see Figure 15.15); (ix) an indoor air blower; (x) four thermostatic expansion and several solenoid valves (which can be individually operated) allowing operating the systems in both heating and cooling modes.

In the heating mode, the system operates with all the three parallel vertical or inclined ground U-tubes, but, in the cooling mode, two of them are cut off after pumping-down the refrigerant in the third U-tube. In this way, the geothermal heat pump's refrigerant charge is optimal in the heating mode when all three vertical U-tubes are operating and, also, correct in the cooling mode when only one vertical U-tube is used, but it contains the full system refrigerant charge. In the heating mode, the direct expansion geothermal heat pump uses all three earth-coupled U-tubes, and solenoid valves SV-A, SV-B, SV-C, and SV-D are open. The cooling expansion valve EX-4 is maintained fully open through connection of its equalizer tube to the refrigerant vapor line entering the condenser that is hot in the heating mode. Liquid refrigerant is allowed to flow into the liquid header through the solenoid valves SV-C and SV-D to heating expansion valves EX-2 and EX-3, respectively. The heating expansion valves EX-1, EX-2, and EX-3 are allowed to control the liquid refrigerant down each of the three vertical borehole U-tubes X, Y, and Z. Vaporized refrigerant flows downward, then to the top of each borehole U-

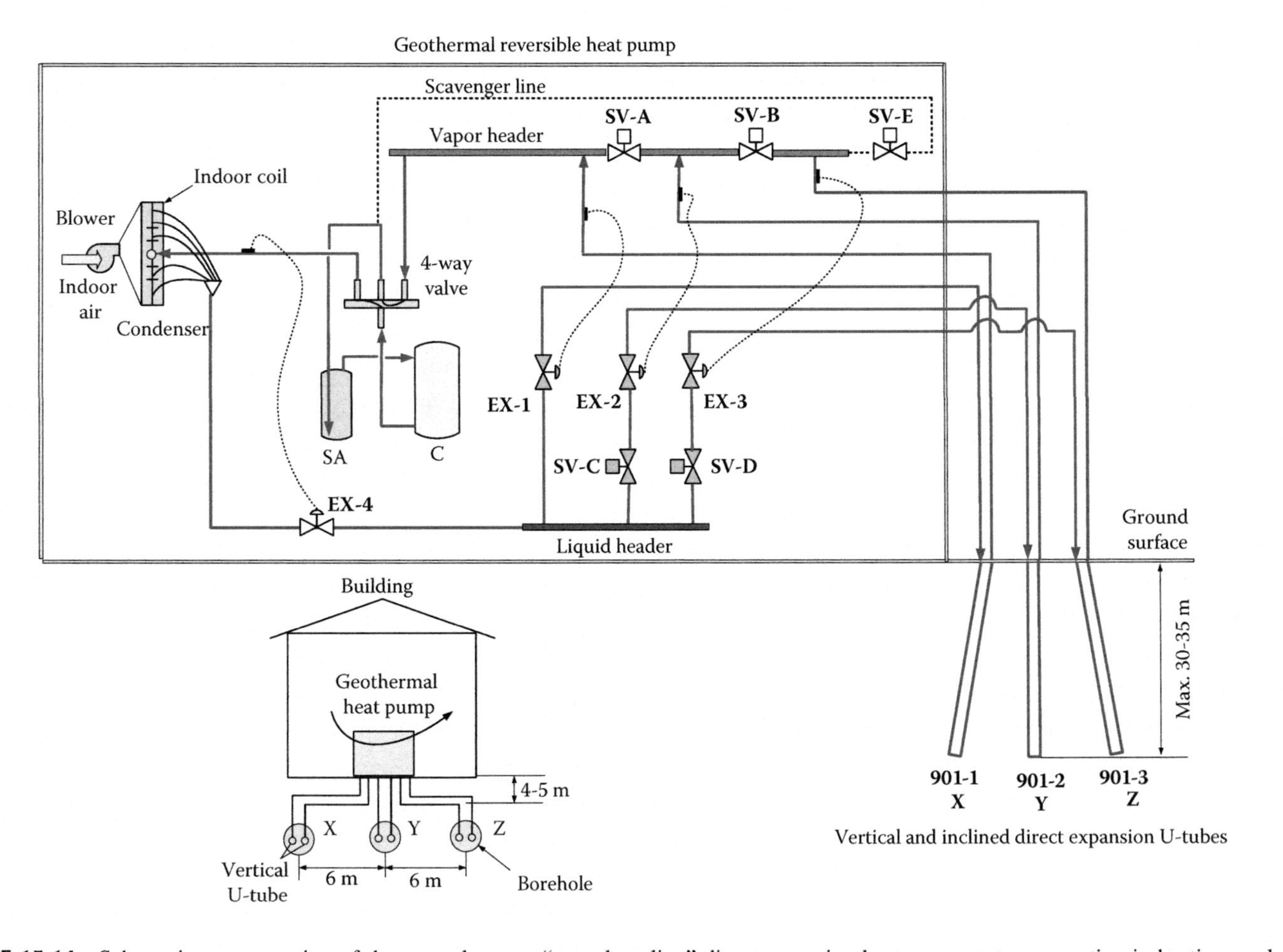

FIGURE 15.14 Schematic representation of the ground-source “staged cooling” direct expansion heat pump system operating in heating mode.

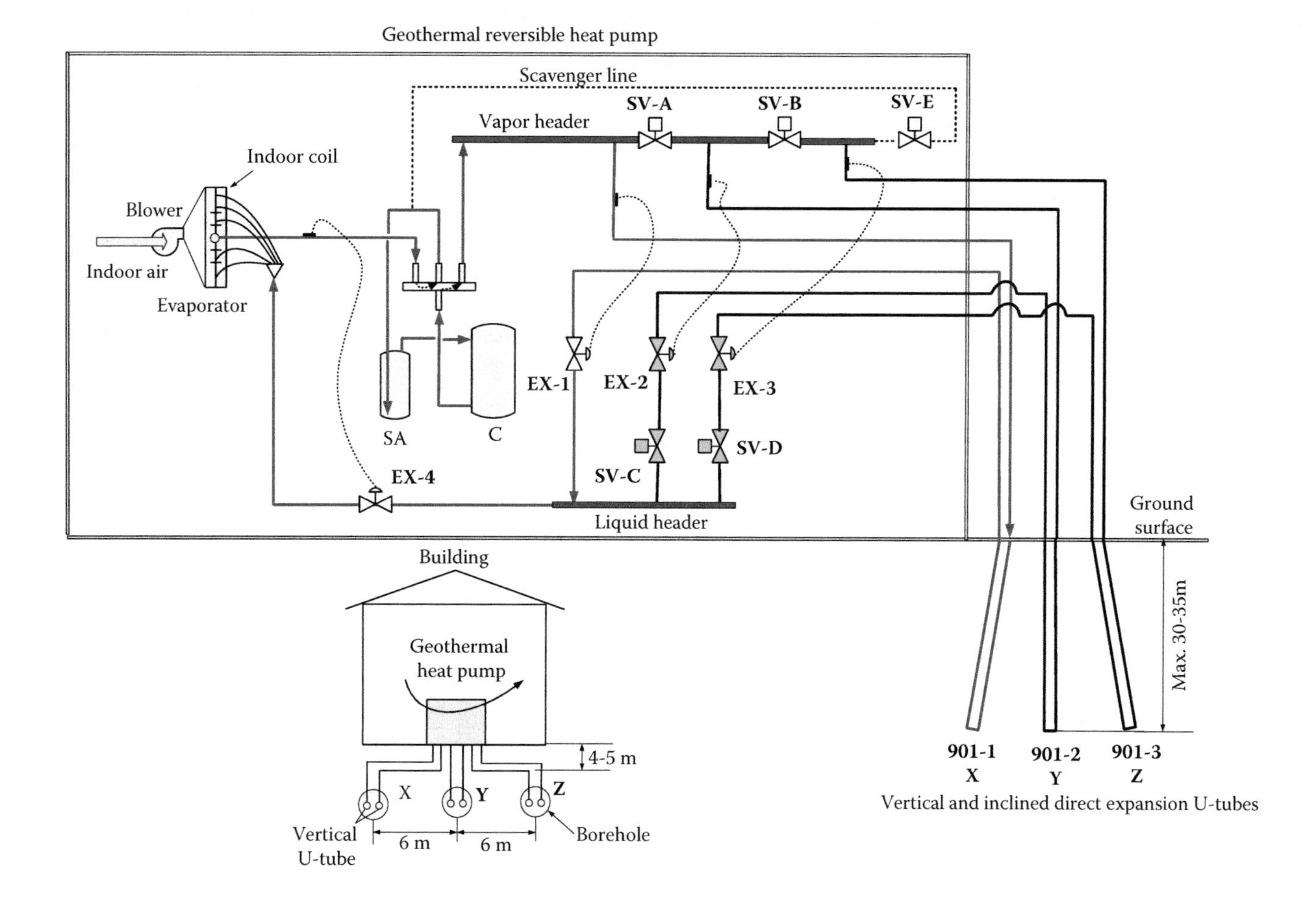

FIGURE 15.15 Schematic representation of the ground-source "staged cooling" direct expansion geothermal heat pump system operating in cooling mode.

tube and, finally, to the refrigerant vapor header. The solenoid valves SV-A and SV-2 are open and, thus, the refrigerant vapor flows through the four-way reversing valve and the suction accumulator to the heat pump compressor.

The compressed superheated refrigerant vapor flows from the compressor through the four-way reversing valve to the refrigerant condenser. The sub-cooled liquid refrigerant is withdrawn from the condenser, passes through the normally open expansion valve EX-4, and the cycle starts again.

The system operation in the cooling mode (Figure 15.4) consists of three stages.

In the first stage: (i) an electric signal from a room thermostat places the heat pump in cooling mode; (ii) solenoid valves SV-A, SV-C, and SV-D are closed; (iii) solenoid valves SV-B and SV-E are opened; (iv) since the solenoid valve SV-A is closed, superheated refrigerant vapor is allowed to enter the vertical direct expansion U-tube X; (v) heating expansion valve EV-1 is maintained open by connection of its controller bulb to the superheated vapor refrigerant line of the vertical direct expansion U-tube that is hot in the cooling mode; (vi) compressed superheated refrigerant vapor passes through the four-way reversing valve and, because the solenoid valve SV-A is closed, it is directed to the vertical direct expansion U-tube X where it condenses at a pressure high enough to force the resulted liquid refrigerant to flow through the open heating expansion valve EV-1 and through the cooling expansion valve EV-4; (vii) because the liquid line solenoid valves SV-C and SV-D are closed, the liquid refrigerant is prevented from entering the vertical direct expansion U-tubes Y and Z which are downstream of the SV-C and SV-D solenoid valves, respectively; (viii) refrigerant liquid is evaporated and scavenged from the liquid refrigerant lines that are downstream of the solenoid valve SV-A, through solenoid valves SV-B and SV-E via scavenger line to the suction accumulator and into the compressor; (ix) the cooling expansion valve EX-4 is allowed to control the correct amount of liquid refrigerant into the air refrigerant-air heat exchanger (evaporator) coil by connection of the equalizer tube thereof to the common suction line ahead of the suction accumulator, and by connection of its controlling bulb thereof a short distance from the outlet of the refrigerant vapor; (x) finally, warmed refrigerant vapor is withdrawn from the refrigerant/air heat exchanger (evaporator) coil into the suction line and through the four-way reversing valve into the accumulator and thence the compressor suction inlet.

In the second stage: (i) all solenoid valves SV-A, SV-B, SV-E, SV-C, SV-D, and individual vertical direct expansion U-tubes X, Y and Y connections to the refrigerant vapor header are actuated by means of a first pressure sensing device placed on the refrigerant vapor header until pressure reaches a predefined point which is sufficient to override an electrical signal sent by the room thermostat and, thereby, to operate a relay (not shown) which activates the solenoid valves SV-A, SV-B, and SV-E; (ii) solenoid valves SV-A, SV-B, and SV-E are opened and solenoid valves 903-2 and 903-5 are closed; (iii) solenoid valve SV-A is then opened and solenoid valve SV-B is then closed to allow the hot gas to flow to vertical direct expansion U-tubes X, Y, and Z; (iv) heating expansion valves EX-1 and EX-2 are maintained open by connexion of the equalizer tubes thereof to the common suction line upstream of the suction accumulator and by connexion of the controller bulbs thereof to the vertical direct expansion U-tubes X and Y, respectively, which are hot in this second-stage cooling mode; (v) hot compressed refrigerant is passed from the compressor through the reversing valve and through the refrigerant vapor header into the hot gas lines which are upstream of the closed solenoid valve SV-B; (vi) the heat pump is allowed to reject its heat to the ground at a pressure which is high enough to force liquid refrigerant through the vertical direct expansion U-tubes X and Y liquid refrigerant lines, through the heating expansion valves SV-1 and SV-2, and on the liquid line header connected to the cooling expansion valve EX-4; (vii) liquid line solenoid valve SV-D is closed, thereby preventing liquid refrigerant from entering the vertical direct expansion U-tube Z which is downstream of solenoid valve SV-D; (viii) residual liquid refrigerant is evaporated and scavenged from the liquid refrigerant lines which are downstream of the refrigerant vapor header solenoid valve SV-2, through the solenoid valve SV-E via scavenger line to the suction accumulator and then to the compressor; (ix) cooling expansion valve EX-4 is allowed to provide the correct amount of liquid refrigerant into the refrigerant/air heat exchanger (evaporator) coil by virtue of the connection of the equalizer tube thereof to such common suction line upstream of the suction accumulator and by connection of the controlling bulb thereof a short distance from the refrigerant vapor outlet; (x) finally, warmed refrigerant vapor is withdrawn from the refrigerant/air heat exchanger coil into the suction line and through the four-way reversing valve into the suction accumulator and, thence, to the suction inlet of the compressor.

In the third stage: (i) all solenoid valves SV-A, SV-B, SV-E, SV-C, SV-D, and individual loop connections to the header are actuated by a second pressure sensing device on the refrigerant vapor header, until the pressure reaches a predefined point which is sufficient to override an electric signal sent by a room thermostat and by the first pressure sensing device, and thereby to operate a relay (not shown) which activates the loop solenoid valves by allowing solenoid valves SV-A, SV-B, SV-C, and SV-D to open while closing the solenoid valve SV-E; (ii) solenoid valves SV-A and SV-B are then open to allow hot vapor to flow to vertical direct expansion U-tubes X, Y, and Z; (iii) heating expansion valves EX-1, EX-2, and EX-3 are maintained open by connection of the equalizer tubes thereof to the common suction line ahead of the accumulator and by connection of the controller bulbs thereof to the vapor lines that are hot in the cooling mode; (iv) hot compressed refrigerant from the compressor passes through the four-way reversing valve and through the refrigerant vapor header into the gas lines which are upstream of the closed solenoid valve SV-E; (v) the heat pump is allowed to reject its heat to the ground at a pressure which is high enough to force the liquid refrigerant lines through the vertical direct expansion U-tubes X, Y, and Z, through the heating expansion valves EX-1, EX-2, and EX-3, through the solenoid valves SV-C and SV-D and on to the liquid line header that is connected to the inlet port of the cooling expansion valve EX-4; (vi) the cooling expansion valve EX-4 is allowed to provide the correct amount of liquid refrigerant into the refrigerant/air heat exchanger (evaporator) coil by connection of equalizer tube thereof to the common suction line ahead of the accumulator and by connection of the controlling bulb thereof a short distance from the outlet of the refrigerant vapor; (vii) finally, warmed refrigerant vapor is withdrawn from the refrigerant/air heat exchanger

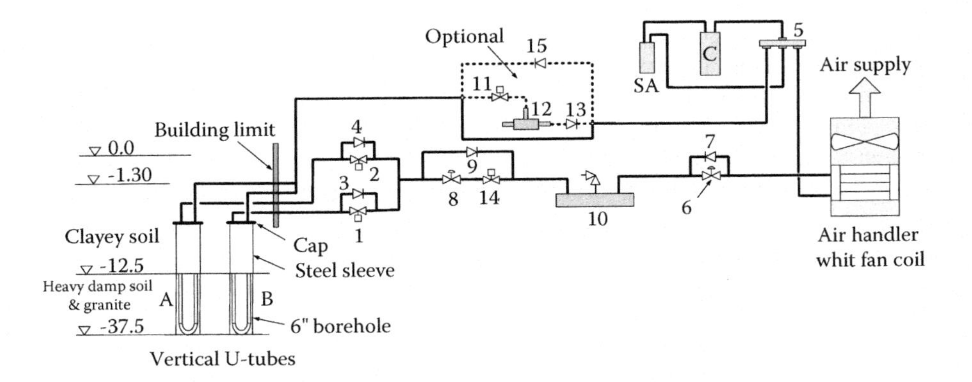

FIGURE 15.16 Schematic of the ground-source heat pump with vertical DX heat exchanger. A, B, vertical ground-coupled U-tubes; C, compressor; SA, suction accumulator.

(evaporator) coil into the suction line and through the reversing valve into the accumulator and to the compressor suction inlet.

15.6.3 Minea's Improved Concept

Minea's improved concept contains new refrigerant circuits, including a bidirectional liquid receiver and control sequences to improve oil return and to make it easier starting up and switching the geothermal heat pump from the heating and/or stand-by modes to the cooling mode (Minea 2008).

The refrigerant circuits (Figure 15.16) contain a number of original features aiming to improve the geothermal heat pump operation and control. Solenoid valves 1 and 2 and check valves 3 and 4 allow the refrigerant to flow down (in the heating mode) and up (in the cooling mode) through the vertical U-tubes. The refrigerant flow direction changes according to the thermostat heating or cooling demands by switching reversing valve 5. Thermostatic expansion valves 6 and 8, based on the refrigerant actual pressures and temperatures at the evaporator outlets in heating and cooling modes, are used as refrigerant control devices. Check valves 7 and 9, installed in parallel, as well as bi-directional liquid receiver 10 let the refrigerant flow in both directions according to the operating mode.

In the heating mode, the refrigerant enters the vertical U-tubes in a saturated two-phase state at low pressure. Convective vaporization takes place with the refrigerant first flowing down and then flowing up through the vertical U-tubes, and forced convective vapor superheating occurs at the end of the vertical tubes. The low-pressure superheated vapor leaving the ground vertical U-tubes (evaporator) flows through the solenoid valve 11 (open) and unidirectional pressure regulating valve 12, check-valve 13, reversible valve 5, and the suction accumulator, and enters the suction line of the compressor C. The sub-cooled liquid from the condenser (fan coil) flows through the check valve 7, bidirectional liquid receiver 10, solenoid valve 14 (open), and expansion valve 8 prior entering the ground vertical U-tubes at low pressure (valves 1 and 2 are open).

In the cooling mode, the refrigerant flow is reversed. The high-pressure superheated refrigerant vapor from the compressor C flows through check valve 15 and enters the vertical U-tubes where it first flows down and then up. Superheated vapor cooling by forced convection followed by convective condensation takes place, and the heat is transferred to the ground. The high-pressure sub-cooled liquid flows through check valves 3, 4, and 9, the bi-directional liquid receiver 10, expansion valve 6, and, finally, evaporates at low pressure inside the fan coil (evaporator) (Minea 2008).

To prevent the ground-coupled heat exchanger from collapsing, it is recommended – if permitted – not to grout the boreholes and, also, not to tightly cap the steel sleeves. These measures may prevent ice expansion, which could be responsible for the refrigerant copper tubes collapsing when very low evaporating pressures occur, and eliminate the risks of copper tube crashing caused by the water and silica grout high static pressure inside the boreholes. Furthermore, these measures may help to reduce the vertical ground-coupled heat exchanger cost, and simplify its installation. As additional measure, an (optional) unidirectional adjustable pressure regulating valve (as shown in Figure 15.16) may be used to keep the refrigerant evaporating temperature above, for example, –10°C, regardless of the refrigerant type, in order to avoid the risk of excessive ground/soil freezing during long running cycles in the heating mode. In this case, the eventual drop in the geothermal heat pump heating capacity must be compensated by supplying additional back-up heating.

To improve the oil return to the compressor in heating mode, a simple control strategy may consist of periodical and alternatively pumping-down of the refrigerant liquid from both vertical U-tubes. At certain intervals (e.g., every 24 or 48 hours), solenoid valve 1 is closed and the refrigerant flow through heat exchanger B is stopped. This valve may be kept closed during a short, fixed period of time (seconds), or until the compressor suction pressure reaches a minimum preset pressure. The velocity of the refrigerant and of the oil eventually trapped at the bottom of U-tube B significantly increase, and they are efficiently pumped to the compressor. At the end of these sequences, the solenoid valve 1 opens and the system continues to operate in the heating mode with both U-tubes open. After a period of time, for example, five minutes, solenoid valve 2 is closed during another short period of time (seconds), or until a minimum preset compressor suction pressure is reached. The velocity of the refrigerant and of the oil eventually trapped at the bottom of U-tube "A" increase, and they are rapidly pumped to the compressor. At the end of these operations, solenoid valve 2 opens, and the system continues to normally operate in the heating mode. Another method for improving oil return may be provided by periodically switching the geothermal heat pump, for example, every 24 or 48 hours, from the heating to the cooling mode. When using two-speed or two parallel compressors to match the heating/cooling capacity with demand, the refrigerant velocity decreases when the compressor speed slows down, or when only one compressor is running. In these cases, to ensure proper oil recovery from the bottom of the vertical U-tubes at lower heating capacities, the compressor may periodically operate, at certain intervals and for short periods of time, at maximum speed.

When the geothermal heat pump is not running, the bulk of the refrigerant within the system naturally migrates to the coldest zones, particularly to the cold ground heat exchanger areas. In this situation, starting up the heat pump in the cooling mode or switching the system from the heating to the cooling mode are critical issues because the temperature of the liquid inside the ground U-tubes is very low. The compressor isn't able to build a sufficient pressure differential to push out the liquid from the ground/soil heat exchanger to the expansion valve. In fact, the refrigerant column, which has a density of about 1 320 kg/m^3 at average evaporating temperatures of –17°C, and a height of 37.5 m, provides a static pressure of about 485 kPa$_r$ at the bottom of the vertical U-tubes. If the compressor has to start immediately in cooling mode, its discharge pressure at such low temperatures will be much lower than the liquid static pressure. Also, when the geothermal heat pump shuts down in the cooling mode for long periods of time (stand-by mode), the refrigerant is trapped inside the vertical tubes at temperatures of maximum 8°C and builds relatively high static pressures. In both cases, the compressor cannot push the liquid up from the vertical U-tubes because the discharge pressures are lower than the static pressure of the vertical liquid column. To efficiently start up the system or switch from the heating to the cooling mode, it is proposed to use bi-directional liquid receiver 10. This liquid receiver lets the liquid flow in both directions and is designed to store almost the entire refrigerant quantity of the system. To facilitate switching the heat pump directly from the heating mode to the cooling mode after having operated for long periods of time in heating mode, or after shut-down periods before cooling demands, the following control strategy is proposed. When the thermostat calls for cooling, the system control first allows the heat pump to start up or to continue running in heating mode during a certain period of time. Solenoid valves 1, 2, and 14 are open and reversible valve 5 is in this heating mode position. The heat pump operates in heating mode, even though the thermostat calls for cooling. After stabilization of the system (5 to 10 minutes), solenoid valve 14 closes to stop the refrigerant liquid flowing towards ground vertical U-tubes A and B. The refrigerant is thus pumped-down and stored inside liquid receiver 10 that has a sufficient capacity to store up to 90% of the system's refrigerant charge. If necessary, the rest of the refrigerant charge is stored inside the condenser (fan coil). Both vertical U-tubes are now almost empty of sub-cooled liquid. After a short period of time, or until a minimum preset suction pressure is detected, the compressor shuts down and the control system switches reversible valve 5 in its cooling position. Simultaneously, solenoid valve 14 opens and the compressor normally starts up in cooling mode (Minea 2008).

15.6.4 Refrigerant-Side Vaporization

The refrigerant-side vaporization thermodynamic process that occurs in vertical U-shaped direct expansion ground-coupled heat exchangers operating in the heating mode is usually by mass and energy conservation equations under the refrigerant two-phase flow.

15.6.4.1 Flow Patterns

The prediction of flow regimes is crucial for an accurate estimations of heat transfer coefficients during condensation of refrigerants in vertical tubes. Heat transfer coefficients for condensation processes depend on the surface geometry and orientation, flow pattern, and condensation rate. The behavior of condensate is controlled by inertia, gravity, vapor-liquid film interfacial shear, and surface tension.

Two major condensation mechanisms in film condensation are as follows: (i) gravity-controlled and (ii) shear-controlled (forced convective) condensation in passages where the surface tension effect is negligible. At high vapor shear, the condensate film may became turbulent.

If the laminar flow direction is downward and gravity controlled, heat transfer coefficient for internal condensation inside vertical tubes can be predicted using the correlations for external film condensation. The condensation conditions usually occur under annular flow conditions (Rohsenow et al. 1998). The flow pattern of the refrigerant condensation process depicts the spatial and temporal distributions of

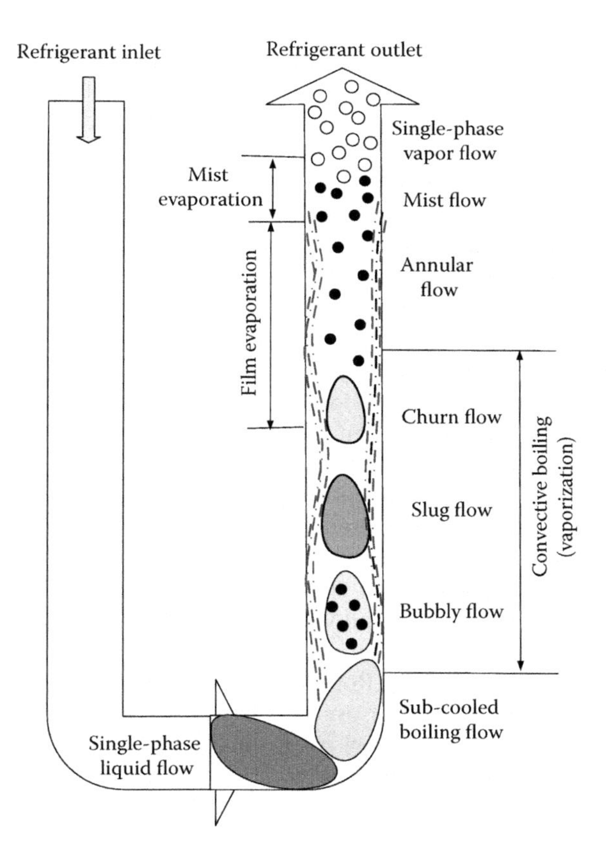

FIGURE 15.17 Schematic of flow patterns and the corresponding heat transfer mechanisms inside a vertical U-shaped tube (Note: upward flow only shown).

vapor and liquid phases of two-phase flows and varies with tube geometry and the fluid physical properties.

At non-adiabatic conditions (i.e., heat transfer with uniform heat addition or removal), the vapor-liquid two-phase flow patterns (Figure 15.17) are very complex physical processes since they combine the characteristics of deformable interface, tube shape, flow direction (upward or downward), and the vapor compressibility, and depends also on refrigerant (liquid and vapor) flow rates, physical properties, and boundary conditions. Depending on the operating conditions, such as pressure, temperature, mass velocity, tube orientation, gravity, and refrigerant properties, various vapor-liquid interfacial geometric configurations occur in two-phase vaporization process (Lawn and Young 1965; Cheng et al. 2008). In addition to inertia, viscous and pressure forces present in single-phase flows, the two-phase flows are affected by the interfacial tension forces, wetting characteristics of the liquid on the tube wall, contact angle, and the exchange of mass, momentum, and energy between the liquid and vapor phases (Carey 1992; Collier and Thome 1996; Rohsenow et al. 1998; Cheng et al. 2008).

As can be seen in Figure 15.17, with single-phase subcooled liquid (below its saturation temperature) formed at the bottom of the vertical U-tube, nucleate boiling begins when the tube wall reaches a temperature exceeding the saturation temperature by a small amount. Bubbles detach from the wall and collapse in the central core of the liquid. When the bulk temperature approaches saturation, bubbles tend to coagulate and form slug and churn flows, which later break down to form an annular flow (Lawn and Young 1965).

This transition occurs at 1% vapor quality (defined as the percentage of vapor in the total mass flow) and it is normally completed between 1% and 2% quality. As the vapor quality increases, slip flow develops between the two phases and small droplets of liquid are picked up from the interface by the high-velocity vapor stream. A large number of droplets forms in the central core continuously and migrate to the liquid stream, forming the annular mist regime. Annular film flow and stratified flow are the two predominant flow patterns in forced convective condensation in vertical tubes. When the vapor quality reaches a certain critical value, the amount of liquid present becomes insufficient to maintain a stable film on the walls. This is the condition sometimes referred to as two-phase burnout (or dry wall condition) that occurs over a region rather than at a point. After the transition region, there is a change over to a mist flow regime in which the walls are dry and the mechanisms of heat transfer are those of forced convection to a vapor stream with migration of droplets to the wall with vaporization on contact. Small droplets have been observed to exist even in superheated vapor.

As can be seen in Figure 15.17 (Rohsenow et al. 1998): (i) at low-vapor qualities and low-mass flow rates, the flow usually obeys the bubbly flow pattern; (ii) at higher vapor qualities and mass flow rates, slug or plug flow replaces the bubbly flow pattern; and (iii) further increase in vapor quality and/or mass flow rates leads to the appearance of the churn, annular, and wispy annular flow patterns.

15.6.4.2 Pressure Drops

Pressure drop process of two-phase flows is relatively difficult to predict with good accuracy because of the complex interaction between the liquid and vapor phases.

The three dominant contributions to the total pressure drop along the two-phase fluid flow path are friction, momentum change, and hydrostatic effects (Rohsenow et al. 1998):

$$\Delta p = \Delta p_{friction} + \Delta p_{momentum} + \Delta p_{gravity} + \Delta p_s \quad (15.44)$$

where

Δp is total pressure drop (Pa).

$\Delta p_{friction}$ is the two-phase friction pressure drop (Pa); it always decreases being a function of the refrigerant velocity, mass flow rate, and density (that depends on the local pressure.

$\Delta p_{momentum}$ is the momentum change loss (caused by acceleration or deceleration of the flow) (Pa); acceleration pressure drop is a function of the change of the velocity caused by the change of density; decreasing density means an increase of the refrigerant velocity and a decrease of the system pressure.

$\Delta p_{gravity}$ is the gravity loss (i.e., the hydrostatic effect) (Pa); pressure drop caused by gravity increases the system pressure in the down-comer and decreases the system pressure in the riser; the pressure change caused by gravity can result in a system pressure increase; depending on the amount of this pressure increase even re-condensation can occur, because temperature differences between refrigerant and ground are relatively low.

Δp_s is the pressure drop due to various singularities along the flow path such as entrance and exit losses, abrupt change in the free flow area, bends, and valves (because this kind of pressure drops is difficult to measure, in two-phase flow, they are often lumped into friction losses).

15.6.4.3 Heat Transfer

The prediction of vaporization heat transfer coefficients is intrinsically related to the refrigerant flow patterns. Because of the complexity of two-phase refrigerant flows, of thermal interference between the vertical U-shaped tubes and of the thermal response of boreholes, simplifying assumptions as the following are generally made to analyze and design the direct expansion heat exchangers: (i) constant temperature of the undisturbed ground/soil; (ii) uniform borehole wall temperature; (iii) uniform U-tube wall temperature and negligible thermal resistance of the tube wall; (iv) one-dimensional homogenous refrigerant flow; (v) two-phase refrigerant flow in thermal equilibrium; (vi) negligible refrigerant pressure drops along the U-tubes; and (vii) negligible axial conduction through the refrigerant.

For upward vaporization (forced boiling evaporation), consider a vertical tube heated uniformly over its length with a low heat flux and fed with sub-cooled liquid at its base at such a rate that the liquid is totally evaporated over the length of the tube (see Figure 15.17). Most of the leading vaporization heat transfer correlations (that consists of convective and nucleate boiling terms) do not contain detailed information on the flow patterns. They are typically accurate for annular flow (Chen 1966; Shah 1982; Gungor et al. 1986; Kandlikar 1990).

A general correlation for in-tube vaporization of refrigerants can be expressed as follows (Kandlikar 1990):

$$\frac{h_{tp}}{h_f} = 0.6683 \cdot Co^{-0.2} + 1058 \cdot Bo^{0.7} \cdot F_{fluid}(1 - x)^{0.8} \tag{15.45}$$

where

h_{tp} is the two-phase heat transfer coefficient (W/m^{2}K).

h_f is the single-phase heat transfer coefficient for the entire flow as liquid (W/m^{2}K).

$Co = (\rho_g/\rho_f)^{0.5}$ is the convection number (–).

$Bo = \dot{q}/G \cdot h_{fg}$ is the nucleate boiling number (–).

ρ_g is the vapor density (kg/m^3).

ρ_f is the liquid density (kg/m^3).

$\dot{q}$ is the heat transfer density (W/m^2).

G is the mass velocity (kg/m^{2}s).

h_{fg} is the evaporation (latent heat) enthalpy (J/kg).

F_{fluid} is the refrigerant–surface parameter that depends on the refrigerant type and the heat transfer surface (–); for example, for HFC-134a, $F_{fluid} = 1.63$.

The vaporization heat transfer rate per unit heat transfer surface area from the wall to the pure vaporizing fluid is given by:

$$\dot{q} = h_{tp}(T_w - T_{sat}) \tag{15.46}$$

where

$\dot{q}$ is the density of heat transfer (W/m^2).

h_{tp} is the two-phase heat transfer coefficient during the vaporization process (W/m^{2}K).

T_w is the tube wall temperature (°C).

T_{sat} is the refrigerant saturation temperature (°C).

15.6.4.4 Refrigerant-Side Condensation

Condensation is a vapor-liquid phase change phenomena that takes place when vapor is cooled below its saturation temperature at a given pressure. Internal condensation processes, consisting of the vapor flowing turbulently in the tube's core and laminar or turbulent film on the tube wall, are complex because, in addition to phase change, a simultaneous motion of both vapor and condensate takes place. The transition from laminar to turbulent film depends not only on the liquid Reynolds number but, also, on the interfacial shear stress (Lawn and Young 1965; Holman 1986; Rohsenow et al. 1998).

Figure 15.18 shows that at a given constant heat flux density (W/m^{2}K) and saturated temperature, the condensation heat transfer coefficients increase with an increase in the vapor quality and mass velocity. Also, the condensation heat transfer coefficients increase with a decrease in the hydraulic diameter at a given saturation temperature.

Because of the complexity of two-phase refrigerant flows, of thermal interference between the vertical tubes and of the thermal response of boreholes, simplifying assumptions as the following are generally made to analyze and design the direct expansion heat exchangers where condensation two-phase changes take place: (i) constant temperature of the undisturbed ground, (ii) uniform borehole wall temperature, (iii) uniform tube wall temperature, (iv) one-dimensional homogenous refrigerant flow, (v) two-phase refrigerant flow in thermal equilibrium, (vi) negligible refrigerant pressure drops along the tubes, (vii) negligible axial conduction through the refrigerant, and (viii) negligible thermal resistance of the tube wall.

Depending on the operating conditions, such as pressure, temperature, mass velocity, tube orientation, gravity, and refrigerant properties, various vapor-liquid interfacial geometric configurations occur in two-phase condensation processes.

Condensation in vertical tubes depends on the vapor flow direction and its flow rate. During downflow of vapor, if the vapor velocity is very low, then the condensate flow is controlled by gravity, and the Nusselt results for a vertical flat plate are applicable (unless the tube inside diameter is very small). The flow may proceed from laminar wave-free to laminar wavy to turbulent conditions, depending on the film Reynolds number (i.e., the heat flux and length of the tube). If the vapor velocity is very high, then the flow is controlled by vapor shear forces, and annular flow models are applicable.

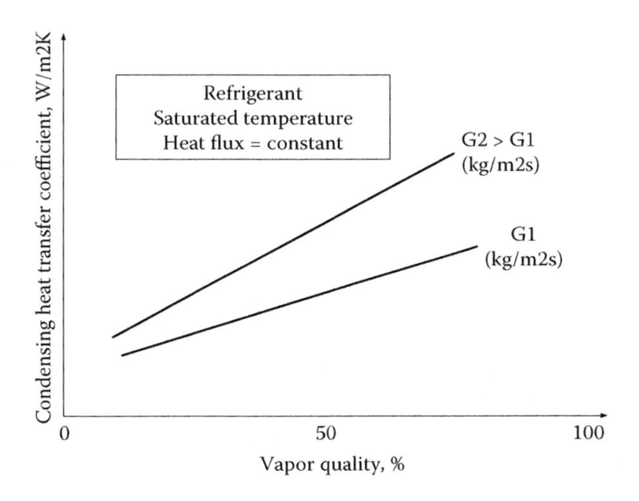

FIGURE 15.18 Variation of heat transfer coefficients of a given refrigerant (condensing at a constant saturated temperature and constant heat flux) with flow velocity.

During up-flow of vapor, interfacial shear will retard the drainage of condensate, thicken the condensate film, and decrease heat transfer. Care must be exercised to avoid vapor velocities that are high enough to cause "flooding," a phenomenon that occurs when vapor shear forces prevent the downflow of condensate. For the interfacial shear-controlled flows, annular film flow pattern is established, and the tube orientation is irrelevant. Consequently, the correlations for annular condensation in horizontal tubes can be applied for vertical internal downward flows as well.

For the upward flow direction, the shear forces may influence the downward-flow of the condensate, causing an increase of the condensate film thickness. Therefore, the heat transfer coefficient under such conditions shall decrease up to 30% compared to the result obtained using the same correlation as the upward-flowing vapor. If the vapor velocity increases substantially, the so-called flooding phenomenon may occur. Under such conditions, the shear forces completely prevent the downward condensate flow and flood (block) the tube with the condensate.

For:

$$7\ \mathrm{mm} \le \mathrm{d_i} \le 4\ \mathrm{mm}$$

$$158\ \mathrm{W/m^2} \le \mathrm{q} \le 1.9 \cdot 10^6\ \mathrm{W/m^2}$$

$$0.002 \le \mathrm{p_{sat}/p_{cr}} \le 0.44$$

$$11 \le \mathrm{G} \le 210\ \mathrm{kg/m^2s}$$

$$21 \le \mathrm{T_{sat}} \le 310\ \mathrm{C}$$

$$100 \le Re_f \le 63 \cdot 10^3$$

$$1 \le Pr_f \le 13$$

$$0 \le \mathrm{x} \le 1$$

the following dimensionless correlation allows to predict with ∓14.4% accuracy the condensation heat transfer coefficient of the annular film condensation flow (Shah 1979):

$$\mathrm{Nu} = h_{cond} \cdot d_i / k_f = 0.023 \frac{\mathrm{k_f}}{\mathrm{d_i}} \mathrm{Re_f^{0.8}} \mathrm{Pr_f^{0.4}} \left[(1 - \mathrm{x})^{0.8} + \frac{3.8 \cdot \mathrm{x}^{0.76} (1 - \mathrm{x})^{0.04}}{(\mathrm{p_{sat}/p_{cr}})^{0.38}} \right] \tag{15.47}$$

where

Nu is Nusselt number (–).
h_{cond} is the condensation heat transfer coefficient (W/m^{2}K).
d_i is the tube inner diameter (m).
k_f is the refrigerant liquid thermal conductivity (W/mK).
$Re_f = Gd_i/\mu_f$ is Reynolds number (–).
G is the refrigerant mass velocity (kg/m^{2}s).
μ_f is the liquid dynamic viscosity (Pa-s).

The filmwise condensation of refrigerants inside tubes represents a vapor–liquid phase-change phenomenon that usually takes place when vapor is cooled below its saturation temperature at a given pressure. The heat transfer density from the pure condensing fluid to the wall is given by:

$$\dot{q} = h_{cond}(T_{sat} - T_w) \tag{15.48}$$

where

$\dot{q}$ is the heat transfer density (W/m^2).
h_{cond} is the condensation heat transfer coefficient (W/m^{2}K).
T_{sat} is the saturation temperature of the condensing refrigerant at a given pressure (C).
T_w is the tube wall temperature (°C).

REFERENCES

ANSI/AHRI Standard 870. 2005. Performance Rating of Direct GeoExchange Heat Pumps (http://www.ahrinet.org/App_Content/ahri/files/STANDARDS/ANSI/ANSI_AHRI_Standard_870_2005_with_Addendum_1.pdf. Accessed March 17, 2020).

Carey, V.P. 1992. *Liquid Vapor Phase Change Phenomena*. Hemisphere, Washington, DC.

Cavallini, A., G. Censi, D. Del Cola, L. Doretti, G.A. Longo, L. Rossetto, C. Zilio. 2003. Condensation inside and outside smooth and enhanced tubes – A review of recent research. *International Journal of Refrigeration* 26:373–392.

Chen, J.C. 1966. A correlation for boiling heat transfer to saturated fluids in convective flow. Industrial and Engineering Chemistry. *Process Design and Development* 5(3):322–329.

Cheng, L., G. Ribatsk, J.R. Thome. 2008. Two-phase flow patterns and flow-pattern maps: Fundamentals and applications. *Applied Mechanics* 61(5):1–28.

Chiriac, F., C. Mihaila, V. Cartas, A.M. Bianchi. 1988. Termotehnica. Institute of Constructions (for students' usage), Bucharest, Romania, 450 pp.

Collier, J.G. 1981. *Convective Boiling and Condensation*, 2nd Edition. McGraw-Hill, New York.

Collier, J.G., J.R. Thome. 1996. *Convective Boiling and Condensation*. Oxford Engineering Science Series, Oxford University Press, New York.

CSA C748-94. Performance of Direct-Expansion (DX) Ground-source Heat Pumps (R2009) (https://webstore.ansi.org/standards/csa/csac7481994r2009. Accessed January 18, 2019).

Dittus, F.W., L.M.K. Boelter. 1930. Heat transfer in automobile radiator of the tubular type. *University of California at Berkley Publications in Engineering* 2:443–461.

Foster, H.K., N. Zuber. 1955. Dynamics of vapor bubbles and boiling heat transfer. *Journal of American Institute of Chemical Engineering* 1:531–535.

Griffith, O. 1998. Two-phase flow, Section 14. In *Handbook of Heat Transfer*, 3rd Edition, Rohsenow, W.M., J.P. Hartnett, Y.I. Cho (Editors), McGraw-Hill, New York.

Gungor, J.K.E., R.H.S. Winterton. 1986. A general correlation for flow boiling in tubes and annuli. *International Journal of Mass and Heat Transfer* 29(3):351–358.

Gungor, K.E., R.H.S. Winterton. 1987. Simplified general correlation for saturated flow boiling and comparisons of correlations with data. *The Canadian Journal of Chemical Engineering* 65(1):148–156.

Halozan, H. 1995. Propane for heat pumps. In Proceedings of the 19th International Congress of Refrigeration, The Hague, Netherlands, 20–26 August.

Halozan, H. 1997. Direct evaporation ground-coupled heat pumps in Austria. *IEA (International Energy Agency) HPC (Heat Pump Centre) Newsletter* 3: 22–23.

Halozan, H., O. Svec. 1993. Heat Pump Systems with Direct Expansion Ground Coils. Final Report of HPC Annex 15, Sittard, The Netherlands.

Halozan, H., R. Rieberer. 2002. *Ground-Source Heat Pumps in Austria*. IIR (International Institute of Refrigeration) – Commission B1, B2, E1 and E2, Guangzhou, China.

Halozan, H., R. Rieberer. 2006. Direct expansion versus secondary loop systems. In Proceedings of the 3rd IEA HPP Annex 29 Workshop, Sapporo, Japan, January 16.

Halozan, H., R. Rieberer, A. Bangheri. 2003. Direct expansion ground-coupled heat pumps. In Proceedings of the 21st International Congress of Refrigeration, Washington, DC, August 17–22.

Halozan, H.R.R. 2002. Direct evaporation ground-coupled heat pumps in Austria. *IEA Heat Pump Centre Newsletter* 20(3):28–32.

Holman, J.P. 1986. *Heat Transfer*, 6th Edition. McGraw-Hill Book Company, New York City.

Kandlikar, S.G. 1987. A general correlation for saturated two-phase flow boiling heat transfer inside horizontal and vertical tubes. In Proceedings of ASME Winter Annual Meeting, Boston, Massachusetts, December 14–18.

Kandlikar, S.G. 1990. A general correlation for saturated two-phase flow boiling heat transfer inside horizontal and vertical tubes. *ASME (American Society of Mechanical Engineers) Journal of Heat Transfer* 112:219–228.

Kaye, G.A. 1995. Staged Cooling Direct Expansion Geothermal Heat Pump. United States Patent 5,388,419, February.

Lawn, J.G., E.H. Young. 1965. Heat transfer to evaporating refrigerants in two-phase flow. *A.I.Ch.E. (American Institute of Chemical Engineers) Journal*, November:124–1132.

Leonachescu, N. 1981. *Termotehnica*. Editura Didactica si Pedagogica, Bucuresti, Romania.

Liu, Z., R.H.S. Winterton. 1999. A general correlation for saturated and subcooled flow boiling in tubes and annuli, based on a nucleate pool boiling equation. *International Journal of Mass Heat Transfer* 34(11):2759–2766.

Martinelli, R.C., D.B. Nelson. 1948. Prediction of pressure drop during forced circulation of boiling water. *ASME (American Society of Mechanical Engineers) Transactions* 70:695.

Mei, V.C., V.D. Baxter. 1990. Experimental study of direct expansion ground coil heat exchangers. *ASHRAE Transactions* 96:634–642.

Minea, V. 2000. Residential DX GSHP with zonal heating and cooling. In Proceedings of the 4th Heat Pump in Cold Climates International Conference, August 17–18, Aylmer (Québec), Canada.

Minea, V. 2005. Development of a vertical direct expansion ground-coupled heat pump (In French). Research report, Hydro-Québec Research Institute, Laboratoire des technologies de l'énergie (LTE), Shawinigan, Québec, Canada.

Minea, V. 2008. Advances in vertical DX ground-source heat pumps design and control. In Proceedings of the 9th International IEA (International Energy Agency) Heat Pump Conference, 20–22 May, Zürich, Switzerland.

Minea, V., G. Kaye. 2008. Residential heat pumps with staged vertical ground-coupled direct expansion heat exchangers. In Proceedings of IEA (International Energy Agency) Heat Pump Program, Annex 29 International Workshop, Zürich, Switzerland, May 19.

Nagano, K. 2012. Current state of ground source heat pump systems. *Journal of Japanese Association of Groundwater Hydrology* 54(2):95–105.

Nusselt, W. 1916. Die Obertlachenkondensation des Wasserdampfes (In German). *Z Ver Dt Ing* 60:541–546, 569–575.

Palen, J.W., G. Breber, J. Taborek. 1979. Prediction of flow regimes in horizontal tube-side condensation. *Journal of Heat Transfer Engineering* 1(2):47–57 (http://dx.doi.org/, Accessed March 12, 2017).

Rayleigh, L. 1880. On the stability, or instability, of certain fluid motions. *Scientific Papers Cambridge University Press* 1:474–487.

Rohsenow, W.M. 1973. Film condensation, in W.M. Rohsenow and J.P. Hartnett (Editors), *Handbook of Heat Transfer, Chapter 12A*, McGraw Hill, New York.

Rohsenow, W.M., J.P., Hartnett, Y.I. Cho. 1998. *Handbook of Heat Transfer*, 3rd Edition, McGraw-Hill, New York.

Shah, M.M. 1979. A general correlation for heat transfer during film condensation in pipes. *International Journal of Heat and Mass Transfer* 22:547–556.

Shah, M. M. 1982. Chart correlation for saturated boiling heat transfer: Equations and further study. *ASHRAE Transactions* 88(1):185–196.

Thome, J.R., J. El Hajal, A. Cavallini. 2003. Condensation in horizontal tubes, part 2: New heat transfer model based on flow regimes. *International Journal of Heat and Mass Transfer* 46:3365–3387.

16 Closed-Loop Vertical Thermo-Syphon Ground-Source Heat Pump Systems

16.1 INTRODUCTION

Conventional closed-loop (indirect, secondary fluid) ground-source heat pump systems with vertical (see Chapter 12) or horizontal (see Chapter 14) ground-coupled heat exchangers need pumps to circulate the geothermal heat carriers (brines or pure water) between the ground/soil buried coils, and geothermal heat pumps.

Because these systems, as well as closed-loop direct expansion ground-source heat pump installations (see Chapter 15), could be harmful to the ground/soil and/or groundwater in the case of brine or accidental refrigerant leakage, thermo-syphons coupled with geothermal heat pumps using, for example, environmentally friendly refrigerants as carbon dioxide, ammonia, or propane as working fluid can be used (Rieberer et al. 2002, 2004).

Closed-loop thermo-syphons (also known as earth probes and self-circulating systems) are essentially heat pipes without capillary wick structures where a two-phase vapor-liquid fluid flows through vertical (or inclined) tubes from condensing to evaporating sections, and vice versa, based on gravity and wetting phenomenon (Kutateladze 1972; Wallis 1969; Imura et al., 1983; Faghri 1989, 1995; Rudolf 2008).

Heat pipes have been manufactured with different sizes and configurations to transfer heat during the evaporation and condensation of working fluids circulating within a sealed cavity. The working fluid circulates inside the heat pipe by either capillary forces in porous wicks, or gravitational forces. The difference between a wick heat pipe and a thermo-syphon is that in the wick heat pipe, the working fluid is returned by capillary forces from the condenser to the evaporator, but the thermo-syphon depends on gravity for the condensate liquid to return to the evaporator. Therefore, the condenser section is always above the evaporator section.

Depending on the purpose of the heat pipe, it is built with different diameters and lengths varying from about three millimeters up to almost a thousand millimeters. There are four main heat pipe configurations: (i) capillary-driven heat pipe or wick heat pipe, (ii) flat-plate heat pipe, (iii) rotating heat pipe, and (iv) wickless heat pipe or two-phase closed thermo-syphon.

DOI: 10.1201/9781003032540-16

Since 1960, 3–10% inclined (sloped) of thermo-syphons (Chato 1962; Hahne and Gross 1981; Wen and Guo 1984; Negishi and Sawards 1983), have been used in various industries as nuclear, space, electronics, automobile, and solar power (Dunn and Reay 1994; Vasiliev 1998, 2005), as well as for recovering low-temperature waste energy (Jeong and Lee 2010), de-icing of roadways, and for preservation of permafrost (e.g., in Canada and Russia) (Hayley 1981).

Since 2000, ground-source heat pump systems based on vertical thermo-syphon principle have been considered as promising alternatives to conventional vertical closed-loop (indirect, secondary fluid) brine, as well as for direct expansion (vertical and horizontal) systems. The first laboratory and in-field applications have been performed in Europe for heating-only purposes especially in low or ultra-low energy single-family houses with limited available ground areas (Rieberer and Mittermayr 2001; Rieberer et al. 2004, 2005; Rieberer 2005; Kruse 2005; Kruse and Russmann, 2005; Bertsch et al. 2005; Halozan and Rieberer 2006; Mittermayer 2006; Acuna et al. 2010; Mastrullo et al. 2014).

The main advantage of thermo-syphons is that no mechanical pumping is needed. As a consequence, they are cheap and reliable. After a heat pipe reached a fully steady-state condition, where the vapor flow is in the continuum state, it is often desirable to determine the temperature and the length of time needed to reach another steady state, for a given new operating condition.

16.2 BASIC CONFIGURATIONS

Ground-coupled thermo-syphon systems consist of one (Figure 16.1) or several (Figure 16.2) vertical closed and pressurized stainless steel or polyethylene-coated copper (except for NH_3) circular tubes, typically of minimum 60–75 m deep (e.g., in Germany) and maximum 100 m deep (e.g., in Austria and Canada) (Rieberer 2005), and inner diameters between minimum 12 mm and maximum 40 mm (Bertsch et al. 2005; Eugster and Sanner 2007). The space between the vertical thermos-syphon tubes and the borehole walls is filled with materials having good thermal conductivity in order to provide low thermal resistance and good contact between the tubes and ground/soil (Florides and Kalogirou 2007). In order to ensure the system sustainability, the maximum number of parallel installed earth thermo-siphons (probes) could be limited to four (Rieberer 2005; Rieberer et al. 2005). As can be seen in Figures 16.1 and 16.2, the vertical thermo-syphon ground-source heat pump systems (in which the geothermal heat pumps are physically decoupled from the earth probes) are similar to classical cascade refrigeration systems.

Their main components are as follows: (i) a first stage consisting of gravity-assisted vertical thermo-syphon (also called earth probe) in which an working fluid (e.g., CO_2, NH_3, or R290) is continuously evaporated and condensed; (ii) an intermediate heat exchanger (also called probe-head located at the top end of each thermo-syphon tube), which is simultaneously condenser for the earth probe and evaporator for the geothermal heat pump's refrigerant (e.g., HFC-410A); and (ii) a second stage consisting of a heating-only geothermal heat pump comprising a compressor, an expansion valve, and a water- or air-cooled condenser.

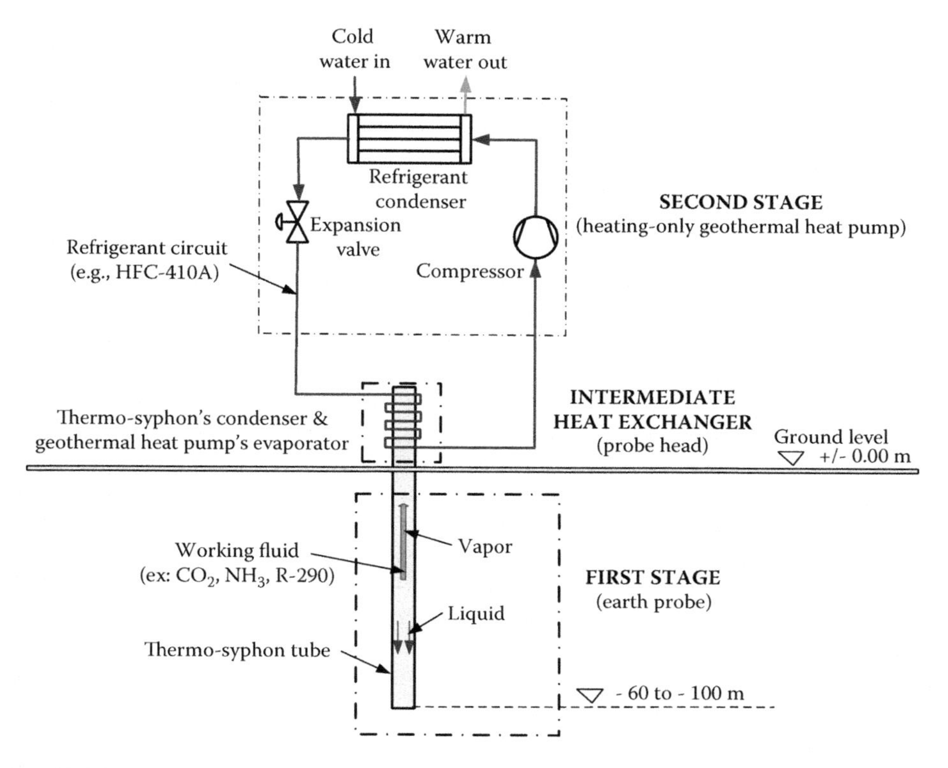

FIGURE 16.1 Schematic of a single-tube, two-phase, closed-loop vertical thermo-syphon ground-source heat pump system (Notes: schematic not to scale; refrigerant condenser may be also air-cooled; not all components represented).

16.3 WORKING FLUIDS

The working fluid is an important part of thermo-syphons for each particular application. Consequently, the design of thermo-syphons must take into account the intended temperature range by selecting a proper working fluid.

The working fluids for thermo-syphon-assisted ground-source heat pump systems should have thermophysical properties such as (Fröba et al. 2000, 2004; Lee et al. 2005): (i) chemically pure, stable, and compatible with the tube material (stainless steel or coated polyethylene); (ii) high density and low viscosity for both liquid and vapor in order to ensure turbulent flow within the thermo-syphons and to enhance the flow return to the evaporating; (iii) be able to change phase at the operating thermal parameters; (iv) high latent heat and good thermal conductivity; (v) high enough surface tension able to lower the risk of liquid entrainment; (vi) high-vapor pressure at low operating temperatures to transport sufficient mass in the vapor phase for adequate heat transfer; it can be seen (Figure 16.3) that, for example, the saturation vapor pressures of CO_2, are quite higher than those of NH_3, and R-290 at the same saturated temperatures; and (vii) compatibility and wet ability with wall materials.

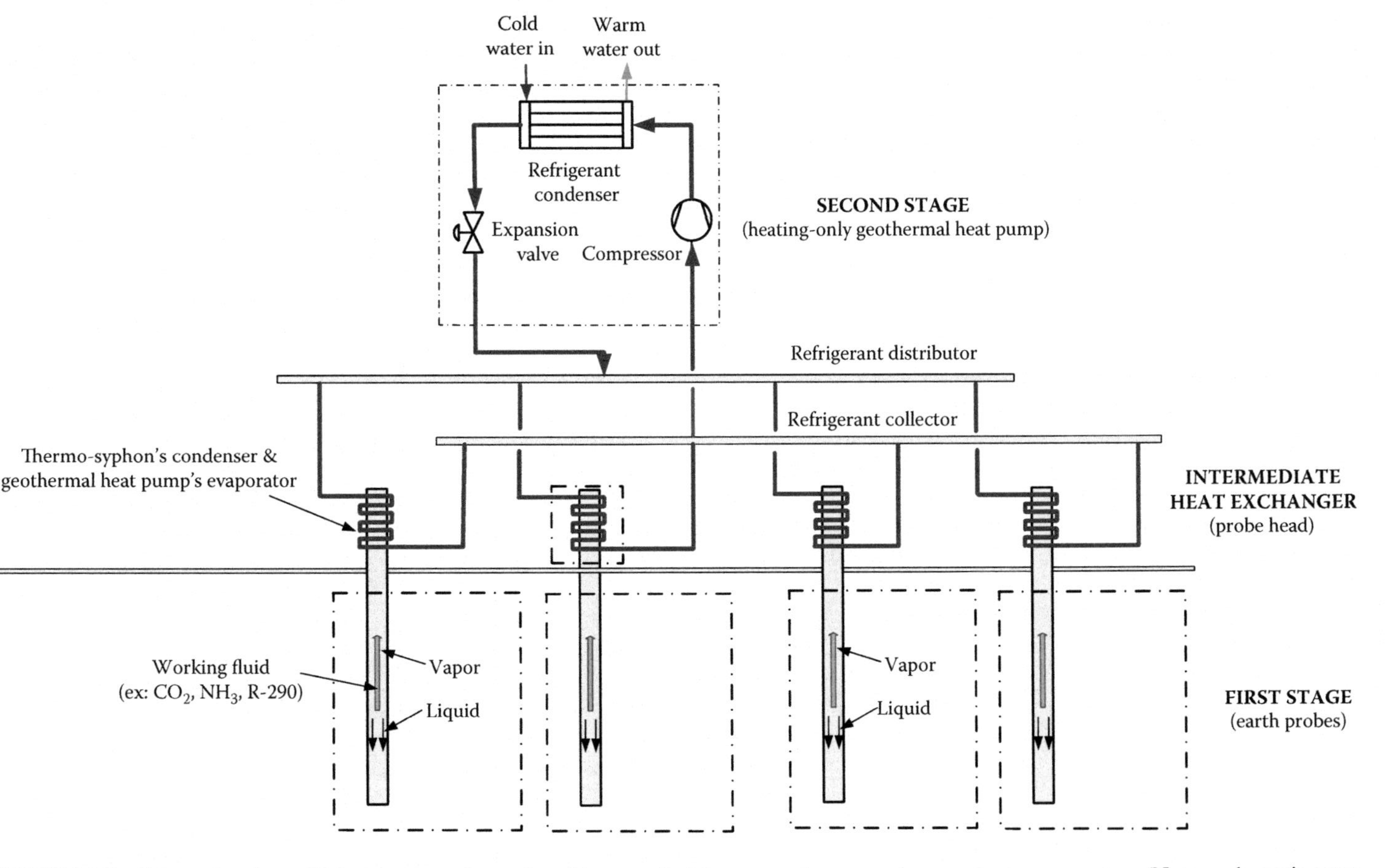

FIGURE 16.2 Schematic of a multiple-tube, two-phase, closed-loop vertical thermo-syphon ground-source heat pump system (Notes: schematic not to scale; refrigerant condenser may be also air-cooled; not all components represented.

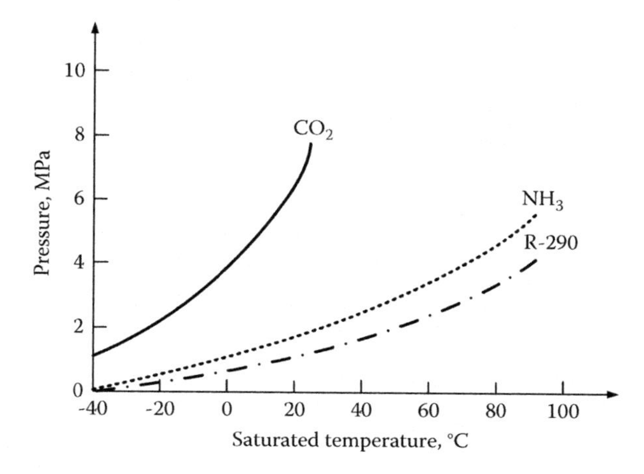

FIGURE 16.3 Saturated vapor-pressure diagram of three natural refrigerants susceptible to be used as working fluid in thermo-syphon heat pump systems.

Depending on local ground/soil properties (e.g., temperature, moisture content, effective thermal conductivity), working fluids such as carbon dioxide (CO_2), ammonia (NH_3), hydrocarbons, such as propane (R-290), or HFC (hydrofluorocarbons) refrigerants as HFC-134a and his replacement HFO-1234yf), and R-410A (a zeotropic mixture, or this replacement R-454B), can be employed in thermo-syphon-assisted ground-source heat pumps.

Among these fluids, carbon dioxide (CO_2) is one of the most desirable because of its favorable thermodynamic and transport properties, such as (Kima et al. 2004; Rieberer et al. 2005; Cheng et al. 2008; Mastrullo et al. 2009, 2012; Kim et al. 2008; Acuna et al. 2010): (i) available, non-explosive, non-flammable, non-toxic, non-reactive, odor- and color-less, and relatively low-price refrigerant; (ii) zero ozone depletion potential and a GWP equal to 1; (iii) harmlessness to the environment since there is no contamination of the ground and/or groundwater in case of leakage; (iv) excellent heat transfer performances during vaporization in vertical configurations; (v) compared to other refrigerants, higher vapor density; for example, the vapor density of CO_2 is seven times higher as the vapor density of HFC-134a; (vi) high thermal conductivity; for example, thermal conductivity of CO_2 saturated liquid and vapor is 20% and 60% higher than those of HFC-134a, respectively; (vii) high vapor density that results in high volumetric refrigeration (cooling) capacity leading to small volumetric flow rates and small pressure losses; (viii) saturation pressure about 4–12 times higher compared to conventional refrigerants (see Figure 16.3); for instance, at 0°C the saturation pressure is about 3,500 kPa (35 bar) for CO_2, and 800 kPa (8 bar) for R-410A; (ix) typical saturated temperatures of CO_2 inside borehole heat exchangers are around 0°C determined by the ground/soil/rocks) temperature levels; this means that both heat absorption and heat rejection take place at sub-critical pressures and temperatures; heat absorption takes place at sub-critical temperatures (i.e., temperatures below 31°C and pressures below 7,400 kPa); in cold climates, normal ground/rock temperatures imply CO_2

operating pressures of around 3,500 kPa (Figure 16.3); (x) owing high pressure levels at 0°C, the CO_2 vapor density is approximately 10 times higher than the density of propane (R-290), and seven times higher than the density of R-134a or HFO-1234yf; and (xi) lower density variations of the fluid during evaporation and lower buoyancy effects.

The shell material of thermo-syphons should be compatible with the working fluid in order to avoid chemical reactions with the working fluid that could produce non-condensable gases and, thus, affect the thermo-syphons' performances.

For low-temperature applications, shell materials as copper and aluminium have been used with refrigerants such as R-134a, HCFC-22, R-410A, and steel, with ammonia.

16.4 OPERATING PRINCIPLE

A two-phase closed thermo-syphon is passive high performance heat transfer device. It is a closed container filled with a small amount of a working fluid. In such a device, heat is supplied to the evaporator wall, which causes the liquid contained in the pool to evaporate. The generated vapor then moves upwards to the condenser. The heat transported is then rejected into the heat sink by a condensation process. The condensate forms a liquid film which flows downwards due to gravity.

The operating principle of vertical single-tube, self-circulating (without fluid circulation pumps), oil-free, two-phase thermo-syphons (that are tubes filled with working fluid at a rated filled ratio to transport heat throughout the system) buried in the ground/soil/rocks is based on the continuous and simultaneous vaporization and condensation of a working fluid (e.g., ammonia, propane, or, preferably, carbon dioxide). As can be seen in Figure 16.4, there are three distinct thermal sections (Rieberer 2005; Bertsch et al. 2005; Wagner 2014): (i) an evaporating ("hot") section at the bottom of the tube where the heat input from the ground/soil (the heat source) undergoes a phase change of the "cold" liquid existing in this section; in the evaporating section, the vapor pressure is higher than the pressure at the condensing section; this pressure difference drives the vapor from the "hot" to the "cold" side of the thermo-syphon; (ii) an adiabatic (thermally insulated) section in the middle of the pipe to separate the evaporation and the condensing regions by a distance suitable to the intended application; this section serves as a flow channel; and (iii) a condensing ("cold") section at the top of the tube; in this section, heat is removed from the system.

The working fluid flow inside a thermo-syphon can be divided into four components: (i) vapor flow in the adiabatic (transport) section; (ii) liquid flow in the adiabatic (transport) section; (iii) vapor flow in the evaporator and condenser; and (iv) liquid flow in the evaporator and condenser.

The principle of heat pipe heat exchange is the application of physical phenomena of water vapor condensation to the two-phase heat transfer. When the water inside the tube absorbs heat, it changes from liquid to gas form, and steam moves along the heat pipe to the cooling side and then condenses. Heat that was previously absorbed through vaporization is released into blast furnace gas or combustion air, completing a thermal cycle and continuing to repeat this heat exchange. The heat

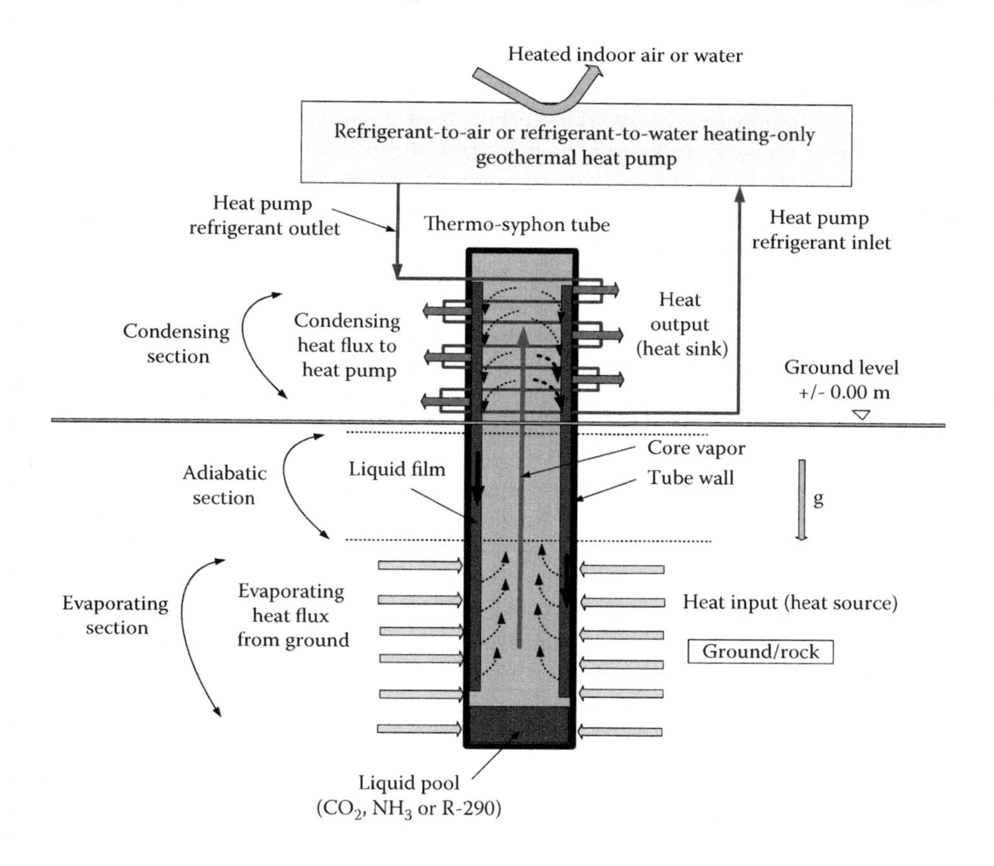

FIGURE 16.4 Operating principle of a thermo-syphon-based ground-source heat pump system.

pipe operating temperature is maintained at the heat absorption of the equilibrium temperature. In order to maximize the efficiency of a heat pipe heat exchanger, some heat pipes are coupled with the outer ring of large radial fins to increase the heat transfer area, and more heat pipes are added to the heat exchanger in a fixed space. The resulted vapor rises upwards because of its lower density.

The temperature of the pressurized liquid refrigerant in the evaporation section increases and a part of the fluid stars to evaporate and rises as a vapor, because of its lower density and density differences, toward the probe head, located at the top of the thermo-syphon, where it condenses. Due to condensation, the pressure decreases and vapor is transported upwards inside the tube through the adiabatic section, and the cycle starts again.

The condensation occurs when the refrigerant temperature is below the saturation temperature of the working fluid vapor. The liquid flows back to the heated section along the wall from the ''cold'' to ''hot'' side of the pipe driven only by gravitation due to the temperature and density gradients occurred because of the fluid phase changes.

The released latent heat (in form of condensation enthalpy) is used to evaporate the refrigerant in the geothermal heat pump cycle.

Figure 16.4 shows a basic working principle of a thermo-syphon. It is a highly efficient heat transfer device where it utilizes the principle of evaporation and condensation of the working fluid. It works by transferring latent heat of evaporation of the working fluid inside the system, which have been transported continuously by changing its phase from liquid to gas. As the amount of the heat absorbed increases, the vapor produced will be transported through the adiabatic section. Its low density causes it to flow upwards to the condenser end of the thermo-syphon, at which it condenses back into a liquid state by releasing the absorbed latent heat to a heat sink. As it condenses and the working fluid takes liquid state due to increasing in density, the liquid is drive back to the evaporator end by gravity. This cycle repeats continuously, resulting in heat transfer to take place between the evaporator end and the condenser end of the thermo-syphon. The basic working principle of a thermo-syphon illustrated in Figure 16.1.

The falling liquid film on the wall is heated up by the neighboring ground/soil of which undisturbed temperature may be initially around 10°C, but decreases over time due to heat extraction.

The heat transfer process may be visualized as a countercurrent two-phase flow between the evaporating and the condensing sections consisting of liquid flowing down through a pipe as it absorbs heat from the surrounding ground and, at some point, evaporating. As this happens, the vapor rises through the same tube due to buoyancy until it arrives to the cooled head probe located at the top of the thermo-syphon tube where it condenses and continuously releases heat to the geothermal heat pump's evaporator.

A heat pipe is a two-phase heat transfer device with a highly effective heat transfer rate through evaporating and condensing a fluid that is circulating in a sealed container.

The heat pipe is filled with a working fluid, such as water. The working fluid will be at a saturation condition because heat pipes operate on a two-phase closed cycle, and pure vapor and liquid phases are present inside the heat pipe. When heat is added to the evaporator section by an external source where a liquid pool exists, the working fluid is heated until it starts boiling and changes to vapor. The resulting vapor pressure cause the vapor to flow through the adiabatic section towards the condenser section, which is the colder section. The wall of the condenser section will be at a lower temperature; therefore, the vapor adjacent to this wall gives up its latent heat. The condensed liquid is then transported back to the evaporator section by the capillary wicks or by gravity in a wickless heat pipe. This cycle will continue transporting heat from the evaporator to the condenser as latent heat. The amount of latent heat that can be transported is much larger than that which can be transported as sensible heat in a conventional system. The temperature difference in a heat pipe needed to transport heat is less than in other systems; thus, it is also referred to as a super conductor as the thermal conductivity for the heat pipe is much larger than any known solid with the same size.

Size, type, material, construction, heat transfer rate, and working fluid are the factors that play an important role in the heat pipe characteristics.

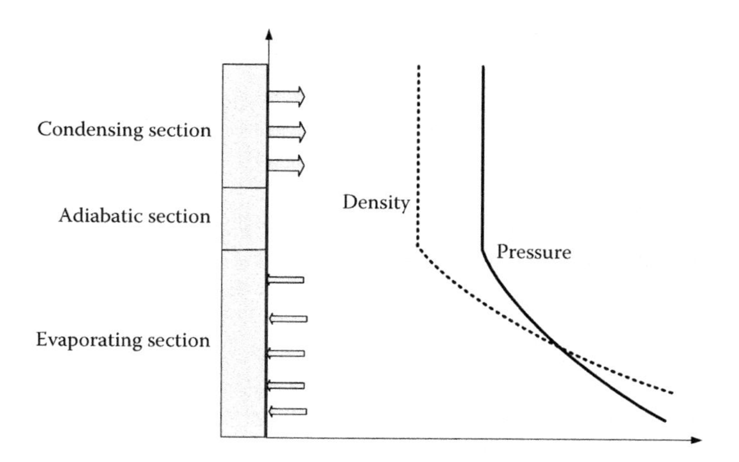

FIGURE 16.5 Schematic behavior of density and pressure along the thermo-syphon working with CO_2 (Note: schematic not to scale).

16.5 DENSITY AND PRESSURE PROFILES

Both density and saturated pressure of the working fluid increase from the top to the bottom of vertical thermo-syphon tubes (Figure 16.5).

Density and pressure profiles of two-phase countercurrent annular fluid flow regime within vertical thermo-syphons (Figure 16.5) affects, among other parameters (Bertsch et al. 2005; Cheng et al. 2008a): (i) the evaporation and the condensation processes; (ii) the optimum filling charge; (iii) the heat transfer rate at a given temperature difference between evaporating and condensing sections; and (iv) the maximum amount of heat transport.

The major part of the pressure drop in two-phase countercurrent annular flow is caused by gravitation, whereas friction losses only need to be considered for small diameter tubes, especially in the adiabatic section where the liquid film thickness and the mass flow rate are the highest (Bertsch et al., 2005), but usually are neglected.

The pressure drop in the tube with the smallest diameter is the largest, and the pressure drops caused by gravity increase as the tube diameter increases.

Depending on thermal characteristics of the ground, the pressure drops of both liquid flowing downward and vapor upwards are caused mainly by (Mastrullo et al. 2014): (i) gravity ($\Delta p_{gravity}$, $Pa = kg/m{\cdot}s^2$); (ii) friction ($\Delta p_{friction}$, Pa); and (iii) acceleration ($\Delta p_{acceleration}$, Pa):

$$\Delta p_{liquid\ or\ vapor} = \Delta p_{gravity} + \Delta p_{friction} + \Delta p_{acceleration} \tag{16.1}$$

where

$$\Delta p_{gravity} = \rho_{liquid\ or\ vapor}{\cdot}g{\cdot}L \tag{16.2}$$

$$\Delta p_{friction} = f/2 \cdot \rho_{liquid\ or\ vapor} \cdot w^2 \cdot L/d \tag{16.3}$$

$$\Delta p_{acceleration} = G^2 \cdot w \tag{16.4}$$

where

$\rho_{liquid\ or\ vapor}$ is the liquid or vapor mass density (kg/m^3).
g is the gravitational acceleration (m/s^2).
L is the length of thermo-syphon pipe (m).
f is the friction factor (–).
w is the liquid or vapor velocity (m/s).
d is the diameter of thermos-syphon pipe (m).
G is the fluid mass flow rate (kg/m^{2}s).
v is the fluid specific volume (m^3/kg).

Due to low-flow velocities, the gravitational pressure drops (almost constant at around 1 kPa/m) has a far greater effect on the pressure profile along the tube than the frictional and acceleration pressure drops, except for the smallest diameter tubes.

16.6 TEMPERATURE PROFILE

If it is assumed that the temperature of the ground/soil is stable, and, because the effect of gravity, the saturated temperature of the working fluid increases along the thermo-syphon being lower at the top than at the bottom.

For example, with CO_2 as a refrigerant, the increase of fluid saturation temperature towards larger depths is about 1 K/65 m (Figure 16.6). That small temperature rise does not influence the system very much. By increasing the temperature difference, the risk of destroying the liquid film trickling downwards increases. Therefore, the temperature difference should be as small as possible (Florides and Kalogirou 2007).

16.7 VELOCITY PROFILE

In thermo-syphons, both liquid and the vapor phases move together under different flow regimes depending on their relative flow rates. The two-phase countercurrent annular film liquid (of which thickness, for a given tube diameter, decrease with the depth and depends also on the vapor velocity) and core vapor can be studied separately by considering or not the liquid droplet entrainment in the vapor core at the wavy liquid-vapor interface (Figure 16.7). As can be seen in Figure 16.7, the velocity of the liquid at the wall is 0, and has a high gradient near the wall. In the middle of the film, the profile is almost flat due to the occurrence of turbulent flow. The velocity in the vapor core is the highest for the smallest diameter. Due to the small mass flow rate of the refrigerant, the velocity is relatively low given for the smaller diameter of thermo-syphons (Acuna et al. 2010). The laminar liquid flow can be reduced to a two-dimensional condensation over a vertical flat-plate (Nusselt

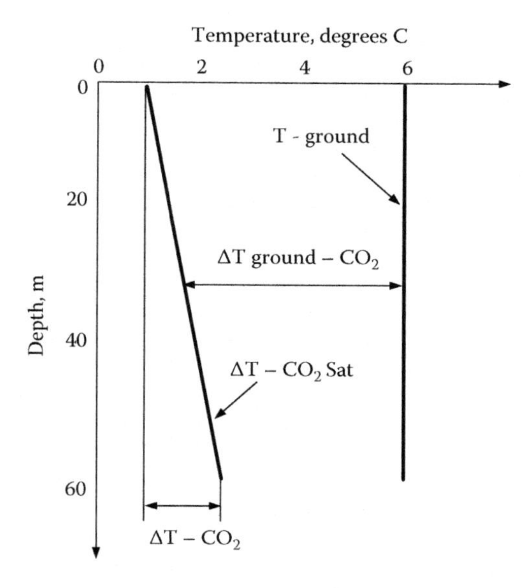

FIGURE 16.6 Example of the temperature profile along a thermo-syphon tube working with CO_2.

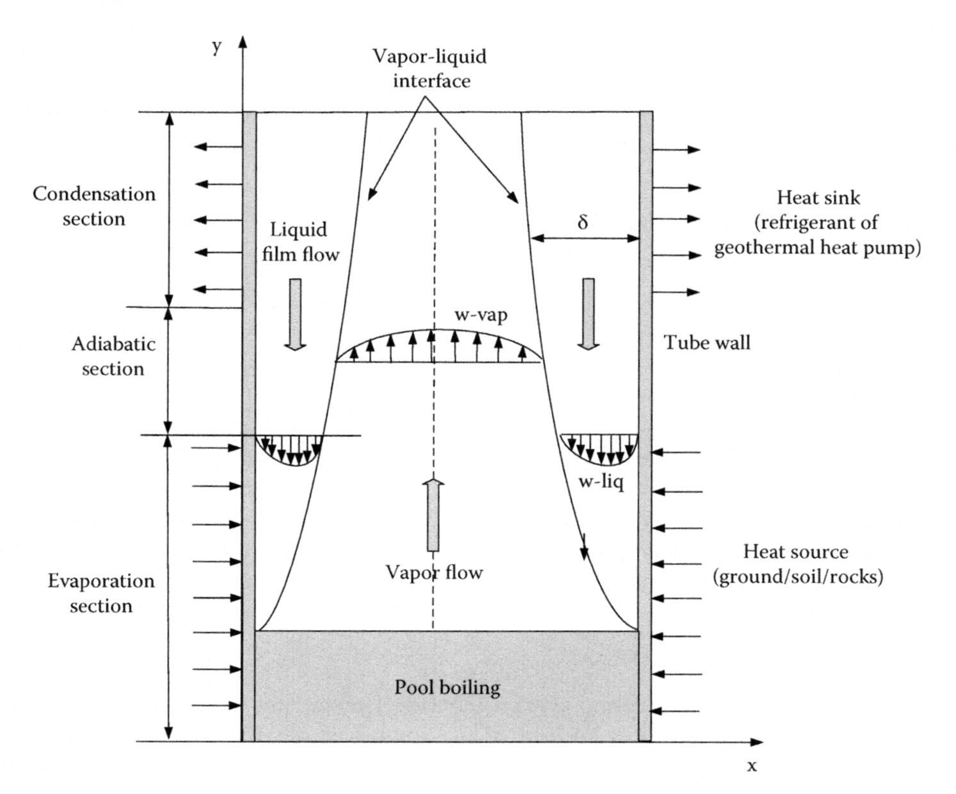

FIGURE 16.7 Schematic representation of velocity profile in a thermo-syphon tube. δ, liquid film thickness (Note: schematic not to scale).

model), accounting or not for turbulent effects and interfacial shear stress. Due to the high shear stress of the liquid at the beginning of the evaporating section on the vapor-liquid contact surface, the smaller the pipe diameter is, the smaller is the liquid velocity.

The velocity reaches a maximum for both phases in the adiabatic section where the working fluid mass flow rate is the highest. This is due to the fact that, immediately after the adiabatic section, the working fluid starts to evaporate and the refrigerant flow rate decreases. The velocities of the liquid film downwards and the vapor film upwards depend on the shear stress of the vapor-liquid interface. The relative velocity of the vapor-liquid interface can be calculated as a difference between the velocity of the upward flowing vapor (considered positive) and the velocity of the downward flowing liquid film (considered negative).

16.8 HEAT TRANSFER

The heat transfer in vertical thermo-syphons (i.e., wickless heat pipe) is a two-phase process that depends on the following (Bertsch et al. 2005; Rudolf 2008): (i) pipe diameter; (ii) working fluid; (iii) flow symmetry; (iv) thickness of the liquid film; (v) temperature (relative small) difference between the condensing and the evaporating sections that varies according to the fluid type; and (vi) thermophysical parameters of the surrounding ground/soil/rocks.

Assumptions of the following are generally made: (i) incompressible fluid; (ii) one-dimensional, steady-state Newtonian fluid flow at saturation temperatures and pressure; (iii) no pressure drop in the liquid film; (iv) axial conduction and viscous dissipation are neglected; and (v) uniform heat fluxes.

16.8.1 Pool Boiling in Evaporator

During pool boiling, that transforms some, or all, of the fluid from the liquid phase to the vapor phase; the heated fluid does not flow by external forces.

If the thermo-syphon is filled with a refrigerant, after applying heat to the pipe, it is heated from a subcooled state to a saturated state. During this process, the refrigerant temperature is raised and chunks of liquid move up and down because of natural convection currents, followed by the formation of vapor bubbles. These bubbles then start rising towards the top of the liquid pool and then smash in the cooler refrigerant above. In this subcooled boiling process, the bulk of the liquid pool is still below the saturation temperature. Eventually, the overall temperature of the liquid pool reaches the saturation (boiling) temperature and the number and size of the vapor bubbles increase and reach the top of the pool section.

16.8.2 Film Condensation

As the condenser in the thermo-syphon involves laminar flow for the liquid film condensation along the inner surface of the condenser section, Nusselt filmwise condensation correlation is used to predict the heat transfer coefficient associated with the film condensation in the condenser section.

The thin liquid film begins to form at the top of the inner surface of the condenser wall of the thermo-syphon and then falls back to the evaporator section under the force of gravity. During the condensation process, heat in the form of the latent heat of vaporization is released by the vapor and transfers through the liquid film to the inner surface of the cold condenser wall.

16.8.3 Thermal Resistances

The thermal resistances of thermo-syphons consist of the following components: (i) thermal resistances between the ground/soil/rocks (heat source) and the external surface of the evaporator; (ii) thermal resistance associated with the evaporation, and consists of two sub-regions: the pool boiling resistance and the nucleate film boiling resistance; (iii) thermal resistance associated with the condensation; (iv) thermal resistances between the external surface of the condenser and the ground/soil/rocks (heat sink); (v) thermal resistance related to the radial heat conduction through the pipe wall of the evaporator and condenser sections; (vi) thermal resistance related to the axial heat conduction through the solid wall of the pipe; (vii) thermal resistance related to the vapor flow inside the pipe from the evaporator to the condenser sections; and (viii) thermal resistances associated with the interface vapor liquid, for the evaporator and condenser sections, respectively, which are usually neglected.

For calculation of heat transfer thermal resistances, assumptions such as the following are usually used: (i) the conduction thermal resistance in the axial direction along the thermo-syphon wall is negligible; (ii) the thermal resistance due to the pressure drop of the vapor as it flows from evaporator to condenser is not considered as it is assumed there is no drop in the saturation temperature between the evaporator and condenser; (iii) the thermal resistance of the radial heat conduction through the thickness of the evaporator and condenser wall is very small, so it is neglected; (iv) since a uniform heat flux has been presented in the evaporator and condenser walls, the thermal resistances associated with the external surface are not considered.

The thermal resistance associated with the evaporation in the evaporator section and the thermal resistance associated with the condensation in the condenser section (two sections thermally connected in series) are the dominant components of overall thermal resistance of thermo-syphons:

$$R_{TS} = R_{evap} + R_{cond} \tag{16.5}$$

The two thermal resistances associated with the evaporation and condensation processes are usually determined with empirical correlations for the heat transfer coefficients.

16.8.3.1 Evaporator

As a result of circulation of the working fluid in the closed-loop thermo-syphons, complex fluid flow and heat transfer behavior occur in the evaporator section, particularly nucleate boiling heat transfer in the pool and liquid film regions. The internal thermal resistance of the evaporator section consists of the thermal resistance of both the nucleate pool boiling in the liquid pool and the nucleate film boiling in the thin liquid film.

The evaporator thermal resistance can be calculated as follows:

$$R_{evap} = 1/(h_{npb} \cdot A_{evap}) \tag{16.6}$$

where

h_{npb} is the nucleate pool boiling heat transfer coefficient (W/m^{2}K).

$A_{evap} = \pi d_{evap} L_{evap}$ is the heat transfer surface area of the evaporator (m^2).

d_{evap} is the evaporator internal diameter (m).

L_{evap} is the evaporator length (m).

One of empirical correlations developed experimentally that can be used to calculate the nucleate pool boiling heat transfer coefficient in the evaporator section is (Rohsenow et al. 1998):

$$h_{npb} = \frac{\dot{q}^{2/3}}{(c_f \cdot h_{fg}/c_g)\left\{(1/h_{fg} \cdot \mu_f)\left[\frac{\sigma}{g(\rho_f - \rho_g)}\right]^{1/2}\right\}^{0.33} Pr_f^n} \tag{16.7}$$

where

h_{npb} is the nucleate pool boiling heat transfer coefficient (W/m^{2}K).

$\dot{q}$ is the heat flux density (W/m^2).

c_f is the liquid-specific heat (J/kg·K).

h_{fg} is the evaporation enthalpy (latent heat) (J/kg).

c_g is the vapor-specific heat (J/kg·K).

μ_f is the liquid dynamic viscosity (kg/m·s).

σ is the wall shear stress (N/m^2).

ρ_f is the liquid density (kg/m^3).

ρ_g is the vapor density (kg/m^3).

Pr_f in the liquid Prandtl number (–).

16.8.3.2 Condenser

For different flow regimes, according to Reynolds numbers, the liquid film condensation heat transfer coefficient at the condenser section can be calculated as follows:

$$h_{film} = Nu_f \cdot k_f / d_{cond} \tag{16.8}$$

where

h_{film} is the liquid film condensation heat transfer coefficient (W/m^{2}K).

$Nu_f = \nu_f/\alpha_f$ is the liquid Nusselt number (–).

ν_f is the liquid kinematic viscosity (m^2/s).
α_f is the liquid diffusivity (m^2/s).
k_f is the liquid thermal conductivity (W/m·K).
The Reynolds number of the liquid film is defined as follows:

$$Re_{film} = 4\dot{Q}_{cond}/\pi d_{cond} \cdot \mu_f \cdot h_{fg} \tag{16.9}$$

where
$\dot{Q}_{cond}$ is the condenser total thermal power (W).
d_{cond} is the inner diameter of condenser section (m).
μ_f is the liquid dynamic viscosity (kg/m·s).
h_{fg} is the condensation enthalpy (latent heat) (J/kg).
For the laminar flow regime, the average heat transfer coefficient in film condensation on a vertical plate was derived as follows (Nusselt 1916):

$$h_{film} = 0.943 \left[\frac{h_{fg} \cdot \rho_f \cdot g \cdot k_f^3 (\rho_f - \rho_g)}{\mu_f \cdot L_c \cdot (T_v - T_{wall})} \right]^{1/4} \tag{16.10}$$

where
h_{film} is the liquid film condensation heat transfer coefficient (W/m^2K).
h_{fg} is the condensation enthalpy (latent heat) (J/kg).
ρ_f is the liquid density (kg/m^3).
g is the gravitational acceleration (m/s^2).
k_f is the liquid thermal conductivity (W/m·K).
ρ_f is the vapor density (kg/m^3).
μ_f is the liquid dynamic viscosity (kg/m·s).
L_c is the height of the condenser section (m).
T_v is the temperature of the saturated vapor (C).
T_{wall} is the temperature of the condenser wall (C).
The thermal resistance of the condenser section is determined by calculating the heat transfer coefficient of the liquid film condensation and then using the following expression:

$$R_{cond} = 1/(h_{film} \cdot A_{cond}) \tag{16.11}$$

where
h_{film} is the heat transfer coefficient of the liquid film condensation (W/m^2K).
$A_{cond} = \pi d_{cond} L_{cond}$ is the heat transfer surface area of the condenser (m^2).
d_{cond} is the condenser internal diameter (m).
L_{cond} is the condenser length (m).

16.8.3.3 Overall Thermal Resistance

The overall thermal resistance of the thermo-syphons can be predicted by adding by adding the thermal resistances at the evaporator (Equation 16.6) and condenser (Equation 16.11) sections as follows:

$$R_{overall} = R_{evap} + R_{cond} = 1/(h_{npb} \cdot A_{evap}) + R_{cond} = 1/(h_{film} \cdot A_{cond}) \quad (16.12)$$

16.8.4 Energy Performance

The performances of thermo-syphon-assisted ground-source heat pump systems are affected by working fluid's parameters and thermo-syphon geometry, such as: (i) heat loads and inner pressures; (ii) working fluid enthalpy (latent heat) of vaporization and condensation; (iii) vapor and liquid density, thermal conductivity, specific heat, and viscosity; (iv) tube diameter and surface treatment; (v) length and heat transfer areas of evaporator, adiabatic section, and condenser; and (vi) tube geometry (shape, length, and aspect ratio) and filling ratio.

Usually, recommended heat extraction rates are of about 55 W/m (in Austria) and 50–80 W/m (in Germany), according to VDI 4640 (1998) (Rieberer 2005; Eugster and Sanner 2007).

For a two-phase thermo-syphon with CO_2 as working fluid, the maximum extraction rate for a normal (wet) ground/soil/rocks structure is approximately 50–58 W per m of thermo-syphon; thus, a thermo-syphon length (depth) of 16 m is required per kW-thermal output for a geothermal heat pump system achieving a Seasonal Performance Factor of 4.5 (Rieberer et al. 2002, 2004; Kruse and Peters 2005). At 0°C, the geothermal heat pump refrigeration capacity is 22,650 kJ/m^3 for CO_2, while for the conventional refrigerants, it is below 6,700 kJ/m^3 (Rieberer and Mittermayr 2001).

At the evaporator and the condenser, the measured overall heat transfer coefficients are 10–40 W/m^{2}K and 20–50 W/m^{2}K, respectively. That means that a house with a heating capacity of 8 kW at the design temperature requires a thermo-syphon length of approximately 130 m, or two thermo-syphons of 65 m each. These numbers were based on heat extraction only, and a maximum operation time of 1,800 hours per year during the heating season (VDI 4640.1998).

According to some authors (Rieberer et al. 2004), the heat efficiency of thermo-syphon ground-source heat pump systems is high because of the natural circulation of the working fluid (e.g., CO_2) instead of a forced circulation in case of brine as the thermal heat carrier (Kruse and Peters, 2008).

16.9 ADVANTAGES AND LIMITATIONS

Among the advantages of self-circulating thermo-syphon ground-source heat pump systems, the following have been provided by several authors (Dunn and Reay 1994; Kruse 2005; Rieberer et al. 2005; Bertsch et al. 2005; Mittermayer 2006; Mittermayer 2006): (i) simple and compact structures with relatively low construction and installation costs (e.g., by using smaller diameter tubes); (ii) compared to closed-loop (indirect, secondary fluid) ground-source heat pump systems, these systems offer a reduction of the electricity consumption and a higher efficiency, respectively, due to the natural circulation of the working fluid; additionally, the application of CO_2 implies favorable heat transfer characteristics and it is absolutely harmless to the ground/soil/rocks; (iii) are very effective, low-cost and reliable heat transfer devices for many thermal and heat recovery applications; (iv) unlike common brine systems, a pump to circulate the working fluid is not required that increases the system reliability, and overall energy performance (by reduction of the electricity consumption) and its durability (by eliminating the potential defects of the fluid circulation pump); (v) the temperature difference between the ground/soil/rocks and the working fluid flowing inside the geothermal heat pump evaporator can be reduced, which may lead to higher overall energy performances; (vi) potentially favorable heat transfer characteristics, such as phase changes at almost constant temperature, high heat recovery effectiveness, and high heat transfer coefficients; and (vii) thermo-syphon systems using non-toxic and non-flammable natural refrigerants as CO_2 and R-290 without lubricant oils address environmental problems with respect to pollution of the ground and groundwater in case of accidental leakage.

The main limitations of thermo-syphon ground-source heat pump systems could be summarized as follows (Kruse 2005; Kruse and Peters 2008; Rudolf 2008; Bertsch et al. 2005): (i) because of natural fluid flow direction and heat transfer from the ground/soil/rocks to the probe head, such systems may work only in the heating mode and not for cooling during the summer; (ii) in other words, can work in only one thermal flow direction, i.e., the heat is always absorbed from the ground/soil/rocks and released at the borehole top; (iii) relatively high initial costs; (iv) require relatively large refrigerant charges; (v) due to the low-temperature differences between ground/soil/rocks and refrigerant, the borehole depth is limited; (vi) depending the fluid properties and the length and diameter of the tube, there is a critical heat flux at which flooding may occur; this means that the vapor flowing out of the lower part of the thermo-syphon into the upper section is able, acting by shear stresses on the downstream fluid film to convert its direction upwards so that no further liquid feeding to the inner side of the tube can occur; this means that dry-out, due to lack of liquid that is stored at the upper end, can occur.

16.10 FURTHER R&D NEEDS

Recommendations for further R&D work such as the following could be formulated in order to guarantee successful market introduction by providing an economical and competitive product able to reduce the first costs and maintenance work: (i) analyze, among others, the number of parallel thermo-syphon pipes per borehole and probe heads, optimum probe length, optimum CO_2 charge, minimum inclination of earth tubes, probe head geometries, and optimum number; (ii) due to the particular thermodynamic properties of CO_2, is important to perform accurate designs to guarantee a correct thermal balancing of working fluid parameters with those of geothermal heat pumps; (iii) two- and three-dimensional mathematical models should be developed and experimentally validated to describe the performance of the system at various operating conditions, such as fluid inlet parameters and flow rate, ground/soil/rocks temperature, and other thermal properties of borehole filling materials, borehole diameter and height, down-comer, riser diameter and thermal properties, and the number of parallel heat pipes per borehole and probe heads; and (iv) further investigation of non-condensable gases (that have a significant influence on the thermal performance of thermo-syphons) should be conducted.

REFERENCES

Acuna, J., B. Palm, R. Khodabandeh, K. Weber. 2010. Distributed temperature measurements on a U-pipe thermos-syphon borehole heat exchanger with CO_2. In *Proceedings of the 9th IIR (International Institute of Refrigeration) Gustav Lorentzen Conference*, April 12–14, Sydney.

Bertsch, S, E.A., Groll, K. Whitacre. 2005. Modeling of CO_2 thermo-syphon for ground source heat pump application. In *Proceedings of the 8th International IEA (International Energy Agency) Heat Pump Conference* "Global Advances in Heat Pump Technology, Applications, and Markets", May 30–June 2, Las Vegas.

Chato, J. C. 1962. Laminar condensation inside horizontal and inclined tubes. *ASHREA Journal* 4(2):52.

Cheng, L., G. Ribatski, J.R. Thome. 2008. New prediction methods for CO_2 evaporation inside tubes: Part II – An updated general flow boiling heat transfer model based on flow patterns. *International Journal of Heat and Mass Transfer* 51:125–135.

Dunn, P.D., D.A. Reay. 1994. *Heat Pipes*, 4th Edition, Pergamon Press, Oxford, UK.

Eugster, W., B. Sanner. 2007. Technological status of shallow geothermal energy in Europe. In *Proceedings European Geothermal Congress*, Unterhaching, Germany, 30 May–1 June.

Faghri, A. 1995. *Heat Pipe Science and Technology*, Taylor & Francis publisher, Abingdon-on-Thames, Oxfordshire, UK.

Faghri, A. et al. 1989. Heat transfer characteristics in two-phase closed conventional and concentric annular thermos-syphons. *Journal Heat Transfer* 111:611–618.

Florides, G., S. Kalogirou. 2007. Ground heat exchangers – A review of systems, models and applications. *Renewable Energy* 32:2461–2478.

Fröba, A.P., P. Pellegrino, L.A. Leipertz. 2004. Viscosity and surface tension of saturated n-pentane. *International Journal of Thermo-Physics* 25:1323–1337.

Fröba, A.P., S. Will, A. Leipertz. 2000. Saturated liquid viscosity and surface tension of alternative refrigerants. In *Proceedings of the 14th Symposium on Thermo-physical Properties*, Boulder, Colorado, June 25–30.

Hahne, E., U. Gross. 1981. The influence of the inclination angle on the performance of a closed two-phase thermos-syphon. *Heat Recovery Systems* 1:267–274.

Halozan, H., R. Rieberer. 2006. Direct expansion versus secondary loop systems. In *Proceedings of the 3rd IEA (International Energy Agency) HPP (Heat Pump Programme) Annex 29 Workshop*, Sapporo, Japan, January 15.

Hayley, D.W. 1981. Application of heat pipes to design of shallow foundations on permafrost. In *Proceedings of the 4th Canadian Permafrost Conference*, Calgary, Alberta, Canada, March 2–6.

Imura, H. et al., 1983. Critical heat flux in a closet two-phase thermos-syphon. *International Journal of Heat Mass Transfer* 26:1181–1188.

Jeong, S.J., K.S. Lee. 2010. An experimental study of a carbon dioxide-filled thermos-syphon for acquisition of low-temperature waste energy. *International Journal of Energy Research* 34:454–461.

Kim, Y.J., J.M. Cho, M.S. Kim. 2008. Experimental study on the evaporative heat transfer and pressure drop of CO2 flowing upward in vertical smooth and micro-fin tubes with the diameter of 5 mm. *International Journal of Refrigeration* 31:771–779.

Kima, M.-H., J. Pettersen, C. Bullard. 2004. Fundamental process and system design issues in CO_2 vapor compression systems. *Progress in Energy and Combustion Science* 30:119–174.

Kruse, H. 2005. Novel CO_2 heat pipe as earth probe for heat pumps without auxiliary pumping energy. In *Proceedings of the 8th IEA (International Energy Agency) Heat Pump Conference*, May 30–June 2, Las Vegas, Nevada.

Kruse, H., S. Peters. 2008. Earth heat pipe working with carbon dioxide – Status of development. In *Proceedings of the 9th International IEA Heat Pump Conference*, 20–22 May, Zürich, Switzerland.

Kruse, H., H. Russmann. 2005. Novel CO_2-heat pipe as earth probe for heat pumps without auxiliary pumping energy. In *Proceedings of the 8th International IEA (International Energy Agency) Heat Pump Conference* "Global Advances in Heat Pump Technology, Applications, and Markets", May 30–June 2, Las Vegas.

Kutateladze, S. 1972. Elements of hydrodynamics of gas-liquid systems. *Fluid Mechanics – Soviet Research* 14:1–8.

Lee, H.S., J.I. Yoon, J.D. Kim, P. Bansal. 2005. Evaporating heat transfer and pressure drop of hydrocarbon refrigerants in 9.52 and 12.70 mm smooth tube. *International Journal of Heat and Mass Transfer* 48:2351–2359.

Mastrullo, R., A.W. Mauro, A. Rosato, G.P. Vanoli. 2009. Carbon dioxide local heat transfer coefficients during flow boiling in a horizontal circular smooth tube. *International Journal of Heat and Mass Transfer* 52:4184–4194.

Mastrullo, R., A.W. Mauro, J.R. Thome, D. Toto, G.P. Vanoli. 2012. Flow pattern maps for convective boiling of CO_2 and R-410A in a horizontal smooth tube: Experiments and new correlations analyzing the effect of the reduced pressure. *International Journal of Heat and Mass Transfer* 55:1519–1528.

Mastrullo, R., A.W. Mauro, L. Menna, G.P. Vanoli. 2014. A model for a borehole heat exchanger working with CO_2. *Energy Procedia* 45:635–644.

Mittermayer, K. 2006. CO_2 – Heat Pipe. In *Proceedings of the 3rd IEA (International Energy Agency) HPP (Heat Pump Programme) Annex 29 Workshop*, Sapporo, Japan, January 15.

Negishi, K., T. Sawards. 1983. Heat transfer performance of an inclined two-phase closed thermos-syphon. *International Journal of Heat and Mass Transfer* 26(8):1207–1213.

Nusselt, W. 1916. Die Oberflächenkondensation des Wasserdampfes (in German), *Z. Ver. Dtsch. Ing.* 60 27:541–546.

Rieberer, R. 2005. Naturally circulating probes and collectors for ground coupled heat pumps. *International Journal of Refrigeration* 28:1308–1315.

Rieberer, R., K. Mittermayr. 2001. CO_2 – heat pipe. *Final Report of the Energy Technology Programme (ETP)-Project supported by the Upper Austrian Government,* Austria.

Rieberer, R., K., Mittermayr, H. Halozan. 2002. CO_2 heat pipe for heat pumps. In *Proceedings of IIR (International Institute of Refrigeration) Workshop*, Commissions B1, B2, E1 and E2. Guangzhou, China, May 2.

Rieberer, R., K. Mittermayr, H. Halozan. 2004. CO_2 two-phase thermos-syphon as heat source system for heat pumps. In Proceedings of the 6th IIR (*International Institute of Refrigeration*) Gustav Lorentzen Conference on Natural Working Fluids, 29 August–1 September, Glasgow, UK.

Rieberer, R., K. Mittermayr, H. Halozan. 2005. CO_2 thermo-syphons as heat source system for heat pumps – 4 years of market experience. In *Proceedings of the 8th IEA (International Energy Agency) Heat Pump Conference*, May 30–June 2, Las Vegas, Nevada.

Rohsenow, W.M., Hartnett, J.P., Cho, Y.I. 1998. *Handbook of Heat Transfer,* 3rd Edition. McGraw-Hill, New York, Chicago, San Francisco, Athens, London, Madrid, Mexico City, Milan, New Delhi, Singapore, Sydney, Toronto.

Rudolf, J. 2008. *Thermosiphon Loops for Heat Extraction from the Ground.* Study Project Energy Technology EGI2008/ETT 451, KTH School of Industrial Engineering and Management Division of Applied Thermodynamic and Refrigeration, SE-100 44, Stockholm.

Vasiliev, L.L. 1998. State-of-the-art on heat pipe technology in former Soviet Union. *Applied Thermal Engineering* 18(7):507–551.

Vasiliev, L.L. 2005. Heat pipes in modern heat exchangers. *Applied Thermal Engineering* 25(1):1–19.

VDI 4640. 1998. Weiterbildung für Ingenieure (German Guidelines for Ground Coupled Heat Pumps) (https://ec.europa.eu/energy/intelligent/projects/. Accessed August 2, 2019).

Wagner, A.M. 2014. *Review of Thermo-Syphon Applications.* Final report. Prepared for US Army Corps of Engineers, Washington, DC.

Wallis, G. 1969. *One Dimensional Two-Phase Flow.* McGraw-Hill, New York.

Wen, Y.P., S. Guo. 1984. Experimental heat transfer performance of two-phase thermossyphons. In *Proceedings of the 5th International Heat Pipe Conference*, Tsukuba, Japan, May 14–18.

17 Open-Loop Groundwater Heat Pump Systems

17.1 INTRODUCTION

In open-loop groundwater heat pump systems, groundwater is extracted from underground aquifers with sufficient flow rates to meet the buildings' thermal needs, brought once through geothermal heat pumps and, subsequently, discharged back into the same or similar geological environment(s).

The selection process of such systems is usually based on criteria such as (DenBraven 2002): (i) availability and temperature of groundwater; in the United States and Canada, for example, groundwater temperatures at depths of about 15 m range from 4.4°C to 21.1°C; (ii) building height and size; and (iii) local regulatory requirements.

Open-loop groundwater heat pump systems can be classified as follows (Figure 17.1): (i) dual- and multiple-well including production (supply) and return (injection, disposal) wells; (ii) single-well (standing column) systems where the groundwater is supplied by and returned to the same well.

Dual-, multiple- and single-well (standing column) groundwater heat pump systems may use groundwater directly or indirectly (via intermediate heat exchangers) by integrating well submersible pump(s) and building internal water (or brine) closed-loop(s) with central or distributed geothermal heat pumps.

17.2 AQUIFERS

Aquifers are underground limited saturated geological formations that hold groundwater in permeable rocks, rock fractures or unconsolidated materials (e.g., sands and gravels, sandstones, limestones, dolomites, basalts, and fractured igneous or metamorphic rocks) in sufficient quantities to allow it to move fast enough at reasonable amounts toward drilled wells (Driscoll 1986).

There are two types of aquifers (Figure 17.2): (i) unconfined (i.e., aquifers of which the upper boundary is the groundwater table) and (ii) confined (where a permeable formation below the groundwater table is covered by a less permeable layer); in other words, these are aquifers overlain by confining layers, often made up of clay with little or no intrinsic permeability with limited or no direct access to recharge or contamination from the surface. In practice, the groundwater wells may penetrate several aquifers separated by aquitard (layers of either clay or non-porous rocks with low hydraulic conductivity, sometimes completely impermeable) that restrict the flow of groundwater from one aquifer to another.

DOI: 10.1201/9781003032540-17

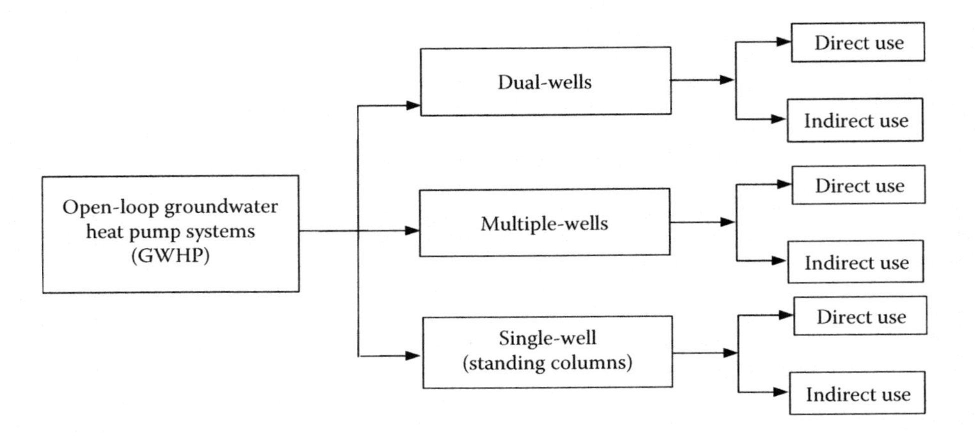

FIGURE 17.1 Classification of open-loop groundwater heat pump systems.

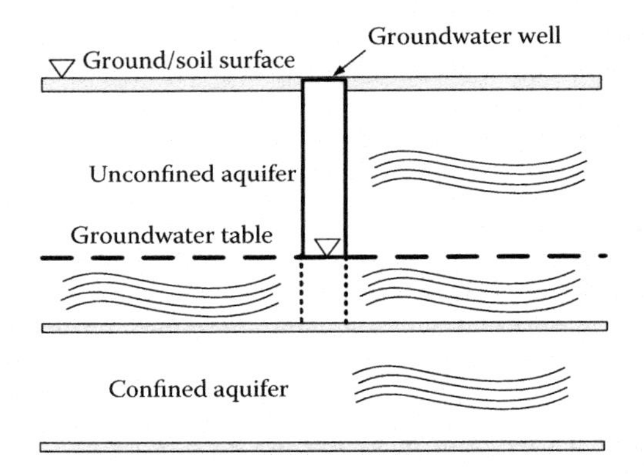

FIGURE 17.2 Types of aquifers.

According to groundwater natural movement, the aquifers are as follows (Dreybrodt 1988): (i) porous (where the groundwater slowly infiltrates through pore spaces between sand grains, typically occurring in sand and sandstone; a groundwater velocity of 0.3 m/day is considered to be a high rate) and (ii) karst (where the groundwater flows through open conduits as underground streams, typically developed in limestone).

17.3 GROUNDWATER QUALITY

The most important quality characteristics of groundwater being used in groundwater heat pump systems are the following: (i) temperature; lowering the groundwater temperature causes an increase in viscosity and, hence, a decrease of hydraulic conductivity; on the other hand, the activity of bacteria-consuming microorganisms drops significantly at temperatures below 10°C; for instance, the mortality rate of pathogens is reduced by about half if the temperature drops from

7 to 2°C; (ii) pH value that expresses the degree of acidity or alkalinity of groundwater; neutral groundwater has a pH of 7.0; groundwater with pH values below 7.0 are acid and may cause corrosion of equipment with which groundwater comes in contact; groundwater containing free mineral acids may have pH values below 4.5; groundwater with pH values above 7.0 or 8.0 are alkaline and, in this case, calcium carbonate scale can be deposited more readily; too intensive cooling of the groundwater, due to excessive energy extraction for heating buildings, increases the *pH* value and reduces calcium solubility, which favors the clogging of pores; (iii) alkalinity, a measure of carbonate and bicarbonates of calcium and/or magnesium, determines the scale-forming tendency of groundwater; (iv) hardness, represented by the sum of calcium and magnesium salts in groundwater; with less than 15 ppm (parts per million) as $CaCO_3$, groundwater is classified as very soft, while with more than 200 ppm as $CaCO_3$, it is considered a very hard; (v) specific conductance (expressed in micro-ohms per cubic centimeter) is a measure of the ability of groundwater to conduct an electric current, indicating whether galvanic corrosion may be a risk; (vi) permeability (or hydraulic conductivity), defined as the resistance for groundwater to flow in the aquifer material, is a main factor for estimating the well capacities at different parts of the aquifer.

17.4 GROUNDWATER TABLE

As can be seen in Figure 17.3, underground water occurs in two zones (Fetter 1994): (i) unsaturated, where ground/soil/rock pores are only partially saturated and the groundwater exists at pressures less than atmospheric pressure; the voids between ground/soil/rocks are mostly filled with air; some water (not groundwater) is held in the unsaturated zone by molecular attraction surrounding the surface of rock particles; this water does not flow toward and doesn't enter the wells; it is held in place by surface adhesive forces and rises above the groundwater table by capillary action to saturate a small zone above the phreatic surface; and (ii) saturated, where all openings in ground/soil/rock pores are fully saturated (i.e., filled with groundwater); groundwater exists at pressures greater than atmospheric pressure and lies below the

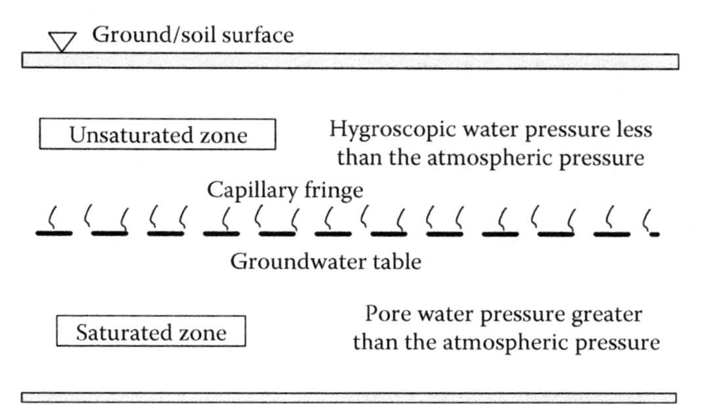

FIGURE 17.3 Unsaturated and saturated zones of underground water.

groundwater table; all the openings in the ground/soil/rocks are full of groundwater that may move through the aquifer or wells from which it is being withdrawn; the term "groundwater" usually refers to groundwater existing in a saturated zone.

The surface separating the unsaturated zone from the saturated zone (the groundwater table) is the level to which groundwater rises and falls freely in wells open to the atmosphere (depending on the amount of available water in the aquifer), and where the groundwater pressure is equal to atmospheric pressure. The groundwater table is the level under which all pores or fractures of ground/soil are water saturated (Figure 17.3). A sloping groundwater table indicates groundwater is flowing.

17.5 PHYSICAL PROPERTIES

The most important parameters and properties of aquifers and surrounding ground/soil/rocks are as follows (Geotrainet 2011): (i) geometry (boundaries and thickness); (ii) static head (groundwater table or hydrostatic level); (iii) groundwater gradient (to detect the natural flow direction); (iv) leakage factor (vertical leakage to the aquifer); (v) boundary conditions (surrounding limits, positive or negative); (vi) ground/soil/rocks effective porosity (i.e., the percentage of solid materials as rocks or ground/soil voids available for groundwater flow); it depends on grain size, shape, and distribution; (vii) aquifer transmissivity, defined as the rate of flow under unit hydraulic gradient through a cross-section of unit width over the whole saturated thickness of the aquifer; (viii) specific storage of saturated aquifer; (ix) aquifer storativity (the volume of groundwater that an aquifer will absorb or expel from storage per unit surface area per unit change in hydraulic head); (x) aquifer density, specific heat, thermal and hydraulic conductivity, and specific retention; (xi) dynamic viscosity; (xii) specific yield (ratio of the volume of groundwater that drains from a saturated rock due to the attraction of gravity to the total volume of the rock).

The aquifer transmissivity can be expressed as the product of the thickness of the aquifer and the average hydraulic conductivity:

$$T = b \cdot K \tag{17.1}$$

where

T is the aquifer transmissivity (m^2/s).

b is the saturated thickness of aquifer (m).

K is the average hydraulic conductivity of the aquifer (m/s).

The specific storage of saturated aquifer is defined as the volume of groundwater that a unit volume of aquifer releases from storage under unit decline in hydraulic head:

$$S_s = \rho_w \cdot g \cdot (\alpha + n.\ \beta) \tag{17.2}$$

where

S_s is the aquifer specific storage (1/m).

ρ_w is the groundwater density (kg/m^3).

g is the gravitational acceleration (m/s^2).
α is the compressibility of bulk porous media ($m{\cdot}s^2/kg$).
n is the aquifer porosity (-).
β is the groundwater compressibility defined as the fractional change in volume per unit increase in pressure ($m{\cdot}s^2/kg$).

For confined aquifers, the aquifer storativity is expressed as follows:

$$S_{confined} = b{\cdot}S_s \tag{17.3}$$

And, for unconfined aquifers, as:

$$S_{unconfined} = S_y + h{\cdot}S_s \tag{17.4}$$

where
$S_{confined}$ is the confined aquifer storativity (-).
b is the saturated thickness of the aquifer (m).
S_s is the aquifer specific storage (1/m).
$S_{unconfined}$ is the unconfined aquifer storativity (-).
S_y is a ratio indicating the volumetric fraction of the bulk aquifer volume that a given aquifer will yield when all the groundwater is allowed to drain out of it under the forces of gravity (1/m).
h is the hydraulic head (m).

17.6 POTENTIAL PROBLEMS

Most groundwater resources contain minerals and other contaminants that could detrimentally affect the equipment of ground-source heat pump systems. Because groundwater always includes impurities that must be reduced and even removed, quality tests should be achieved prior to groundwater heat pump installation.

The impurities in groundwater can be classified as: (i) dissolved solids (as sodium chloride in solution); (ii) liquids or gases; and (ii) suspended matter which can be removed by filtration of mud, clay, or silt.

Open-loop groundwater heat pump systems are in general more sensitive for operational problems compared to closed-loop ground-source heat pump systems, mainly because of operational problem such as clogging or bio-fouling that may occur in the wells and heat exchangers.

Without regular chemical treatment, different contaminants outside recommended standard limits (including iron bacteria, dissolved minerals as calcium carbonate and inorganic constituents as dissolved solids, liquids, and gases, and suspended materials) can cause scaling, corrosion, clogging, and organic growth that can have serious impacts on the mechanical components as circulation pumps and heat exchangers and, finally, on the economic performances of once-through groundwater heat pump systems.

In order to protect the rest of the ground-source heat pump systems from such operational problems related to water chemistry, the groundwater circuits are normally separated from other fluid loops by intermediate heat exchangers.

17.6.1 Scaling

Precipitation of carbonates (often referred to as scaling), may take place if carbon dioxide is allowed to be stripped out from the groundwater. This happens if the draw down in the well exceeds the bubble point for CO_2. For this reason, large drawdowns should be avoided for groundwater types that are scaling sensitive.

For groundwater heat pump systems that use groundwater directly (as in small residential units), the primary groundwater quality problem is scaling, mainly with: (i) mud (soft, sticky matter resulting from the mixing of earth and water); (ii) algae (simple, nonflowering, and typically aquatic plant); and (iii) lime (a white, caustic, alkaline substance consisting of calcium oxide).

Scale formation, that usually occurs when minerals and salts are precipitated and deposited on the inside surface of the pipes when the groundwater is heated, reduces the heat transfer rates and increases the pressure drops through the heat exchangers.

The tendency of calcium carbonate scaling, partially temperature driven, may be estimated using the Ryznar stability index (DOE 1995):

$$Ryznar\ Stability\ Index = 2pH_s - pH_g \tag{17.5}$$

where

pH_s is the pH above which calcium carbonate will precipitate (-).

pH_g is the measured pH of the groundwater (–).

Table 17.1 shows the groundwater scaling tendency according to Ryznar stability index.

17.6.2 Corrosion

Corrosion is caused partly by absorption of gases from the air or by using salted groundwater, creating conditions to aggressively attack the exposed metal surfaces. Corrosion problems may require using expensive corrosion resistant materials, groundwater conditioning devices, and frequent cleaning of the equipment.

Equipment components sensitive for corrosion are as follows: (i) carbon steel casing with threaded or welded joints, (ii) screens, and (iii) heat exchangers.

TABLE 17.1
Ryznar Stability Index

Ryznar index	Scaling tendency
≤4.0	Extreme
4.0–5.0	Heavy
5.0–6.0	Moderate
6.0–7.0	Light
≥7.0	None

The main corrosive species commonly found in groundwater, and their corrosive effects, are as follows(DOE 1995): (i) oxygen: extremely corrosive to carbon and low-alloy steels; concentrations of more than 50 ppb (parts per billion) cause serious pitting; (ii) hydrogen ion reduction (pH) (a figure expressing the acidity or alkalinity of a solution on a logarithmic scale on which 7 is neutral, lower values are more acid and higher values more alkaline); the pH is equal to $-\log_{10}c$, where c is the hydrogen ion concentration in moles per liter; low pH (≤ 5) promotes sulfide stress cracking of high-strength, low-alloy steels and some other alloys coupled to steel; low pH may cause breakdown of passivity of stainless steels; primary cathodic reaction of steel corrosion in air-free brine; corrosion rate decreases sharply above pH 8; (iii) carbon dioxide species (dissolved carbon dioxide, bicarbonate ion, carbonate ion): dissolved carbon dioxide lowers pH, increasing carbon and low-alloy steel corrosion; dissolved carbon dioxide provides alternative proton reduction pathway, further exacerbating carbon and low-alloy steel corrosion; may exacerbate sulfide stress cracking; there is a strong link between total alkalinity and corrosion steel; (iv) hydrogen sulfide species (hydrogen sulfide, bisulfide ion, sulfide ion): potent cathodic poison, promoting sulfide stress cracking and low-alloy steel corrosion; highly corrosive to alloys containing both cupronickels; (v) ammonia: causes sulfide stress cracking of some copper-based alloys; (vi) chloride ion: strong promoter of localized corrosion of carbon, low-alloy steel, stainless steel, and other alloys.

Groundwater quality factors that increase the corrosion potential are preferably a low pH, a high content of salts, and dissolved gases like oxygen and hydrogen sulphite.

Common types of corrosion that may take place and the components that may be corroded are: (i) chemical, mainly caused by the presence of O_2, CO_2, H_2S, and Cl in the groundwater; (ii) electrochemical pitting caused by a local attack on a "weak" point or a passive area; and (iii) galvanic corrosion (also called bimetallic corrosion) is an electrochemical process in which one metal corrodes preferentially when it is in electrical contact with another, in the presence of an electrolyte as the groundwater, the driving force for corrosion is a potential difference between the different materials; galvanic corrosion can be prevented by selecting materials with similar corrosion potentials or by breaking the electrical connection by insulating the two metals from each other.

Corrosion problems can be limited by (i) using corrosion-resistant material, such as plastics and/or more noble metals/alloys; (ii) not mixing materials with different electro chemical potential; and (iii) using cathode protection for wells with carbon steel casing; use coated casing and pipes done by a thin plastic or epoxy layer on the inside of the metal pipes. However, such films are sensible for mechanical damages and fits badly together with threads (or welding). It is therefore recommended not to use coated protections, especially not for well components that are easily damaged during construction.

In most commercial applications, groundwater-source geothermal heat pumps use copper groundwater-to-refrigerant heat exchangers. If, in rare cases, groundwater corrosion properties are outside the standard contaminant limits of copper heat exchangers, alternate heat exchangers with inner surfaces constructed of cupronickel should be used (Table 17.2).

TABLE 17.2
Property Limits for Using Copper or Cupronickel Heat Exchangers

	Cu	Cu-Ni
Water velocity (m/s)	≤3.6	≥12
Filtration (μ)	≤25	≥25
Uninhibited chloride (ppm)	≤200	≥200
Alkalinity (ppm as $CaCO_3$)	80–100	≥100
pH	7.2–7.6	≤7.2
Free residual chloride level (ppm)	1.0–1.7	≤1.7

17.6.3 Clogging and Fouling

Operational problems of 25–30% of the groundwater geothermal heat pump systems are usually related to clogging and fouling due to high contents of materials as humus, iron bacteria, manganese, or carbonates.

The most common technical problem is clogging of wells, a process defined as an increased flow resistance for groundwater to enter the well or be disposed through the well. The clogging (that can easily be traced and dealt with in an early stage by monitoring flow rates and drawdown) normally gets more and more evident with time and will result in a decreased flow rate with time (i.e., the wells lose their original capacity most commonly due to clogging); detection of clogging requires monitoring over long period of time.

Main causes for particle clogging are: (i) invasion of unbreakable mud into the porosity; (ii) migration of fines towards gravel packs; (iii) bridging that may form bridges in the well vicinity which will increase the groundwater flow resistance; however, these bridges can be broken down by a reversed flow and the fines may be flushed to the surface by a further well development; for this reason, groundwater wells with potential bridging should be constructed so they can easily be flushed; (iv) under certain conditions, the groundwater wells may be clogged by hydrochemical processes involving solid chemical precipitates as iron, hydroxide, calcium carbonate, and siderite; (v) precipitation of dissolved minerals caused by temperature changes and precipitation of iron manganese hydroxides; it increases with temperature variations in the aquifer and with air entering the wells or pipework; the latter can be avoided by operating the system with a slight overpressure; furthermore, such wells need submersible pumps that can be lifted for maintenance.

Clogging by iron is by far the most common one. Occasionally, iron is mixed or even replaced by manganese oxides forming a soft and sticky black layer on the screen. Stripping of CO_2 will often cause iron to precipitate, normally as iron hydroxide.

In groundwater that is close to be oversaturated of carbonates and with iron in solution, the wells are sometimes clogged by siderite, the worst form of clogging since siderite is almost impossible to remove. For prevention of hydrochemical clogging, measures as the followings should be adopted: (i) design the system in a way that entrance of air to the groundwater loop is prohibited; (ii) place wells so mixing of different water types is avoided; and (iii) operate the wells with a restricted drawdown.

Some of the clogging prevention methods, or at least to minimize the effects, are as follows: (i) wells should be designed, based on representative sampling and other geo documentation obtained by test drillings; (ii) the drilling and flushing method should be applied to minimize the disturbance of permeable layers in which the well screen are placed; (iii) at the completion of wells, the screens and gravel packs should be carefully placed and the development should proceed until the back-flushed water is perfectly clean; (iv) at the operational mode, a proper pumping schedule should be applied in order not to have the wells working above its capacities or designed flow rates; (v) wells should be equipped with devises that allow monitoring of well behavior over time, to detect potential clogging in an early stage; (vi) clean the screen and restore a groundwater well; (vii) use of chemicals to break down encrusting materials; and (viii) use wire brushing, swabbing, jetting, and surging can be used to assist the chemical treatment.

High-pressure jetting may be the most effective mechanical treatment method, preferably to get rid of softer encrustations, bio-films, and iron slime. This type of jetting includes a self-rotating nozzle head through which water is pumped with a high velocity. The pressure would typically be in the order 10–14 kPa at a flow rate of 5–10 kg/s. By pumping the well during the jetting, the removed material is brought to the surface. Acids are commonly used to chemically dissolve encrustation formed in a well. Acids can dissolve a number of mineral deposits. Although acids typically are not effective at killing bacteria, they are in general used to dissolve iron and manganese oxides formed because of bacteriological growth. Acid would also dissolve carbonate scales. For iron bacteria and slime, a liquid hypochlorite would be effective. This strong oxidizer will also kill any bacteria. For clogs with carbonate scale, sulfuric acids are commonly used. For clogs with iron or manganese oxides, muriatic acid or hydroxyacetic acid may be a proper choice. The chemicals are pumped into the well and agitated frequently for 24–72 hours before it is recovered. The use of dispersants, often polyphosphates, has sometimes been successful in improving the well performance due to mechanical blockage. Fine-grained materials can be dispersed by the dispersant and allow "sticky" particulates to disperse and move more freely in the formation pore space. In this way, the pores can be cleaned from fine grains and eventually be pumped out of the well.

All chemicals used during a well cleaning process must be carefully removed from a well and properly disposed. Water should be pumped from the well until the water quality is essentially the same as prior to treatment.

Implementation of rehabilitation procedures for individual wells should generally include the following steps: (i) perform mechanical cleaning of the well screen; (ii) apply the proper quantity and type of chemical treatments; (iii) allow sufficient chemical reaction time; (iv) remove spent chemicals from the well; and (v) conduct a performance pumping test.

17.6.4 Organisms

Organism growth of slime (a thin, glutinous mud) and algae that form under certain conditions can reduce the heat transfer rate by forming an insulating coating on the inside tube surface. Algae can also promote corrosion by pitting.

Groundwater quality tests (e.g., gas chromatograph/mass spectrograph open scan for volatile organic compounds) should be required to assess the potential for bacteria growth. Such quality analysis is required because flow blockage and pitting corrosion attributed to the growth of nuisance bacteria (e.g., iron and sulfur bacteria) are supported by conditions such as the following (DOE 1995): (i) pH (6.0–8.0); (ii) dissolved oxygen ($O_2 \leq 5\frac{mg}{L}$); (iii) ferrous iron ($\geq$ 0.2 mg/L); (iv) groundwater temperature (7.7–16.1°C); (iv) redox potential (≥ -10 mV) (also known as oxidation/reduction potential that is a measure of the tendency of a chemical species to acquire from or lose electrons to an electrode and thereby be reduced or oxidized, respectively).

Groundwater contains hundred thousands of bacteria with different species. The one that frequently makes clogging occur is the iron bacteria.

Iron bacteria are microorganisms that use iron as a source of energy. In the northern part of the United States and in Canada, iron bacteria are naturally present in the ground/soil and in surface water. This surface water may reach the groundwater well and enter to geothermal heat pump piping system.

Although iron bacteria are not harmful, they can cause troublesome, persistent, and expensive well problems, including reduced well efficiency and excessive corrosion of well and plumbing components.

17.6.5 Sand

In practice, a groundwater well producing just 10 ppm (particles per million) of sand, operating a total of 1,000 hours/year at 19 kg/s, will produce 680 kg of sand.

If it is not possible to complete the groundwater well in such a way as to limit sand production, some form of surface sand separator will be necessary, except, however, open tanks that allow oxygen to enter the groundwater and alter the ferrous iron (if present) to a ferric state with a much lower solubility leading to fouling of heat exchangers. The most effective surface sand removal devices are the strainers that ensure effective removal at any flow rates.

17.7 THERMAL STORAGE

Depending on the aquifer's hydraulic conductivity value (a physical property that measures the ability of the material to transmit fluid through pore spaces and fractures in the presence of an applied hydraulic gradient), according to Darcy's law, the hydraulic conductivity is the ratio of the average velocity of groundwater through a cross-sectional area to the applied hydraulic gradient, which is dependent on the size and shape of media pores, very low-temperature thermal (heat) energy can be stored in underground aquifers with negligible groundwater flow rates when

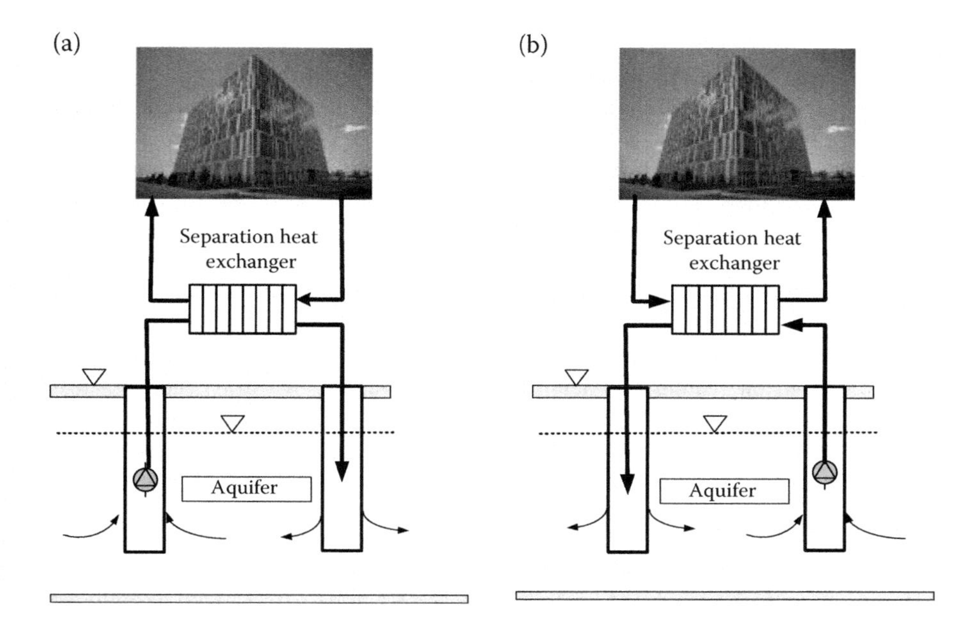

FIGURE 17.4 Principle of an aquifer thermal energy storage; (a) winter heating dominated mode; (b) summer cooling dominated mode (Building image retrieved from https://unsplash.com/photos/e0uCDHd19U4).

they are hydraulically connected to groundwater heat pump heating and cooling systems via at least two (or more) wells (Holmberg 2018).

As can be seen in Figures 17.4a and 17.4b, a reservoir can be created around each well or group of groundwater wells. In periods with a predominant heating demand (in winter), groundwater from the "warm" wells is cooled by the geothermal heat pump evaporator(s) and returned to the "cold" wells (Figure 17.4a). In periods with a predominant cooling demand (in summer), the cycle is reversed and groundwater from the cold wells is used for cooling purposes, and the heated water is returned to the "warm" wells (Figure 17.4b).

Such a heat storage concept can be generally successful if (Hall and Raymond 1992; Stene 2004): (i) there are suitable aquifers for groundwater supply; (ii) low-grade thermal energy users as geothermal heat pumps are available; and (iii) there are temporal mismatches between thermal energy availability and thermal energy use.

17.8 MAINTENANCE ASPECTS

Open-loop ground-source heat pump systems need to be systematically controlled, monitored, and visually inspected in order to trace operational problems in an early stage, the earlier the better. The control should consist of visual inspections, monitoring of operational parameters, and analyses of the groundwater; a simple way to check the function of wells is to monitor parameters that show the behavior of the wells. This could of course be done manually by measuring the drawdown, flow rate, and injection pressure now and then, but the modern computerized controlling and monitoring systems makes it much simpler.

Maintenance activities should: (i) keep records of the specific drawdown and injection pressure to check the status of the groundwater wells; (ii) keep record of temperature and pressure drops over heat exchangers connected to groundwater loop; (iii) in geothermal heat pump systems with reversed flow, the wells used for injection must be always back flushed before changing the flow direction; (iv) survey the leakages, corrosion damages, and status of monitoring meters; and (v) take water samples for chemical, bacterial, and physical analyses, in order to detect changes causing operational or environmental problems.

The visual inspections (should be regular executed, once every quarter of a year or so) should focus on visiting of groundwater well houses in order to (i) check for leakages, (ii) the function and status of monitoring and controlling devices, (iii) potential corrosion problems on equipment, and (iv) the indoor status of well houses or cellars.

17.9 ADVANTAGES

Under suitable conditions, the main advantage of groundwater heat pump systems with respect to ground-coupled heat pump systems can be summarized as follows: (i) relative design and installation simplicity; (ii) achieve better energy performances due to their constant heat source temperature year-around; (iii) aquifers' temperatures are generally advantageous for geothermal heat pump applications that are very close to the undisturbed ground/soil/rocks temperature; (iv) in regions with abundant resources, the groundwater extracted and returned using relatively inexpensive wells requiring small (compact) amount of ground/soil land area; in some cases, a single pair of high-volume wells can serve an entire building; (v) therefore, the initial installation costs required for wells' drilling per unit capacity are much lower; (vi) groundwater well drilling contractors are widely available; (vii) properly designed by selecting reliable components, and using well-developed groundwater wells, maintenance costs could not be higher than those of conventional central HVAC systems; (viii) negligible impact(s) on the environment, when groundwater is injected back into the same aquifer by a second well, net groundwater use is zero; (ix) the space required for the water well is very compact; (x) in cold and moderate climates, possible direct use of the groundwater for cooling; and (xi) possibility of coupling with water uses (e.g., with potable water supply or irrigation).

17.10 LIMITATIONS

Among the limitations of open-loop groundwater heat pump systems in comparison to more closed-loop (indirect, secondary fluid) ground-source heat pump systems can be mentioned: (i) even more attractive because of the efficient direct use of groundwater, are less frequently used because they are more demanding due to hydro-geological requirements and frequent maintenance requirement of wells; (ii) the feasibility and the sustainability of such HVAC systems depend on the availability of adequate bacteriologic/chemical groundwater quality, flow rate, and groundwater static level; (iii) for small residential applications, the cost of groundwater well drilling can be relatively high; (iv) both in direct and indirect use, fouling, corrosion, scaling, and clocking may cause problems at the level of circulating pumps and heat pumps' heat exchangers; (v) groundwater presents risks for corrosion, fouling and blockage of wells, heat exchangers and other components; (vi) presence of sand in the groundwater stream can plug injection wells if screens or strainers aren't included in construction of the production wells to minimize the production of sand; (vii) in some regions (localities), restrictive regulations over groundwater withdrawal and re-injection are restrictive because of environmental concerns; and (viii) in systems drawing groundwater from deep aquifers, the energy cost of pumping may be relatively high.

While open-loop systems can be very effective in zones with a high groundwater table and in the presence of groundwater flow, there are two major concerns that must be addressed when considering such systems. Due to the fact that the circulation fluid is discharged directly into the ground, for environmental and practical reasons, the only fluid that can be used is water.

Moreover, open-loop systems cannot be used in frozen ground or in sites with a very deep groundwater table. If the extraction well is unable to draw water from the ground aquifer, the system cannot operate.

REFERENCES

DenBraven, K.R. 2002. Regulations for open-loop ground-source heat pumps in the United States. *ASHRAE Transactions* 108(1):962–967.

DOE. 1995. *Commercial/Institutional Ground-Source Heat Pump Engineering Manual*. Prepared by Caneta Research Inc. for U.S. Department of Defense U.S. Department of Energy, Oak Ridge National Laboratory, Tennessee, TN.

Dreybrodt, W. 1988. *Processes in Karst Systems: Physics, Chemistry, and Geology*. Springer Series in Physical Environment. Springer, Berlin, Germany.

Driscoll, F.G. 1986. *Groundwater and Wells*, 2nd Edition. Johnson Filtration Systems Inc., Amarillo, TX.

Fetter, C.W. 1994. *Applied Hydrogeology*. 3rd Edition. Macmillan, New York.

Geotrainet. 2011. *Training Manual for Drillers Shallow Geothermal Systems. Geo-Education for a Sustainable Geothermal Heating and Cooling Market Project*. IEE/07/581/SI2.499061. Compiled and Edited by European Geothermal Energy Council ISBN: 978-2-9601071-1-1. Brussels, Belgian (www.geotrainet.eu. Accessed May 17, 2021).

Hall, S.H., J.R. Raymond. 1992. Geo-hydrologic characterization for aquifer thermal energy storage. In *Proceedings of the 27th Intersociety Energy Conversion Engineering Conference*, San Diego, CA, pp. 101–107.

Holmberg, H. 2018. *Analysis of Geo-Energy System with Focus on Borehole Thermal Energy Storage*. Thesis for the Degree of Master of Science. Division of Thermal Power Engineering Department of Energy Sciences, Faculty of Engineering, Lund University, Sweden. ISRN/TMHP-09/5192-SE (https://lup.lub.lu.se/luur/download?func=downloadFile&recordOId=1479061&fileOId=1479064. Accessed September 22, 2018).

Stene, J. 2004. *IEA HPP Annex 29 – Ground-Source Heat Pumps Overcoming Technical and Market Barriers*. Norway Status Report. SINTEF Energy Research, December.

18 Open-Loop, Dual and Multiple-Well Groundwater Heat Pump Systems

18.1 INTRODUCTION

Open-loop and dual- and multi-well groundwater heat pump systems remove groundwater from one (as usual in residential and small commercial/institutional buildings), or two or more well(s) (as in large-scale commercial/institutional/industrial buildings) and deliver it directly or indirectly (via intermediate heat exchangers) to one or several geothermal water-to-air and/or water-to-water heat pumps to serve as a heat source (for space and domestic/process hot water heating) or sink (for space cooling/dehumidifying).

Successful application of such systems (that are among the most energy-efficient and cost-effective heating and cooling systems when groundwater is available in sufficient quantities and appropriate quality in order to avoid fouling, clogging, and corrosion at given geographic sites) depends on the building size and height and national and/or local regulatory requirements (Bose et al. 1985).

18.2 BASIC CONFIGURATIONS

18.2.1 Residential and Small-Scale Commercial/Institutional Buildings

In residential and small-scale commercial/institutional buildings, open-loop groundwater heat pump systems are generally designed for direct "use" of groundwater (i.e., without intermediate heat exchangers) by the aid of at least two wells drilled at a certain distance away from one another.

In such systems, in both heating and cooling operating modes, groundwater is pumped from one production well by the aid of a well-immersed circulating pump, passed once through a strainer provided with a bypass to allow its replacement and service, and at least one geothermal heat pump usually provided with cupronickel heat exchangers and discharged into a suitable receptor, such as an aquifer via one return (injection) well (Figure 18.1a), or drained to a surface water body (lake, river, pond) (Figure 18.1b).

The groundwater disposal to surface bodies, a less expensive solution, is possible if the receiving body is capable of accepting the groundwater over long periods of time according to local hydrological and regulatory context. The diaphragm expansion tank, kept under pressure by the submersible pump, is allowed to absorb the groundwater

DOI: 10.1201/9781003032540-18

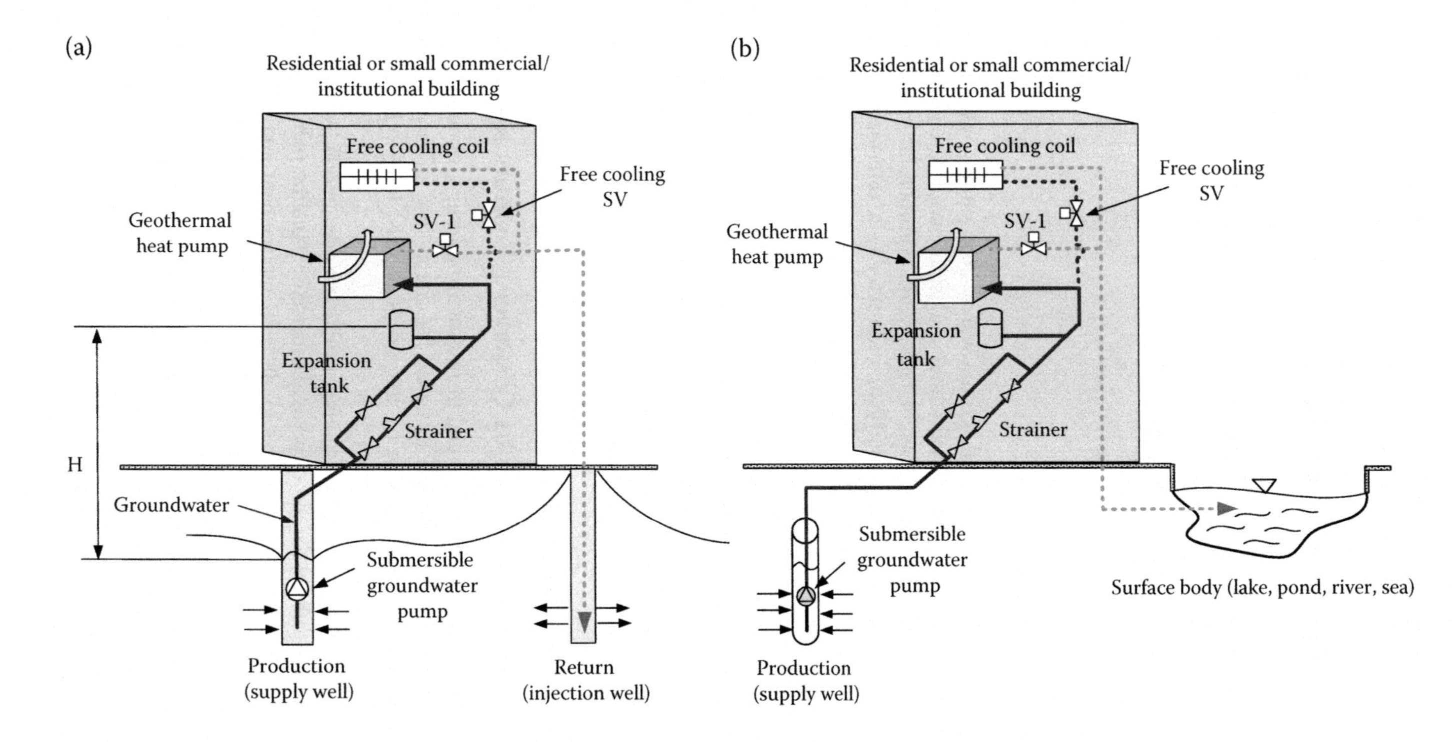

FIGURE 18.1 (a) Direct use (without intermediate heat exchanger) of groundwater; (a) dual (supply and return) wells; (b) one supply well and surface water body disposal; SV, solenoid valve (Notes: schematic not to scale; not all component shown).

volume variations. A motorized (solenoid) valve, placed downstream of the geothermal heat pump to keep it under pressure, opens and closes on call for heating or cooling to provide groundwater flow.

In cold-climate countries, such as Canada, Sweden, and Norway, where the temperature of groundwater is usually between 5 and 10°C, by closing the solenoid valve SV-1 and opening the solenoid valve SV-2, it can be circulated through water-to-air coils in order to provide to the building space free-cooling/dehumidification (see Figures 18.1a and 18.1b).

18.2.2 Large-Scale Commercial/Institutional Buildings

In cold-climate countries (such as Canada, Sweden, and Norway where huge quantities of groundwater are usually available at relatively good qualities), groundwater heat pump systems are frequently installed in large-scale commercial/institutional buildings (Reiley 1990; Minea 2005; Lo Russo et al., 2011) as well as in district heating and cooling systems (Stene 2004).

Large-scale groundwater heat pump systems consist of the following main components (Figure 18.2): (i) one or several production (supply) groundwater well(s), each provided with a submersible circulating pump; (ii) a strainer with a bypass to allow its replacement and service; (iii) an intermediate frame plate heat exchanger (generally of stainless steel materials) to isolate the groundwater from the building water closed-loop; it protects the building closed loop from exposure to, for example, iron bacteria and other undesirable impurities contained in the groundwater, in addition to allowing a groundwater loop to be operated at different flow rates than those of building internal water closed loop; even the intermediate heat exchanger can be subject to corrosion, scaling, fouling, and blockage, and it can be easier to clean and/or replace; inside the intermediate heat exchanger, the groundwater rejects (in the dominated heating mode) or absorbs (in the dominated cooling mode) heat to or from the building internal water closed loop connected to distributed (or central) water-to-air or water-to-water geothermal heat pumps; the groundwater is then dumped back into the re-injection well(s), preferably in the same aquifer to ensure adequate thermal recharge of the resource; (iv) a diaphragm expansion tank between the intermediate heat exchanger and the geothermal heat pumps; (v) distributed (or central) reversible geothermal heat pumps (each with or without desuperheater); in the heating mode, heat is absorbed by the geothermal heat pump's evaporator from the groundwater, while in the cooling mode, heat is rejected by the geothermal heat pump's condenser in the same groundwater flow; (vi) solenoid (or motorized) valves downstream each geothermal heat pump; (vii) at least one groundwater return (rejection) well; and (viii) internal building thermal distribution system that delivers space heating and cooling via, for example, air ducts or underfloor/ceiling heating/cooling pipes.

18.3 GROUNDWATER WELLS

The design and construction of production (supply) and return (injection, recharge) wells, usually with ID diameters between 150–200 mm and 10–40 m deep (occasionally deeper), must be achieved after a hydro-geological investigation of the site of which

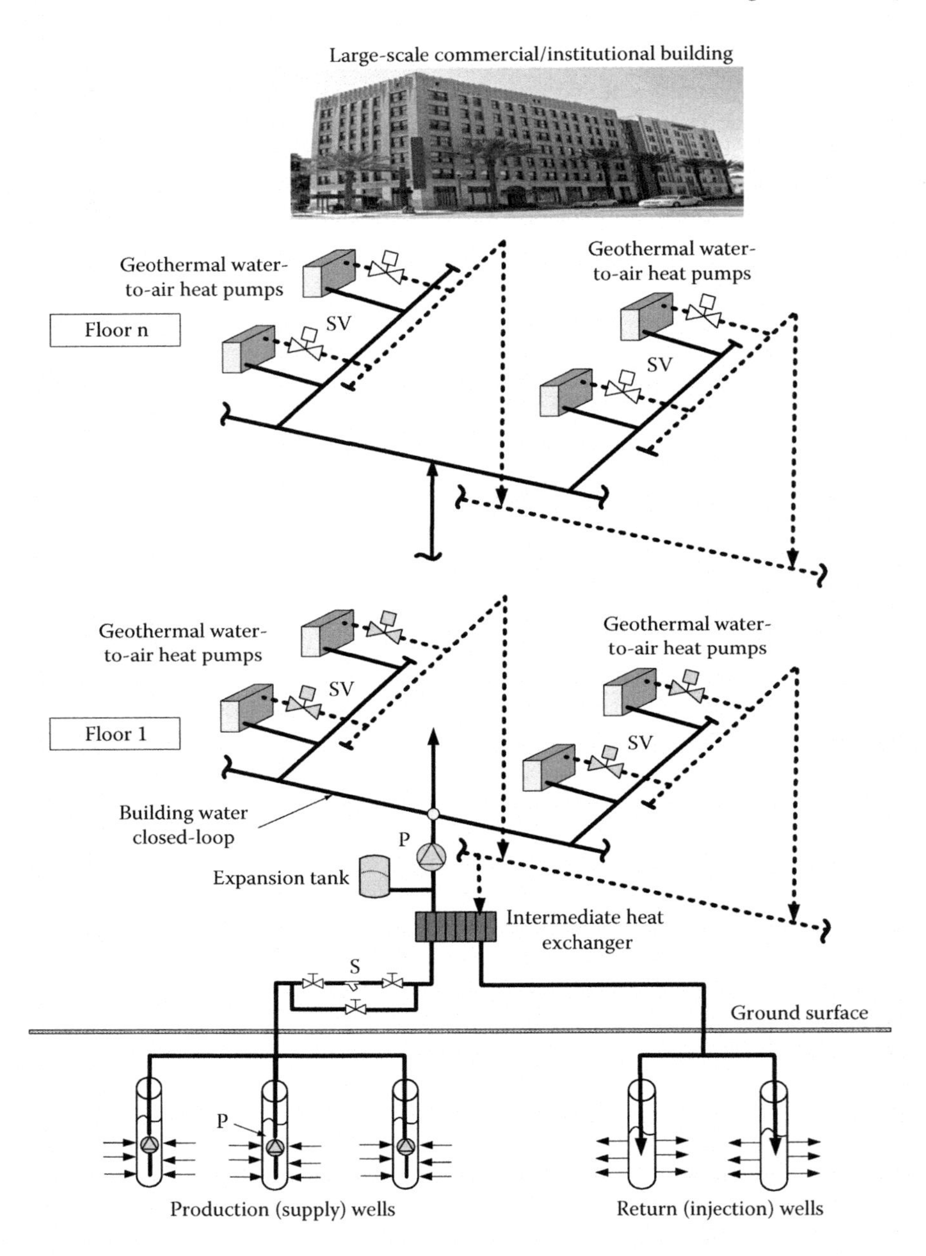

FIGURE 18.2 Schematic configuration of a typical open-loop, indirect, multi-well groundwater heat pump system for large-scale commercial/institutional buildings; P, groundwater pump; S, strainer; SV, solenoid valve (Notes: schematic not to scale; not all components shown) (Building view retrieved from: https://www.nps.gov/tps/how-to-preserve/briefs/14-exterior-additions.htm, accessed March 9, 2020).

main objectives are: (i) identify the ground/soil/rock types and their thermos-physical properties; (ii) determine the groundwater quality, temperature profile, hydraulic gradient, dominant movement direction and velocity; (iii) estimate the aquifer thickness, porosity, static level, specific capacity (for confined aquifers), average flow rate, and drawdown (for unconfined aquifers); (iv) review all local and national regulations affecting or restricting the withdrawal and return of groundwater, and establish what is permissible at the site in terms of drilling, excavation, allowable fluids, and legal rights to use properties; (v) respect all requirements of local and national authorities concerning casing, grouting, and cementing materials, placement procedures, disposal of drilling cuttings, fluid, and mud, and the protection and conservation of fresh groundwater supplies; (vi) review of published geological and/or hydrogeological reports (including well pump tests already achieved in the vicinity of the proposed site) and maps, if available, and assess the impact, if any, on existing wells; and (vii) recommend the most appropriate drilling methods and their approximate costs, well (production/supply and return/injection/recharge) locations, number, diameter(s), depth (s), and spacing aiming to achieve required groundwater flow rates and ensure return of the discharge groundwater to the same aquifer from which it was drawn; (vii) recommend well testing method(s).

18.3.1 Production (Supply) Wells

The diameter and key components of production (supply) wells depend on the diameter of the submersible pump bowl assembly (impeller housings and impellers, always placed sufficiently below the expected pumping level to prevent cavitation at the peak production rates) necessary to provide the required groundwater flow rate (Figure 18.3).

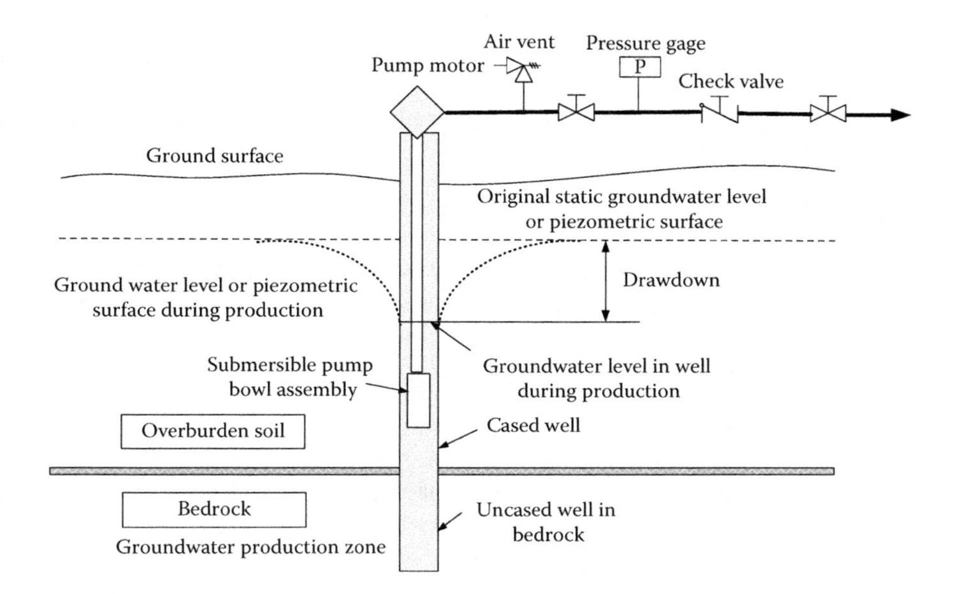

FIGURE 18.3 Schematic cross sections of production (supply) well and aquifer drawdown.

The pumping of groundwater creates a cone of depression around the production well of which size and shape are mainly related to the pumping rate and the hydraulic conductivity of the aquifer. In the case of confined aquifers, the natural recharge and flow is more restrained compared to unconfined situations. Depending on climate, the groundwater table will vary with seasons under a year according to the amounts of local precipitations.

In addition, the design and construction of groundwater production (supply) and return (injection) wells must consider the aspects as the following: (i) be deep enough to reach the groundwater table(s); (ii) groundwater delivery system should eliminate potential oxygen intrusion and induced turbulence in the production (supply) well(s); (iii) a small, positive pressure on the groundwater discharge piping should be maintained to prevent air from entering into the piping, and reduce the potential for groundwater hammer by using check valves; (iv) account for occurrences such as failure of submersible circulation pump(s); (v) use adequate groundwater pumping equipment; (vi) provide the best well spacing (i.e., the distance between the production and injection wells) that may eliminate siphoning from one well to the others, through the connecting piping, when the groundwater heat pump system is not operating; and (vii) provide an air vent valve at each return well head, downstream of the pressure-sustaining valve, to allow air to be vented instead of being forced into the injection aquifer.

In competent rock formations, often the bottom of the production wells is uncased. However, in formations in which there is a tendency to cave, sand removal should be controlled by careful well construction (by placing screens and/or gravel packs around the screens to provide additional filtering and to increase the permeability of the near well materials) as a primary strategy, and surface strainers or separators as a secondary strategy.

The submersible pumps should be equipped with check valves located at the base of the groundwater column near the bowl assembly in order to maintain the column full of water and, thus, prevent the reverse thrust during the start up. In addition, submersible motors should be equipped with thrust bearings to resist the down thrust developed in normal operation.

18.3.2 Return (Injection) Wells

The return (injection) wells aim to dispose of the groundwater after it has passed once through the intermediate (separation) heat exchangers or directly through geothermal pump system in order to stabilize the aquifer from which the groundwater is withdrawn by reducing or eliminating long-term drawdowns, and helps to ensure long-term productivity. Return (injection) wells, particularly those likely to be subject to positive injection pressure, should be fully sealed from the top of the injection zone to the ground/soil surface (EPA 2018) (Figure 18.4).

Return (injection) groundwater wells may present the risk of hydraulic (Figure 18.5a) and thermal (Figure 18.5b) feedback among wells (Banks 2009; Milnes and Perrochet 2013; Casasso and Sethi 2015).

In order to avoid the imbalance between the amount of groundwater extracted and rejected and, thus, protect groundwater resource, design of return (injection)

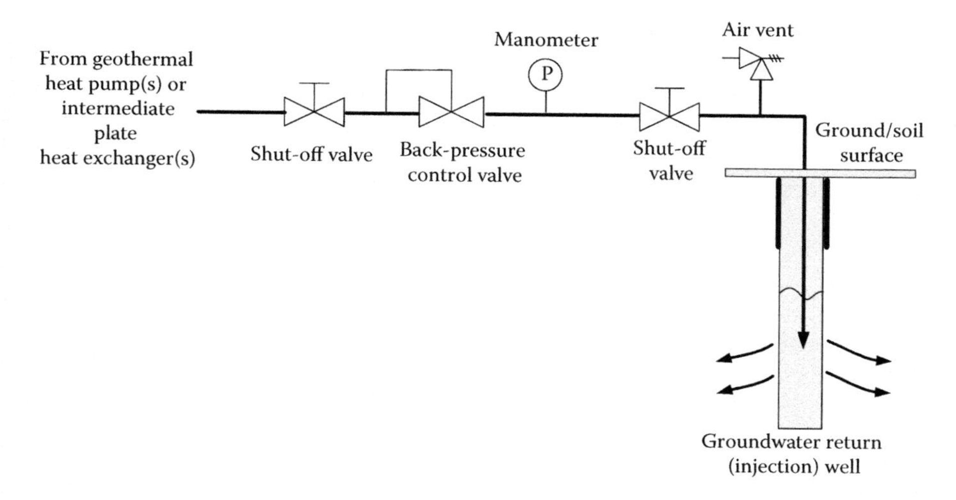

FIGURE 18.4 Schematic of a typical groundwater return (injection) well.

wells must consider aspects as: (i) groundwater entrance velocity through the screen should be limited to a maximum of 0.03 m/s to avoid incrustations; (ii) because it is likely that the groundwater level in the well will be higher than the static water level when the groundwater heat pump is in operation, it is important that the seal (grout placed between the borehole and the outside of the casing) be carefully placed and that it extends from the top of the aquifer to the ground/soil surface; this prevents the injected groundwater from finding a path up around the outside of the casing to the surface; (iii) injected groundwater should always be released below the static water impede groundwater flow; (iv) an air release valve should be provided to avoid the air entering the injection well; (v) the provision for pressure (or groundwater level) monitoring is important as a means of determining any accumulation of particulate; (vi) successful injection requires clean, particle-free (e.g., sand < 1 ppm) groundwater; and (vii) poor production well performance in terms of sand content coupled with the lack of surface removal devices inevitably means that this material will be deposited in the injection well(s).

18.4 GROUNDWATER PUMPING

Under non-pumping conditions, the groundwater level indicates of the groundwater table level in unconfined aquifers; when the pump is started, groundwater level drops to a new, lower level referred to as the pumping level that is a function of the groundwater pumping rate; the difference between the static water level and the pumping level is referred to as the drawdown (see Figure 18.3). At a given pumping rate, the size and shape of the cone of depression that forms in the aquifer around the well during pumping, and the depth of the drawdown are functions of the aquifer and its ability to deliver groundwater; the specific capacity of a groundwater well (a value indicating the ease with which the aquifer produces groundwater) may be quoted in L/s per meter of drawdown. For example, for a well with a static level

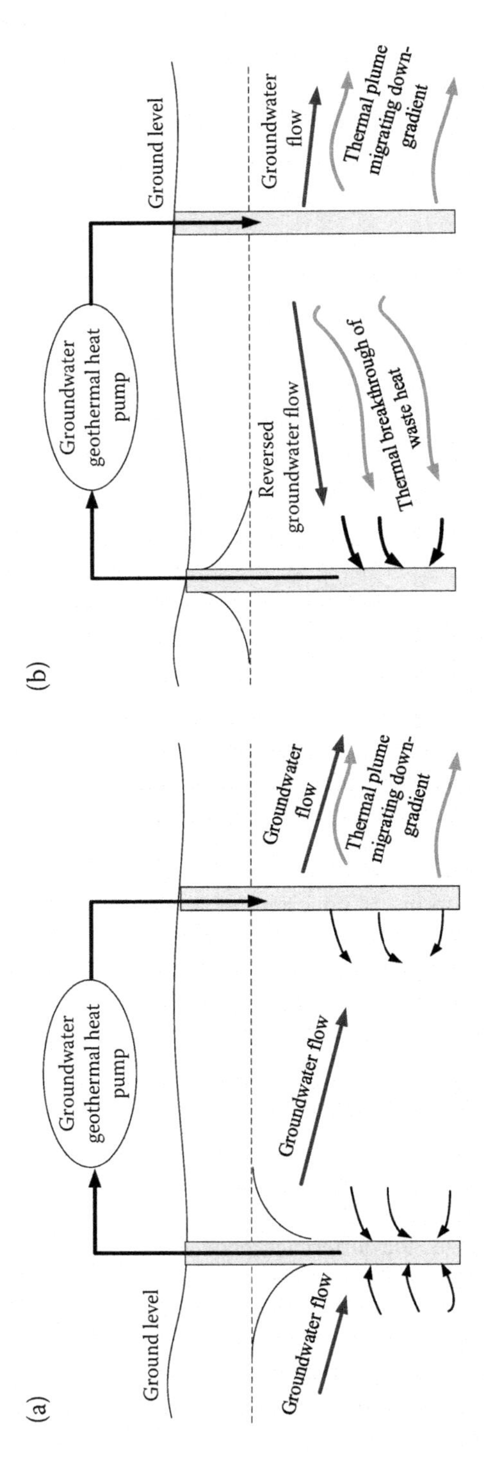

FIGURE 18.5 Well doublet with: (a) hydraulic feedback and (b) thermal feedback.

of 15 m that produces 30 L/s at a pumping level of 25 m, the drawdown is (25–15 m) 10 m, and the specific capacity (30/10), 3.0 L/s·m; a high value (e.g., 2.1 L/s·m) would indicate a good groundwater well, whereas a value of 0.1 L/s·m would indicate a poor well. The specific capacity diminishes as pumping rates increase.

The head on the groundwater well pumps is equal to the vertical rise (H in Figure 18.1) from the seasonal low pumping water level in the supply wells to the highest water level in the recharge wells during operation plus the vertical submerged distance of the discharge in the recharge well plus the friction head of the piping system from the well pump to the down-comer discharge into the recharge well plus the head of the backpressure valve minus the siphon head.

The head equals the vertical rise from the groundwater table level and the expansion tank, plus the greatest friction pressure losses from the expansion tank to the outflow in the return well, plus the pressure drop through the geothermal heat pump.

The total pump head pressure of submersible groundwater pumps is composed of the following components: (i) static groundwater level plus the drawdown at the design flow rate (or the vertical distance over which the groundwater must be pumped to reach the surface, e.g., 25 m); (ii) column friction losses between the bowl assembly and the ground/soil surface, usually in the order of 3–9 kPa); (iii) surface friction pressure losses through piping from the production well to the building, through the mechanical room piping and equipment (heat exchangers, controls, etc.) and through the piping from the building to the disposal point; unless there are significant elevation considerations or distances involved, surface losses normally amount to less than 78 kPa assuming a 35 kPa loss in the intermediate heat exchanger (typically, between 78 and 108 kPa); and (iv) injection pressure requirements, a function of well design, groundwater quality, and/or the type of disposal means; injection pressure may be 10–40% higher than production one; for poor groundwater quality, high sand production, or poor well construction, injection pressure may be 30–60% higher compared to theoretical (calculated) values; in surface discharge applications, often a pressure sustaining valve is used to maintain a small (less than 35 kPa back pressure on the system to keep it full of groundwater.

18.4.1 Groundwater Submersible Well Pumps

The electric, multi-stage, centrifugal submersible pumps assembled with hermetically sealed motors submerged in vertical position in vertical drilled wells with nominal 3,600 rpm motors are cost-effective options for groundwater ground-source heat pump systems since they are able to produce a higher flow rates per unit diameter than line surface shaft pumps which typically operate at speeds of 1,800 rpm or less.

The main components of multiple-stage submersible groundwater pumps are: (i) in the sub-surface: submersible pump, seal electric cable, gas separator; the pump shaft is connected to the gas separator or the protector by a mechanical coupling at the bottom of the pump; groundwater enters the pump through an intake screen and is lifted by the pump stages; other parts include the radial bearings (bushings) distributed along the length of the shaft providing radial support to the pump shaft;

and (ii) at the ground/soil surface: motor controller (or variable speed controller), transformer and electric cable.

To select a groundwater circulation pump it is necessary to know: (i) the performance of the production well(s) in terms of the pressure head; and (ii) groundwater static level recovery data after pumping ceases.

The main disadvantages are as follows: (i) the pump seals can become corroded with time, allowing the groundwater to seep into the motor, rendering it useless until it is repaired; and (ii) must be fully submerged since the groundwater around a submersible pump actually helps to cool the motor and thus avoid the overheat.

Typical problems leading to pump failures are listed in the following: (i) up-thrust wear occurs when the pump is operated at flow rates greater than the maximum recommended pumping rate; in floating-type pumps, the impellers are forced up against the diffusers, the up-thrust washers may be overloaded and the stage eventually destroys itself; abrasives in the well groundwater accelerate the process; (ii) down-thrust wear in pump stages is typical at rates lower than the minimum recommended rate and is exaggerated by the presence of sand or abrasive solids; down-thrust washers are destroyed first before the stages fail; (iii) radial wear caused by well groundwater loaded with abrasives increases the clearances in the pumps journal bearings. This usually leads to increased vibrations of the shaft and a further acceleration of wear in the bearings; (iv) erosion in pump stages occur when abrasive-laden fluids are produced and can cut through pump housings; and (v) scale buildup can plug or even lock pump stages.

Groundwater circulation (well) pumps can be controlled according to the type of ground-source heat pump system they serve. For unitary geothermal heat pump systems, water flow is typically on/off with the loop operating between upper and lower temperature limits. For central geothermal heat pump systems, the groundwater flow is generally modulated to maintain a specific chilled- or hot water loop temperature.

Typically, one of the following methods can be employed: (i) dual set-point; (ii) staged, multiple-well pumping; and (iii) variable-speed pumps.

The dual set-point control approach, most common in systems with a single production well, energizes the well pump above a given temperature in the cooling mode and below a given temperature in the heating mode. Between those temperatures, the temperature of building water closed loop varies between the set points. To control well pump cycling, it is necessary to establish a temperature range (difference between pump-on and pump-off temperatures) over which the pump operates in both the heating and cooling modes. The size of this range is primarily a function of the building closed-loop water volume. To avoid the short cycling on the well pump, the range around each set point (difference between cut in and cut out temps) must be sufficient to result in an acceptable pump cycle time.

In systems in which multiple wells are required due to aquifer hydrology or redundancy, it is possible to employ a staged groundwater pumping control.

The variable-speed control varies the pump speed in response to some temperature or thermal load signals, but energy savings are a minimum mainly because a large portion of the well pump head is a static head. In addition, at higher speeds, the pump wear increases, particularly if the pumped groundwater contains abrasive solids, like sand.

The groundwater pump electrical power can be calculated as follows:

$$\dot{P}_{pump} = \rho \cdot g \cdot \dot{V} \cdot h / \eta \tag{18.1}$$

where

$\dot{P}_{pump}$ is the groundwater pump shaft electrical power (W).

ρ is the groundwater density (kg/m^3).

g is the local gravitational acceleration (m/s^2).

$\dot{V}$ is the groundwater volumetric flow rate (m^3/s).

h is the total differential head (m).

η is the groundwater circulating pump efficiency (-) that can be assumed to be constant (i.e., 65%).

18.4.2 Groundwater Flow Testing

The capacity of groundwater production wells should be established by pumping tests that are intended to assess the groundwater resource by providing useful information on (Osborne 1996; Kruseman and Ridder. 1990; ASHRAE 2011): (i) ground/soil/rocks stratigraphy and temperature; (ii) groundwater static level(s) and drawdowns, physical characteristics (temperature at various depths, quality, etc.), and available quantities and/or potential flow rates.

For groundwater heat pump systems serving commercial/institutional buildings of less than approximately 2,800 m^2, one test well is recommended, while for larger such buildings, at least two test wells should be drilled and operated.

To perform groundwater flow rate tests, the following elements are generally required: (i) obtain legal permits for drilling wells and using groundwater on the property; (ii) locate the groundwater test wells in such a way to be able later to convert them into production wells; (iii) the diameters of testing wells must be sufficient to allow installation of an appropriate submersible groundwater circulating pumps; (iv) have each well pump tested for water yield for at least 24 hours and for recharge rate up to the maximum recharge capability required, to determine the combined withdrawal and recharge capability of the wells; (v) monitor nearby wells to gather additional information on the aquifer and assess the potential for interference with these wells; and (vi) measure the groundwater temperature during the pumping tests, and analyze the water samples collected for chemical and microbiological analyses.

The length of the test is governed by the time required for the groundwater to become clean.

Depending on the time allowed, groundwater flow tests with wells as those shown in Figure 18.1 can be (DOE 1995; ASHRAE 2011): (i) a short-term pumping test during 4–24 hours with gradually increased flow rates by using a drilling rig onsite; the larger large flow rates must coincide with the nominal capacity of the groundwater well; (ii) these tests that generally run with a temporary electric

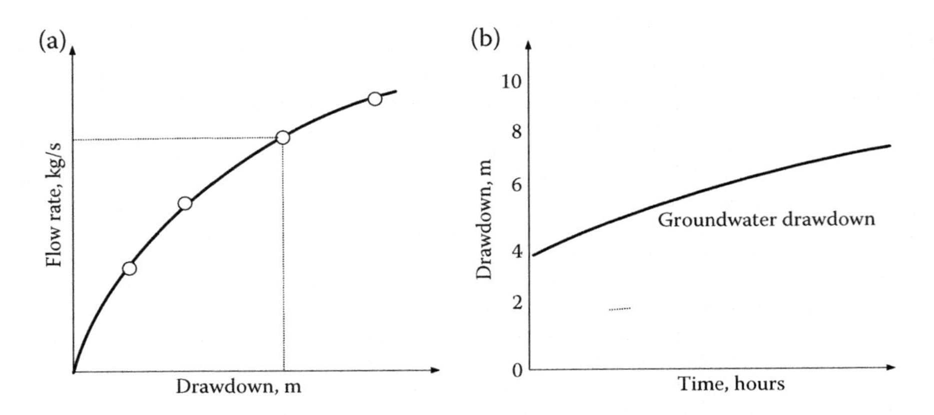

FIGURE 18.6 Typical curves from constant rate pumping tests: (a) flow rate and (b) drawdown.

submersible pumps, are used most frequently for direct use for small groundwater heat pump applications; (iii) long-term tests of up to 30 days may provide information on the aquifer reservoir helping to evaluate the interference effects between nearby wells, to calculate the hydraulic transmissivity and storage coefficient, reservoir boundaries, and recharge areas.

For short periods of time, the well drawdown can be calculated as follows (Banks et al. 2012):

$$s_w = B\dot{Q} + C\dot{Q}^2 \tag{18.2}$$

where

s_W is the well drawdown (m).

B is a constant that depends on the aquifer's hydraulic resistance (-).

$\dot{Q}$ is the groundwater flow rate (kg/s).

C is a constant related to the hydraulic resistance of the well structure and several fluid dynamics mechanisms (e.g., turbulence) that are constant for short time-scales (–).

In order to calculate the necessary pumping electrical power and the desired flow rate, a well productivity curve (Figures 18.6a and 18.6b) can be drawn by plotting the well pumping data. Such a curve evaluates the maximum groundwater flow rate that could be extracted with a reasonable electrical energy consumption.

REFERENCES

ASHRAE. 2011. *HVAC Applications – Geothermal Energy*. ASHRAE (American Society of Heating, Refrigerating and Air-Conditioning Engineers), Atlanta, GA.

Banks, D. 2009. Thermo-geological assessment of open-loop well-doublet schemes: a review and synthesis of analytical approaches. *Hydrogeology Journal* 17(5):1149–1155.

Banks, D., H. Scott, C. Ogata. 2012. From Fourier to Darcy, from Carslaw to Theis: The analogies between the subsurface behaviour of water and heat. *Groundwater – Italian Journal of Groundwater AS03013* (10.7343/AS-013-12-0025), pp. 9–18.

Bose, J.E., J.D. Parker, F.C. McQuiston. 1985. *Design/Data Manual for Closed-Loop Ground-Coupled Heat Pump Systems*. Oklahoma State University. American Society of Heating, Refrigerating and Air-Conditioning Engineers, Inc., Tullie Circle, NE, Atlanta.

Casasso, A., R. Sethi. 2015. Modelling thermal recycling occurring in groundwater heat pumps (GWHPs). *Renewable Energy* 77:86–93.

EPA. 2018. Ground Water Discharges (Underground Injection Control Program) (https://www3.epa.gov/region1/eco/drinkwater/pc_groundwater_discharges.html. Accessed July 29, 2018).

Kruseman, G.P., N.A. Ridder. 1990. Analysis and evaluation of pumping test data. ILRI (International Livestock Research Institute) Publication 47, pp. 377.

Lo Russo, S., G. Taddi, G. Baccino. 2011. Different design scenarios related to an open-loop groundwater heat pump in a large building: Impact on subsurface and primary energy consumption. *Energy and Buildings* 43(2–3):347–357.

Milnes, E., P. Perrochet. 2013. Assessing the impact of thermal feedback and recycling in open-loop groundwater heat pump (GWHP) systems: A complementary design tool. *Hydrogeology Journal* 21(2):505– 514.

Minea, V. 2005. Large heating and cooling GSHP applications in the Canadian cold climate. In *Proceedings of IEA (International Energy Agency) HPP (Heat Pump Program) Annex 29 Workshop (Ground Source Heat Pumps – Overcoming Market and Technical Barriers)*. Las Vegas, May 30 (http://www.annex29.net).

Osborne, P.S. 1996. *Suggested Operating Procedures for Aquifer Pumping Tests*. EPA (Environmental Protection Agency) Environmental Assessment Sourcebook. Technical Report, p. 191.

Reiley, J.S. 1990. Applications of closed-loop water source heat pumps for space conditioning in commercial buildings. *Journal of the Association of Energy Engineering* 87(5) (1990) 6–16.

Stene, J. 2004. *IEA HPP Annex 29 – Ground-Source Heat Pumps Overcoming Technical and Market Barriers. Norway Status Report*. SINTEF Energy Research, December.

19 Open-Loop Single-Well (Standing Column) Groundwater Heat Pump Systems

19.1 INTRODUCTION

Compared with fossil energy used in conventional HVAC (heating, ventilating, and air conditioning) systems, earth energy is a reliable and stable heat/sink source for providing building space and domestic hot water heating, and cooling by using geothermal heat pumps with relatively low electricity consumption and high energy efficiency.

In cold-climate average houses, typically equipped with 10.5-kW (nominal cooling capacity) brine-to-air geothermal heat pumps, usually require one 160-m deep, 152.4-mm ID diameter borehole in which one 31.75-mm ID high-density polyethylene U-shaped tube is inserted. This means that, depending on the thermal properties of ground/soil/rocks, the required drilling depth is of about 15.2 m per kW of nominal cooling capacity. The total cost of such vertical U-shaped ground-coupled heat exchangers, including drilling, U-tubes, grout, brine, filling, and purge, could be as high as US$12,000 (2019), a relatively high cost that, without subsidies, may discourage consumers from installing vertical closed-loop (indirect, secondary fluid) ground-source heat pumps.

One of the potentially effective and environmentally acceptable alternatives consists of using groundwater drawn and returned from/to wells drilled in areas with abundant resources and highly permeable ground/soil/rock formations.

Standing column well systems present an interesting potential for energy savings, especially in dense urban areas with suitable geological conditions where lack of land constitutes an impediment to the use of horizontal and even vertical closed-loop systems. Advantageous hydrological and geological conditions exist, for example, in the northeastern United States, as well as in Canada, Norway, and Sweden, where most of the territories are underlain by near-surface bedrocks at relatively low mean ground/soil/rocks temperatures (Orio 1994; Orio et al. 2005; Rees et al. 2004).

Open-loop single-well groundwater heat pump systems are commonly known as standing column systems (also referred as "energy wells," "recirculating wells," "geo-wells," "thermal wells," or "pumping and recharging wells"). The wells are usually drilled in hard rock geological formations, generally where it is not possible to produce sufficient groundwater for groundwater heat pump systems. In such

DOI: 10.1201/9781003032540-19

systems, the groundwater is circulated once from the wells through the geothermal heat pump(s) (see Figure 19.3a) or through intermediate heat exchangers (see Figure 19.3b) in open-loop pipe circuit(s) and then it is returned to the same well in totality or as a proportion if emergency bleed is applied (see Section 19.5). In other words, standing column wells combine supply and injection wells into one, pumping and recharging that occur simultaneously.

Several operational successes observed with standing column systems in the United States have generated a growing interest for these systems among the scientific and commercial communities in the last two decades. Financial simulations carried out over a 20-year life cycle showed that, despite higher maintenance costs, the use of standing column systems may result in overall costs between 27% and 51% lower than for conventional closed-loop systems. Beyond the financial gains associated to standing column systems, their real potential lies in their ability of being integrated to existing buildings located in dense urban areas where the land required for a wide, closed-loop ground-coupled heat exchanger is not available. Standing column wells can provide energy savings at lower first costs than conventional vertical ground-coupled heat exchangers while having a higher potential in dense urban areas. Unfortunately, operating these open wells with groundwater near the freezing point has limited so far their use in northern climates and studies illustrating their successful operation in heating mode are limited.

The operation of standing column wells modifies the temperature and pressure of the groundwater circulating in the wells and in the surrounding aquifers, and promotes its oxygenation through gaseous diffusion. Consequently, the chemical signature of the groundwater changes over time, which can produce undesirable effects such as scaling and growth of bacteria or algae in the system's components. Scaling problems are well known for open-loop systems and can also jeopardize operation of standing column systems. If concerns exist regarding the quality, groundwater can be transferred to an intermediate heat exchanger to prevent scaling and clogging within the geothermal heat pumps and ease maintenance. Many private users use their well groundwater both for standing column systems and for drinking. Since the operation of standing column wells could promote development of pathogen microorganisms in groundwater, such dual use of standing column systems should be permitted only after proving the local drinking water standards are met all year-round.

Iron bacteria is another possible nuisance of standing column systems that can clog the wells and thus affect their productivity. If groundwater has naturally low iron concentration, designers should be aware that using steel casing and steel pipes introduce sources of iron that can foster growth of iron bacteria and accumulation of ocherous precipitates. If needed, one can avoid the use of steel casing or cover the casing with appropriate paints to reduce bio-accessibility of iron.

19.2 STANDING COLUMN WELLS

Standing columns are open-loop, single wells generally drilled in hard rocks, creating columns of groundwater from the static groundwater table level down to the bottom of the borehole(s). The differences between different configurations of standing columns consist of well types (uncased or partially cased) parameters (depth and diameter), type of pipes (e.g., two open end or concentric), location of groundwater circulation pumps (bottom or top). In cold and moderate climates, typical standing column boreholes have at least 150 mm ID diameters, and depths ranging from 70 m to 450 m, mostly drilled in bedrock. When boreholes are deeper than 150 m, there is a substantial increase in thermal energy extraction available for building heating, and in heat rejection capacity for space cooling/dehumidification.

Standing column systems may supply groundwater from the well(s) to geothermal heat pump(s) through (Oliver and Braud, 1981; Yavuzturk and Chiasson 2002): (i) two open end tubes consisting of a dip tube extending, for example, to very near the bottom of the well, and a discharge tube from the geothermal heat pumps ending near the well's top (Figure 19.1) (mostly found in small-size residential and light commercial/institutional buildings), where the groundwater is recirculated from one end of the well to the other with centrifugal pumps; the placement of groundwater circulation pumps generally depend on the depth of standing column boreholes; they can also be located at the ground/soil surface, or inside the building; for relatively small-scale standing columns (e.g., 75–150 m deep), as those for residential applications, the circulation pumps are placed at the bottom of vertical wells, the groundwater entering the geothermal heat pump(s) is drawn from the bottom of vertical wells, while groundwater leaving the geothermal heat pump(s) is returned to the top of the same well; for large commercial/institutional applications, the submersible groundwater circulation pumps can be placed at the tops of boreholes by using lightweight plastic tubes (dip or tail pipes) inserted to the bottom of the borehole, a solution that uses shorter electrical cables and facilitates both installation and service works; (ii) concentric pipes consisting of a conduit to draw up groundwater and an annulus to return it downward (Figure 19.2) (mostly found in large-scale commercial/institutional buildings); in the case of concentric pipes, the groundwater circulation pump is installed at ground/soil level, and the suction tube at the bottom of the well; some thermal "short circuiting" could occur between the inner and outer flow channel, but this can be reduced with use of low thermal conductivity inner pipes.

As can be seen in Figure 19.1, the space inside the standing column wells can be divided into three zones): (i) low pressure (production), (ii) sealed section, and (iii) high pressure (injection).

To prevent the surface unconsolidated ground/soils falling down into the open hole, accommodate the installation of submersible pump(s), and to prevent surface water from draining down in the borehole, the overburden should be cased from the surface to the rock basement with steel or PVC pipes (see Figures 19.1 and 19.2). Standing column wells can be partially cased (at the top of the wells), totally cased (over the entire length of the wells), and even totally uncased.

Partially cased standing column wells consist of 7.5–15 m casings in steel or other material until competent bedrock is reached. Bedrock sealing requirements vary from a geographic region to another. The remaining depth of the well is then self-supporting through bedrock. As the borehole wall is porous, the groundwater is able to flow from the borehole uncased wall into and out of the ground/rock's porous formation. Boreholes drilled in unconsolidated materials, such as sand, gravel, clay, or

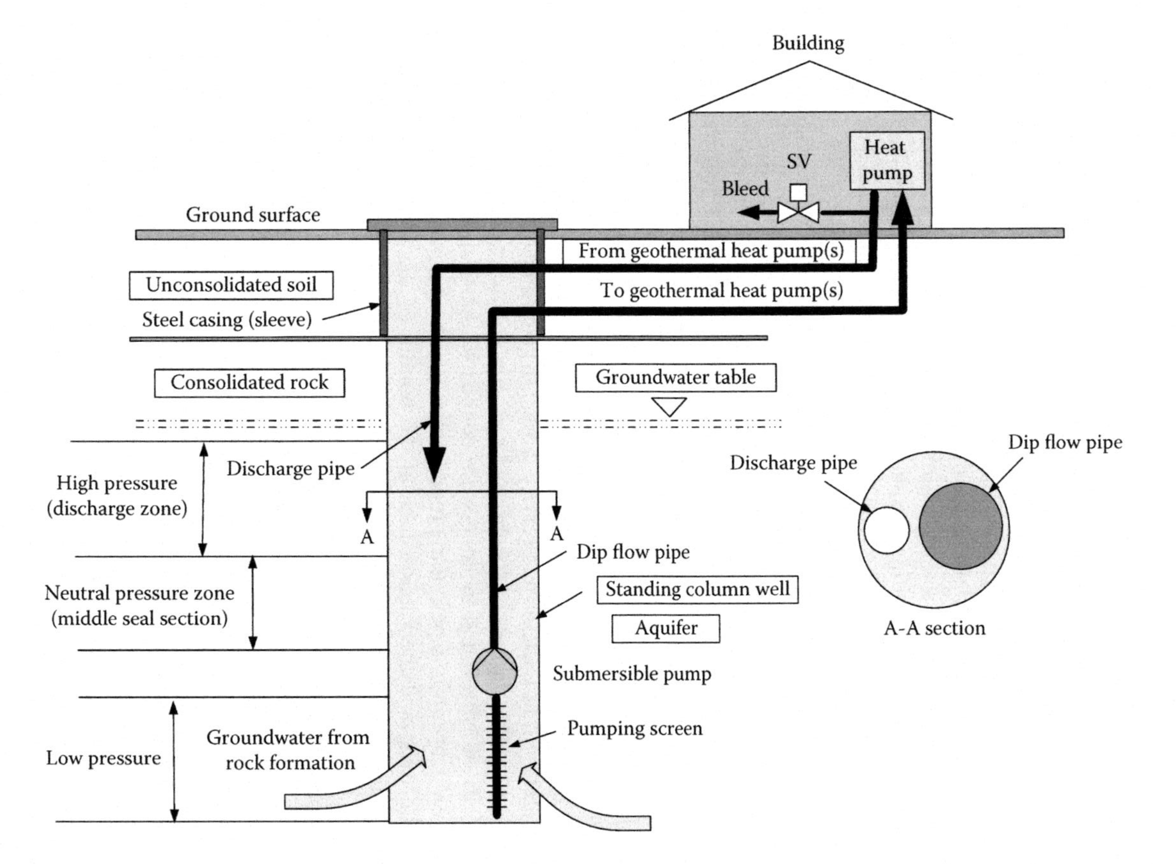

FIGURE 19.1 Simplified representation of a standing column two open-end pipes (Notes: schematic not to scale; not all components shown).

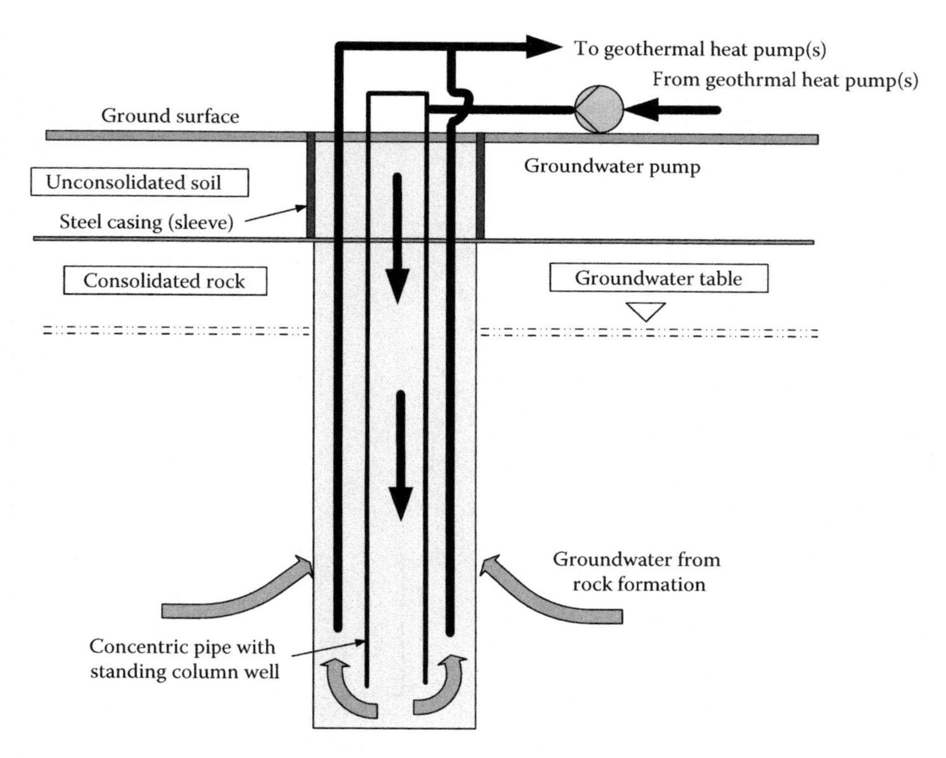

FIGURE 19.2 Simplified representation of a standing column with well concentric pipe (Notes: schematic not to scale; not all components shown).

soil, must be totally cased to stabilize the borehole due to the unstable nature of these geological formations. Totally uncased standing column boreholes, limited to geologic regions (such as those found in Sweden, Norway, and Canada) with competent rocks and good groundwater quality are drilled in hard rock formations in which groundwater coming from rock fractures infiltrates over the entire length of the borehole. The direct contact with the earth eliminates the thermal conductive resistances of plastic pipes and grout associated with conventional closed-loop (indirect, secondary fluid) U-tube heat exchangers. In addition, the rough uncased borehole walls induce turbulence producing higher heat transfer coefficients, and the substantial groundwater infiltration/exfiltration enhances the system overall energy performance.

19.3 BASIC CONCEPTS

Open-loop, single-well groundwater heat pump systems for residential and commercial/institutional ground-source heat pump systems generally consist of the following main components (see Figure 19.3a and 19.3b): (i) at least one standing column well; (ii) at least one geothermal heat pump with or without desuperheater for domestic hot water preheating; (iii) circulating submersible pump installed within a riser pipe screened or slotted at its base; submersible variable-speed pumps

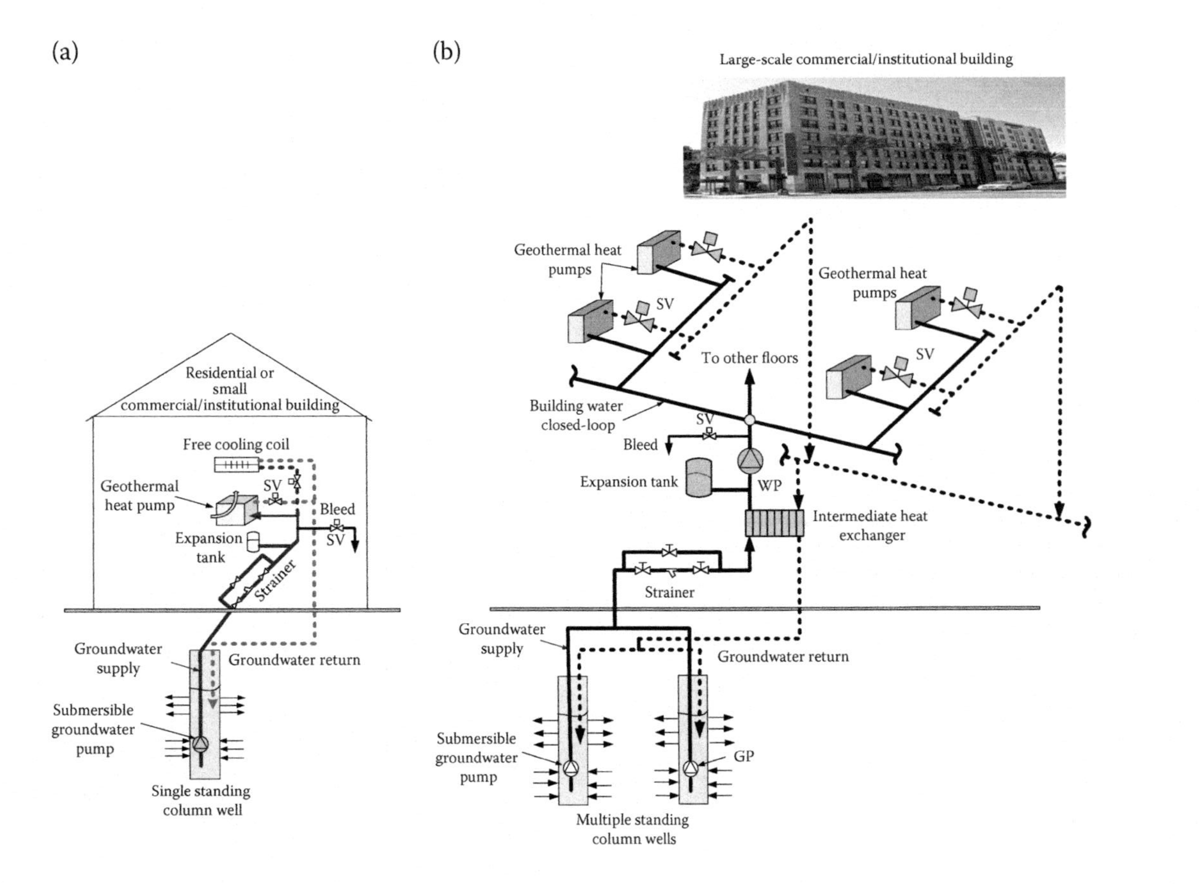

FIGURE 19.3 Schematic representation of open-loop single-well (standing column) groundwater heat pump systems with: (a) direct use; (b) indirect use of groundwater as heat and sink for geothermal heat pumps; GP, groundwater pump; WP, water pump; SV, solenoid valve (notes: schematics not to scale; not all components shown) (Building view retrieved from: https://www.nps.gov/tps/how-to-preserve/briefs/14-exterior-additions.htm, accessed March 9, 2020).

can be used to modulate the total pumping rate and reduce pumping energy when ground-source heat pump systems are operating at part load; (iv) an intermediate (separation) heat exchanger; and (v) internal building heating/cooling distribution systems that delivers space and domestic hot water heating, and cooling via, for example, air-handling units, underfloor heating pipes, or radiators.

Depending on groundwater quality and physical properties, it could flow from standing column wells through the geothermal heat pumps directly (Figure 19.3a) or indirectly (via intermediate heat exchanger) (Figure 19.3b).

The hydrological conditions in standing column wells vary in time and space and, also, depend on the mode of operation of the well (e.g., with or without bleed). Therefore, both groundwater flow and heat transfer must be simultaneously considered (Rees et al. 2004).

19.4 GROUNDWATER FLOW

The groundwater flow is a non-linear process that should be studied in both inside (where the groundwater is re-circulated from the bottom to the top of the column, or vice-versa) and outside (through the surrounding ground/rocks) regions of the standing column wells. The scope is to identify the effect of groundwater flow on the system energy performances evaluated in terms of groundwater minimum and maximum temperatures, and annual energy consumption and operating costs.

19.4.1 Flow Outside Standing Columns

Due to pumping, buoyancy, and, eventually, bleed of groundwater from the standing columns, a recirculating radial (horizontal) flow in the surrounding ground/rock formation with associated temperature and hydraulic gradients, is induced and superposed over the local natural groundwater movement. Because the standing column depths are much larger than their diameters, the vertical groundwater flow around standing columns may be neglected.

The immediate effect of groundwater flow around standing columns is the enhancement of the aquifer "effective" thermal conductivity defined as:

$$k_{eff} = n \cdot k_{gw} + (1 - n) \cdot k_{ground} \tag{19.1}$$

where

k_{eff} is the aquifer "effective" thermal conductivity (W/m·K).

n is the ground/soil/rock porosity (the ratio of the pore area to cross-sectional area) (–).

k_{gw} is the groundwater thermal conductivity (W/m·K).

k_{ground} is the ground/soil/rock thermal conductivity (W/m·K).

Outside standing column wells, the governing steady-state, one-dimensional equation for groundwater flow giving the specific discharge can be determined based on Darcy's law (that describes the relationship among the instantaneous rate of discharge through a porous medium and pressure drop at a given distance) (Chen and Jiao 1999):

$$\dot{V} = -(K \cdot A)/\mu \cdot \frac{\Delta p}{L} \tag{19.2}$$

where

$\dot{V}$ is the volumetric flow rate (m^3/s).
K is the ground/soil/rock permeability (m^2).
A is the cross-sectional area of ground/soil/rocks (m^2).
μ is the groundwater dynamic viscosity (Pa.s).
Δp is the pressure drop across the sample (-).
L is the sample length (m).

It can be seen from Equation 19.2 that the flow of groundwater around standing column wells is proportional to hydraulic conductivity and pressure gradient along the borehole (according to Darcy's law) in the same way that the heat flux is proportional to thermal conductivity and temperature gradient (according to Fourier's law).

The magnitude of induced groundwater flow around standing columns also depends on the relative resistance to flow along the borehole compared to the resistance to flow through the ground/soil/rocks. If the dip tube is arranged to draw groundwater from the bottom of the well, groundwater will be induced to flow into the ground/soil/rock in the top part of the borehole and will be drawn into the borehole lower down. At some distance down the borehole, there will be a balance point (i.e., with no net head pressure gradient) at which there will be no groundwater flow either into or out of the ground/soil/rock.

By applying the law of mass conservation to a control volume, and using Darcy's law (Equation 19.2), the governing equation defining the hydraulic head distribution in a porous medium as the ground/soil/rock can be derived as (Deng 2004):

$$S_z = \frac{\partial h}{\partial t} = \frac{\partial}{\partial x}\left(K \cdot \frac{\partial h}{\partial x}\right) + R \tag{19.3}$$

where

$S_z = \rho_w \cdot g \cdot (\alpha + n\beta)$ is the specific storage (the amount of water per unit volume of a saturated formation that is stored or expelled from storage (1/m) (Fetter 1994).
h is the hydraulic head (m).
t is the time (s).
K is the hydraulic conductivity (m/s).
R is the source/sink (1/s).
ρ_w is the density of groundwater (kg/m^3).
g is the acceleration of gravity (m/s^2).
α is the compressibility of aquifer ($1/(N/m^2)$.
n is the ground/soil/rock porosity (-).
β is the compressibility of water ($1/N/m^2$).

19.4.2 Flow Inside Standing Columns

The groundwater flow inside standing column wells is a non-Darcy laminar or turbulent flow (i.e., the relationship between the flow rate and the pressure gradient is non-linear). It can be described just by replacing K (for Darcy flow in the aquifer) with the "effective" hydraulic conductivity (K_{eff}) defining the hydraulic head distribution (Chen and Jiao 1999; Deng 2004):

For laminar flow in the standing column well:

$$K_{eff} = \frac{d^2 \cdot \rho \cdot g}{32\mu} \tag{19.4}$$

For transient and turbulent flow in the standing column well:

$$K_{eff} = \frac{2g \cdot d}{w_{gw} \cdot f} \tag{19.5}$$

where

K_{eff} is the "effective" ground/rock hydraulic conductivity (m/s).
d is the well diameter (m).
ρ is the groundwater density (kg/m^3).
g is the gravitational acceleration (m/s^2).
μ is the groundwater dynamic viscosity (Pa.s).
w_{gw} is the average groundwater velocity in the standing column well (m/s).
f is the friction factor (-).
For the dip tube walls inside the well, K = 0.

Because of rough surfaces of standing column wells, the following correlation for friction factor can be used for turbulent flow ($10^4 < Re < 10^7$) and all values of ϵ/r (Chen 1979):

$$1/\sqrt{f} = 3.48 - 1.7372 \cdot ln\left[\varepsilon/r - \frac{16.2426}{Re} lnA_2\right] \tag{19.6}$$

where

f is the Fanning friction factor (-).
ϵ is the height of the surface roughness (m).
r is the borehole radius (m).
Re is the Reynolds number ($Re = (D_{eq} \cdot w_{gw})/\nu$).
D_{eq} is the borehole equivalent hydraulic diameter (m).
w_{gw} is the groundwater velocity (m/s).
ν is the groundwater kinematic viscosity (m^2/s).

$$A_2 = (\varepsilon/r)^{1.1098}/6.0983 + (7.145/Re)^{0.8981} \tag{19.7}$$

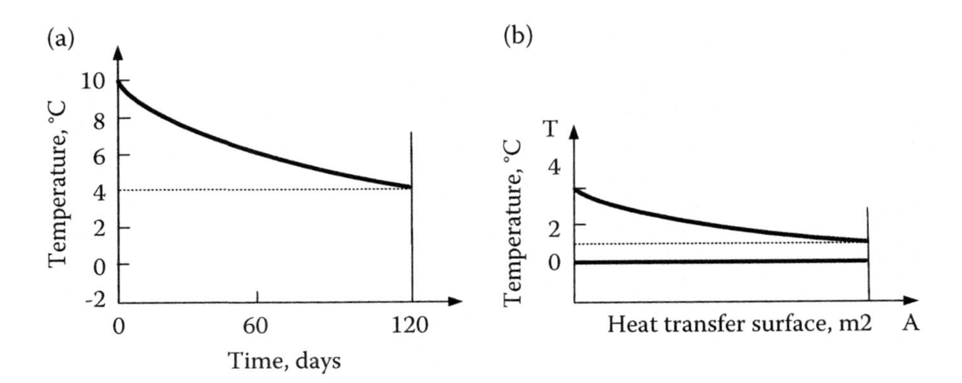

FIGURE 19.4 (a) Example of groundwater temperature variation inside a standing column operating in the heating mode without bleed; (b) T-A diagram of a geothermal heat pump's evaporator running in heating mode.

19.5 GROUNDWATER BLEEDING

Generally, during much of the year, most standing column systems may operate in heating and cooling modes by returning recirculating full groundwater flow rate to the wells after being cooled and heated, respectively, by the geothermal heat pump(s). To avoid freezing (in the heating mode) or overheating (in the cooling mode) of groundwater at the geothermal heat pumps' outlets, the temperature of the groundwater is usually controlled, for example, by an on/off control strategy. However, as can be seen in Figure 19.4a, if the initial temperature of groundwater is, for example, 10°C, after a certain period of operation in the heating mode (let's say, 120 days), the temperature of groundwater could drop up to 4°C. In such a case, groundwater may enter the evaporator(s) of geothermal heat pump(s) at temperatures as low as 4°C and leave at 1°C (or even less) (Figure 19.4b) presenting the danger of freezing if the refrigerant evaporating temperature attains, for example, –2°C or less.

This hypothesis suggests that standing columns without bleed are more feasible for the heating mode of operation in locations where the initial groundwater temperature are, let's say, higher than 12–13°C. In the absence of a bleed, conduction is the main heat transfer mechanism for standing columns. In that sense, increasing the borehole depth can have a significant, almost linear, effect on the performance of the system. When bleed is activated, however, heat advection becomes the primary driving force and the effect of well depth becomes significantly less important. A 100-m deep standing column operating with a maximum bleed ratio of 30% can present a performance similar to a 200-m deep standing column operating with a maximum bleed of 22.5%. Nonetheless, an increase of the well depth can significantly increase the volume of water contained in the standing columns, which can be beneficial to damp the peaks in fluid temperature. Similarly to the well depth, increasing the well diameter also increases the overall conductive heat transfer and borehole thermal capacity, resulting in better performances. However, increasing the drilling diameter from 150 mm to 200 mm has a limited influence on the energy consumption, especially for high bleed ratios. Note also that, depending on local drilling contractors, drilling costs can significantly increase with well diameter.

If, during peak heat extraction (winter) or rejection (summer) periods, the well groundwater temperature drops too low or climbs too high, respectively, the standing column wells may bleed, an operation that brings the groundwater temperature closer to the far-field ground/soil/rock temperature by discharging a certain percentage of the flow rate to another well, a watercourse, an infiltration ditch, or toward a surface water or a storm sewer, if permitted by local regulations, used for domestic water consumption or simply reinjected in the origin aquifer with a relatively small injection well rather than reinjecting the entire quantity of groundwater in the same well. Increasing the bleed flow rate helps stabilize the temperature at the ground-source heat pump inlets but increases the volume of water discharged outside the well. This is a simple control strategy to heat the standing columns during heat extraction (in the winter) and cools them during heat rejection (in the summer) by allowing warmer (or colder) groundwater getting into the well in heating (or cooling) mode. In other words, the scope of bleed operation is to induce an additional fresh groundwater flow inside the standing columns from/to the surrounding aquifer(s) and, in both heating and cooling modes, reestablish and stabilize (i.e., increase and decrease, respectively) the groundwater temperature within optimum limits in order to avoid groundwater freezing or overheating. When a bleed is activated, the induced groundwater flow instantly boosts the thermal efficiently of the standing column as advection becomes the dominant heat transport mechanism. The bleeding strategy is therefore a key feature for maintaining the heat pump's entering brine temperature within its operational limits, especially during heating and cooling peak load periods.

According to dead-band and the groundwater temperature difference, an intermittent control strategy could be applied by fixing the conditions at which bleed must start and stop by using solenoid valves, as shown in Figures 19.3a and 19.3b. These solenoid valves must open when the groundwater temperature is close to the freezing point, or is too high, exceeding the geothermal heat pumps' extended operating limits in heating and cooling modes, respectively. For example, in the winter, when the groundwater temperature leaving the standing column is lower than, let's say 5°C, a bleed is started to prevent groundwater freezing in the geothermal heat pump evaporator. When the groundwater temperature leaving the standing column well is higher, for example, than 10°C, a bleed is stopped. In summer, a bleed can be started when the groundwater temperature leaving the well is higher than, for example, 30–35°C, and stopped when the water temperature leaving the well is lower than 25°C.

In practice, the maximum amount of bleed flow rate achievable may be limited by well-pumping capacity and practical difficulties in disposing of the bleed groundwater. The bleed flow rate is typically between 5% and 25% of the total pumping rate and can be modulated to reduce the volume of discharged water. This means that, after passing through the geothermal heat pumps, approximately 85–90% of the groundwater is returned to the well.

19.6 HEAT TRANSFER AROUND AND INSIDE STANDING COLUMNS

The complex, non-linear heat transfer process consisting of heat extraction (in the winter, when the groundwater inside standing column is cooled), and heat rejection (in the summer, when the groundwater is heated), mostly affected by the transient building thermal loads, should be considered in both inside (where the groundwater is recirculated from the bottom to the top of the column, or vice versa), and outside (through the porous surrounding ground/soil/rock formations).

The heat transfer analysis aims to identify the most significant design parameters and their effect on standing column energy performances, usually expressed in terms of required well depth, minimum and maximum groundwater entering and leaving temperatures in/from the geothermal heat pumps, optimum bleed rates, and the system annual energy consumption and costs (Ni et al. 2007).

Most of two-dimensional (to analyze the coupled radial thermal and hydraulic energy transfer around the standing column wells) and three-dimensional models (to study the heat transfer both inside and outside of standing column wells with and without groundwater bleed) are based on assumptions, as the following (Rees 2001; Rees et al. 2004): (i) constant temperature of undisturbed ground/soil/rocks formations; (ii) homogenous (i.e., same permeability from point to point), isotropic (i.e., same permeability and thermal conductivity in all directions) and porous aquifers; (iii) groundwater flow independent of ground/soil/rocks average density; (iv) groundwater flow and energy transfer in the aquifer in the vertical direction are ignored; (v) only the local flows caused by standing column pumping are considered; (vi) regional/local groundwater flow (such as pumping or recharge from local rivers) and seasonal groundwater table movement are ignored; (vii) groundwater flow in the lateral direction (due to head gradients induced by adjacent rivers, local pumping, and changes in topology and geology on a larger scale) are usually not considered; consequently, in this case, it is assumed that the groundwater flow and heat transfer are symmetrical about the centerline of the borehole; (viii) all physical parameters of the aquifer are independent of time, location, pressure, and temperature; (ix) temperature of groundwater is constant inside the standing column supply and return pipes; (x) temperature difference between the groundwater flowing in the annular area and the earth is the driving force for the heat transfer processes; (xi) temperature difference in the two pipes is the driving force for the crossover heat transfer; (xii) groundwater-rock thermal exchange has an insignificant effect on the total process of heat transfer, and can be neglected because of the short duration of the process; therefore, the thermodynamic equilibrium is considered completely instantaneously, with the same temperature of aquifer matrix and groundwater around matrix; (xiii) in the aquifer, the effect of natural vertical convection due to the temperature difference between the cold and the warm groundwater is ignored; (xiv) the aquifer thickness is so thin that vertical temperature gradients can be neglected; (xv) the change of air temperature has no effect on the groundwater temperature; (xvi) the aquifer and its cap-rock and bedrock remain at thermal equilibrium (i.e., the temperatures are same between the surface of aquifer and the bedrock) before the operation of a geothermal heat pump; (xvii)

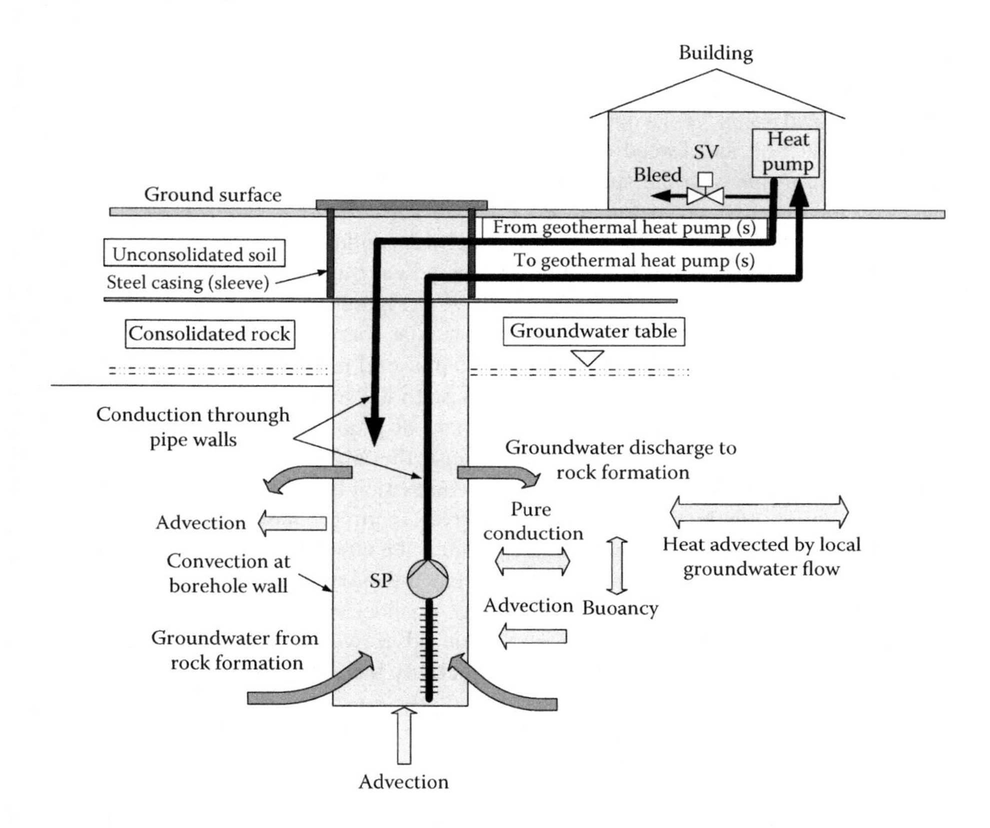

FIGURE 19.5 Main heat transfer mechanisms around and inside a standing column well; SP, submersible groundwater pump (Notes: schematic not to scale; not all components shown).

the specific volume heat and thermal conductivity of the same aquifer are assumed to be uniform and constant in time; and (xviii) in aquitards of leaky aquifers, the forced thermal convection and thermal dispersion are ignored, and only the thermal conduction is considered due to low velocity of groundwater. Figure 19.5 shows the main heat transfer mechanisms inside and around a typical standing column well (Rees 2001; Rees et al. 2004).

19.6.1 Heat Transfer Around Standing Column Wells

The heat transfer in Earth is a result of the interaction of energy transfer mechanisms, primarily near the ground/soil surface, and of dynamic thermal properties of the ground/soil/rock body as mass and volumetric specific heat and thermal conductivity, and the ratio of thermal conductivity to thermal diffusivity.

Open-loop, one-well (standing column) groundwater heat pump systems, where the groundwater is pumped and injected in the same well, induce changes in groundwater flow and temperature of surrounding formations that consist of a solid phase and interconnected void spaces totally filled with groundwater. The groundwater flow occurs in the interconnected voids and the heat transfer through both the solid matrix and voids form a coupled, more complex, heat transfer process consisting of pure conduction and forced convection, and advection/diffusion (that introduces further heat transfer with the surroundings). These processes are significantly influenced by the thermal and hydraulic properties of surrounding geological formations (as minimum and maximum temperatures, thermal conductivity, and specific heat). Around the standing column wells, the following heat transfer processes may occur simultaneously (Domenico and Schwartz 1990) (Figure 19.5): (i) pure heat conduction (i.e., the energy transport due to molecular thermal diffusion) through an aquifer (the fluid phase) and bedrocks (the solid phase) if not affected by groundwater flow; forced heat convection from the fluid phase to the solid phase, and through the fluid phase (advection), caused by the movement of groundwater with different velocities and temperatures gradients; (iii) because the advection augments the conductive heat transfer, the combined advection-conduction heat transfer process can be treated as a pure conduction by using "effective" thermal conductivity values; (iv) however, when a bleed occurs at higher flow rates, the advection process can become significant, and it is no longer feasible to treat the process as a pure conduction process with an increased "effective" thermal conductivity; (v) increased thermal conductivity of ground/soil/rocks increases the advective heat transfer around the borehole; (vi) although higher hydraulic conductivity increases the groundwater flow rate to and from the borehole into and out of the surrounding ground/rocks, the groundwater flow along (up and down) the borehole is correspondingly reduced; this, in turn, reduces the convective heat transfer at the borehole wall; hence, there is some trade-off between increased advective heat transfer and reduced convective heat transfer; (vii) heat dispersion, a process present at relatively high velocities in the aquifer, depends on the properties of a solid and fluid matrix, as well as on the structure of the aquifer pores and channels; thermal dispersion results in a huge intensification of heat transfer, especially near the standing column walls.

19.6.2 Heat Transfer Inside Standing Column Wells

Inside the standing column wells, the groundwater velocity near the wall is high enough to produce turbulent flow. The complex heat transfer process in the radial direction (enhanced by groundwater pumping and thermal short-circuiting due to temperature differences between the groundwater flowing up in the dip tube and down in the annulus) consists of (Figure 19.5) (Deng 2004; Rees et al. 2004): (i) forced convection along the surfaces of the dip tube and at the borehole wall and casing in the direction of the borehole radius; (ii) conduction through groundwater and advection induced by the groundwater flow at the borehole well surface in the vertical direction; advection is beneficial for standing column energy performances, especially in systems with bleed whether the groundwater is withdrawn from the top or the bottom of the well; without bleed, the heat transfer by advection is negligible, and forced convection at the borehole wall and casing becomes the dominant heat transfer mode.

The energy balance can be formulated at each annular horizontal plane in any standing column well as follows (Rees et al. 2004):

$$\dot{Q} = \dot{Q}_{conv,\ i} + \dot{Q}_{adv,\ j} \tag{19.8}$$

where

$\dot{Q}$ is the total heat transfer rate (W).

$$\dot{Q}_{conv,i} = \frac{1}{R_i}(T_i - T_{annular}) \tag{19.9}$$

where

$\dot{Q}_{conv,\ i}$ are the convective heat transfer rates in suction tube, discharge tube, and ground/rock, respectively (W).

i is an index referring to suction tube, discharge tube, and ground/rock (-).

R_i is the thermal resistance of suction tube, discharge tube, and ground/rock, respectively (K/W).

T_i is the temperature of suction tube, discharge tube, and ground/rock, respectively (°C).

$T_{annular}$ is the groundwater temperature in the annular region (°C).

$$\dot{Q}_{adv,j} = \dot{m}_{gw} c_p (T_j - T_{annulus}) \tag{19.10}$$

where

$\dot{Q}_{adv,\ j}$ are the advection heat transfer rates in ground/rock and annulus, respectively (W).

j is an index referring to the rock and annulus groundwater (-).

$$\dot{Q}_{adv,\ rock} = \dot{m}_{gw} c_p (T_{rock} - T_{annulus}) \tag{19.11}$$

$$\dot{Q}_{adv,\ annulus} = \dot{m}_{annulus} c_p (T_{annulus} - T_{tube}) \tag{19.12}$$

where

$\dot{m}_{gw}$ is the groundwater mass flow rate through the rock (kg/s).

c_p is the specific heat of groundwater (J/kg·K).

T_{rock} is the rock average temperature (°C).

$T_{annulus}$ is the groundwater temperature in the annular region (°C).

$\dot{m}_{annulus}$ is the groundwater mass flow rate through the annulus (kg/s).

$T_{tube}$is the groundwater temperature in the tube (discharge tube or suction tube) (°C).

The borehole thermal resistance based on the well inside area is:

$$R_{borehole} = \frac{1}{A_{in}}\left[\frac{1}{h_{in,\ wall}} + \frac{d_{in}}{2 \cdot k_{pipe}} \ln\left(\frac{d_{in}}{d_{Out}}\right) + \frac{d_{in}}{d_{out}}\left(\frac{1}{h_{out}}\right)\right] \tag{19.13}$$

where

$R_{borehole}$ is the borehole thermal resistance (m^{2}K/W).

A_{in} is the borehole (standing column) inner heat transfer area (m^2).

$h_{in,wall}$ is the equivalent tube interior convection heat transfer coefficient (W/m^{2}K) that depends on the Reynolds number:

$$Re_{d_{eq}} = w_{gw} \cdot d_{eq} / \nu_{gw} \tag{19.14}$$

where

w_{gw} is the groundwater velocity (m/s).

d_{eq} is the borehole hydraulic equivalent diameter (m).

ν_{gw} is the groundwater cinematic viscosity (m^2/s).

The interior convection heat transfer coefficient can be calculated using the Gnielinski correlation for the Nusselt number pertaining to smooth tubes:

$$Nu_{d_{eq}} = h_{in,\ wall} \cdot d_{eq} / k_{gw} = \frac{(f/8)(Re_{d_{eq}} - 1000)Pr_{gw}}{1 + 12.7(f/8)^{1/2}(Pr_{gw}^{2/3} -)} \tag{19.15}$$

where

$Nu_{d_{eq}}$ is the Nusselt number (-).

$h_{in,wall}$ is the interior convection heat transfer coefficient (W/m^{2}K).

d_{eq} is the the borehole hydraulic equivalent diameter (m).

k_{gw} is the groundwater thermal conductivity (W/mK).

$Re_{d_{eq}}$ is the Reynolds number (-).

$Pr_{gw} = \nu_{gw} / \alpha_{gw}$ is the groundwater Prandtl number (-).

ν_{gw} is the groundwater kinematic viscosity (m^2/s).

α_{gw} is the groundwater thermal diffusivity (m^2/s).

f is the standing column well wall friction factor (-); for smooth surfaces, f depends on the Reynolds number and can be approximated by the following equation:

$$f = 0.31 Re_{d_{eq}}^{-0.25} \tag{19.16}$$

Because the friction factor increases with the surface roughness of uncased wells, it was considered approximately 1.5 times higher than that of a smooth surface (0.0564). For turbulent flow, the wall heat transfer coefficient increases with wall roughness. As a first approximation, it may be computed using Equation 19.16 with friction factors obtained from the Moody diagram. However, although the general trend is one of increasing $h_{in,\ wall}$ with increasing f, the increase in f is proportionately larger, and when f is approximately four times larger than the corresponding value for smooth surfaces, $h_{in,\ wall}$ no longer changes with additional increases in f (Bergles et al. 1983).

d_{in} is the tube inner diameter (m)
k_{pipe} is the equivalent tube thermal conductivity (W/mK)
d_{out} is the equivalent tube outer diameter (m)
h_{out} is the equivalent tube exterior convection heat transfer coefficient (W/m^{2}K)

REFERENCES

Bergles, A.E., V. Nirmalan, G. H. Junkhan, R.L. Webb. 1983. *Bibliography on Augmentation of Convective Heat and Mass Transfer – II. American Society of Mechanical Engineers (ASME).* Iowa State University, Engineering Research Institute, Heat Transfer Laboratory Ames, Iowa, December.

Chen, C., J.J. Jiao. 1999. Numerical simulation of pumping tests in multiplayer wells with non-Darcy-an flow in the well bore. *Groundwater* 37(3):465–474.

Chen, N.H. 1979. An explicit equation for friction factor on pipe. *Industrial and Engineering Chemistry Fundamentals* 18:296–297.

Deng, Z. 2004. *Modelling of Standing Column Wells in Ground Source Heat Pump. Submitted to the Faculty of the Graduate College of the Oklahoma State University in partial fulfillment of the requirements for Ph.D. degree*, Stillwater, OK, US, December.

Deng, Z., S.J. Rees, J.D. Spitler. 2005. A model for annual simulation of standing column well ground heat exchangers. *HVAC&R Research* 11(4):637–655.

Domenico, P.A., F.W. Schwartz. 1990. *Physical and Chemical Hydrogeology*. John Wiley & Sons Inc., New York.

Fetter, C.W. 1994. *Applied Hydrogeology*, 3rd Edition. Macmillan, New York.

Ng, B.M., C. Underwood, S. Walker. 2009. Numerical modeling of multiple standing column wells for heating and cooling buildings. In *Proceedings of the 11th International Building Performance Simulation Association (IBPSA) Conference*, Glasgow, Scotland, July 27–30.

Ni, L., Y. Jiang, Y. Yao, Z. Ma. 2007. Thermal modeling of groundwater heat pump with pumping and recharging in the same well. In *Proceedings of the 22nd International Congress of Refrigeration*, Beijing, China, 23–26 August.

Oliver, J., H. Braud. 1981. Thermal exchange to earth with concentric well pipes. *ASAE (American Society of Association Executives) Transactions* 24(4):906–910.

Orio, C.D. 1994. Geothermal heat pumps and standing column wells. *Geothermal Resources Council Transactions* 18:375–379.

Orio, C.D., A. Chiasson, C.N. Johnson, Z. Deng, J.D. Spitler, S.J. Rees. 2005. A survey of standing column well installations in North America. *ASHRAE Transactions* 111(2):109–121.

Rees, S.J. 2001. Advances in modeling of standing column wells. In *Proceedings of International Ground Source Heat Pump Association Technical Conference & Expo*, Stillwater, OK, US, July 14–17.

Rees, S.J., C.D. Orio, J.D. Spitler, Z. Deng, C.N. Johnson. 2004. A study of geothermal heat pump and standing column well performance. *ASHRAE Transactions* 109(1):1–14.

Yavuzturk, C., A.D. Chiasson. 2002. Performance analysis of U-tube, concentric tube, and standing column well ground heat exchangers using a system simulation approach. *ASHRAE Transactions* 108(1):925–938.

20 Surface Water Ground-Source Heat Pump Systems

20.1 INTRODUCTION

Surface water ground-source heat pump systems use very low-temperature water from nearby stationary (e.g., lakes, ponds, or seas, all referred here as lakes) or moving surface water (e.g., rivers with reliable flow rates and modest currents), as heat/sinks sources for one or more reversible geothermal heat pumps providing heating and cooling to residential as well as to light and large-scale commercial/institutional/industrial buildings located in cold and moderate climates.

In such systems, the conventional ground-coupled heat exchangers and groundwater boreholes are replaced by multiple parallel-circuited heat exchangers submerged in the body of surface water and connected directly or indirectly to geothermal heat pumps located inside the buildings.

Because, unlike the ambient air temperatures, the temperatures at the bottom of deep surface water bodies are warmer in winter and colder in summer, they provide valuable heat/sink sources to geothermal heat pumps able to achieve higher thermal efficiency than air-source heat pumps.

Successful application of surface water ground-source heat pump systems mainly depends on: (i) surface water type, size, and affordability; (ii) building location, type, size, and height; and (iii) national and regional regulatory requirements concerning the surface water usage and environmental issues.

Surface water ground-source heat pump systems utilize lakes, ponds, rivers, seas, oceans, etc. (all referred here as lakes) to provide heat sources and heat sinks for geothermal heat pumps in order to heat and cool/dehumidify residential and commercial/institutional buildings. In winter, the lakes are warmer than the ambient air, mainly because of the thermal inertia of water from the previous summer and the added solar energy. Therefore, the surface water geothermal heat pump systems could be more efficient than conventional air-source heat pumps in (winter) heating mode of operation. In addition, in cold and moderate climates, the lakes may provide free (direct) cooling to buildings' spaces if, at depths where the water could be withdrawn, their temperatures are not higher than 8–10°C in summer. In such systems, if large volumes of lakes are available, the water of surface waters passes once through air cooling/dehumidifying coils prior to returning to the lakes or downstream of the abstraction points in rivers with temperature increases not more than 3–4°C.

DOI: 10.1201/9781003032540-20

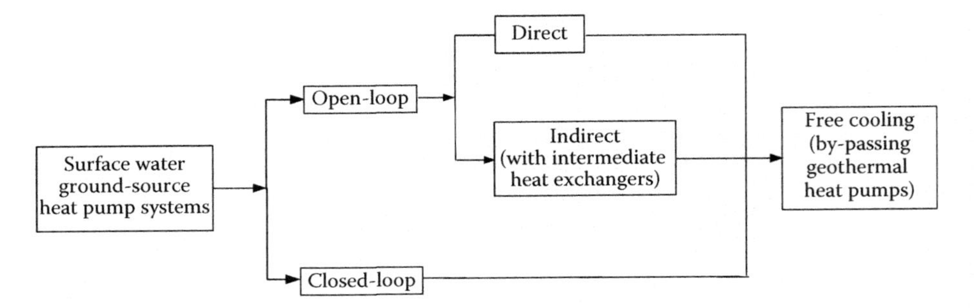

FIGURE 20.1 Classification of surface water ground-source heat pump systems.

20.2 BASIC CONCEPTS

Depending on whether the water is directly pumped from the lake body or indirectly, via submerged heat exchangers, the surface water ground-source heat pump systems can be classified as open loop (similar to groundwater heat pump systems) and closed loop (similar to ground-coupled heat pump systems) (Figure 20.1).

Residential as well as light- and large-scale commercial/institutional surface water ground-source heat pump systems generally consist of the following main components (see Figures 20.2a, 20.2b, 20.3, and 20.4): (i) a surface water source (lake, pond, river, sea); (ii) in the case of closed loops, submerged heat exchangers; (iii) for both open- and closed-loop systems, water or brine circulating pumps; (iv) at least one water (or brine)-to-air or water (or brine)-to-water reversible geothermal heat pump with or without a desuperheater for domestic hot water preheating; (v) an internal building heating/cooling distribution system that delivers space heating and cooling via, for example, air-handling units, underfloor heating pipes and ceiling cooling panels, or radiators.

20.2.1 Open-Loop Systems

In open-loop once-through surface water ground-source heat pump systems, the surface water can be directly supplied to geothermal heat pumps (Figure 20.2a) or indirectly, via intermediate heat exchangers (Figure 20.2b). Water at temperatures above of at least 4°C (to prevent freezing in the heating mode) is withdrawn from a natural body of water (lake, pond, sea, or river) through screened intakes at adequate depths to prevent fish or any debris from being entrained, pumped through geothermal heat pump(s), or isolation heat exchanger(s) for building heating or cooling, and returned (discharged) generally to the same surface water body at some distance away from the intake. The groundwater circulation pumps must be located at or below the surface water body level in order to achieve net positive suction head requirements and thus avoid cavitation that will occur at the circulation pump inlet (Figures 20.2a and 20.2b).

An advantage of open-loop systems consists of overcoming surface water heat exchanger thermal resistance, which can allow achieving better systems' coefficients of performance than those of closed-loop systems.

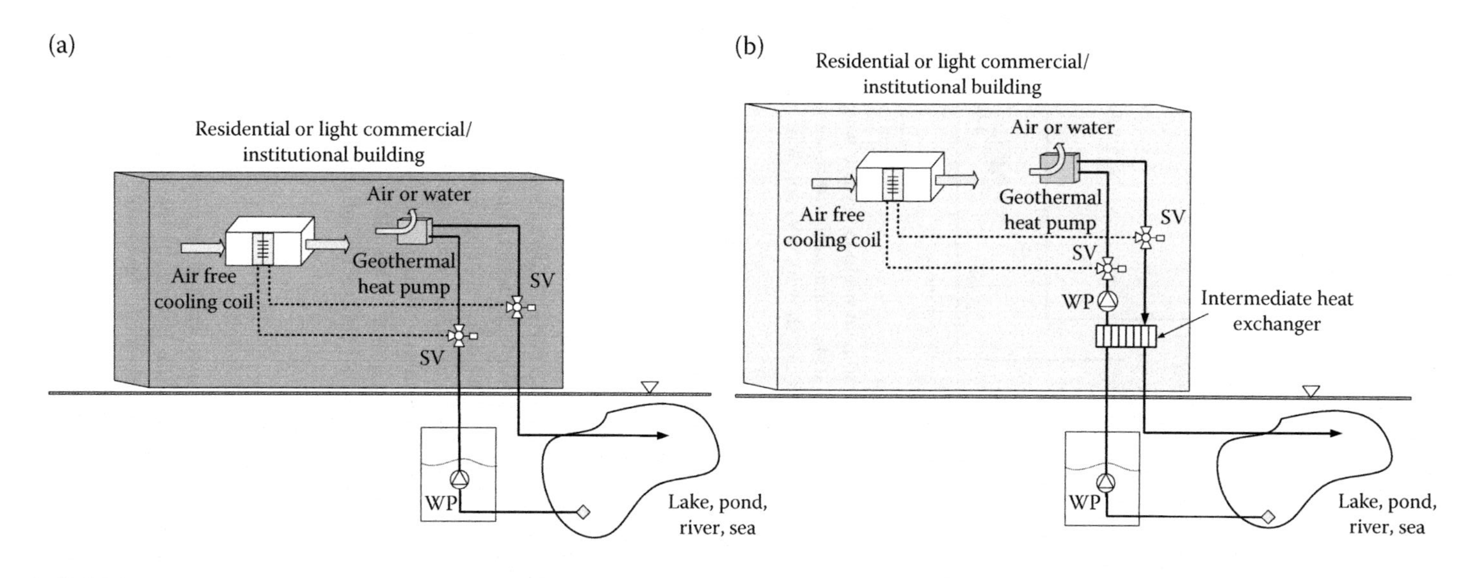

FIGURE 20.2 Schematics of open-loop, once-through surface water ground-source heat pump systems with free (direct) air cooling capability; (a) direct; (b) indirect (with intermediate heat exchanger) (Notes: schematics not to scale; not all components are shown); WP, water circulating pump; SV, solenoid valve.

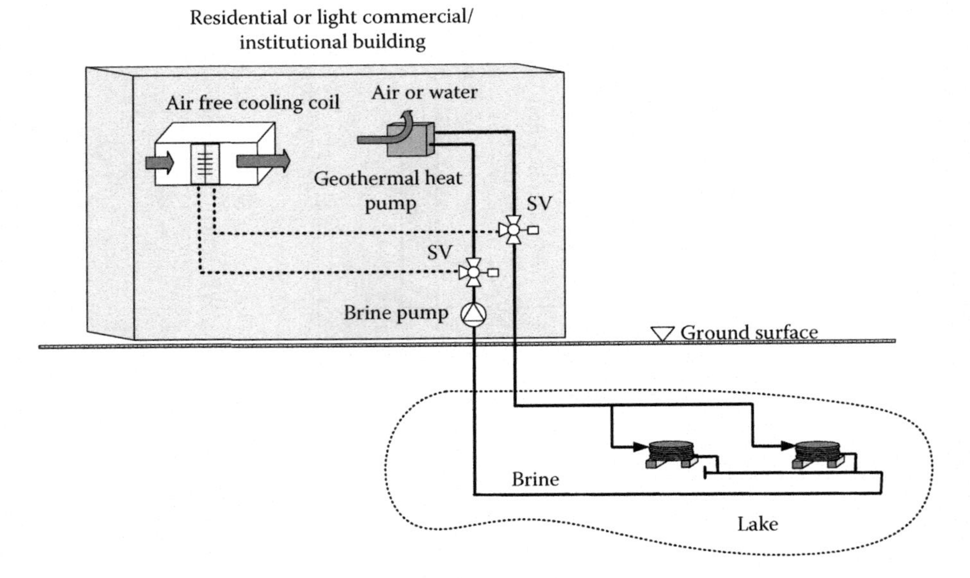

FIGURE 20.3 Schematic representation of a closed-loop surface water ground-source heat pump system for residential and small-scale commercial/institutional buildings (Notes: schematic not to scale; not all components are shown); SV, solenoid valve.

In deep surface bodies (12 m or more) with enough thermal stratification and water temperature below 10°C throughout the year, it is possible to directly (sensibly) cool and even dehumidify the building indoor air in order to substantially reduce the energy consumption. In this case, the water, pumped from the bottom of deep lakes, bypasses the geothermal heat pump(s) and then passes through air cooling/dehumidifying heat exchangers placed in the return air duct(s) of building HVAC system (see Figures 20.2a and 20.2b). If the entering temperature of surface water is greater than what is required to meet the sensible or latent cooling load, the geothermal heat pump can make up the difference, and allow the building thermal loads to be met. The most prominent advantage of such free (direct) cooling systems, where the primary energy consumers are the water circulation pumps, is that no mechanical cooling equipment is required. This allows these systems to operate at exceptionally high values of system coefficients of performance (e.g., as high as 25).

20.2.2 Closed-Loop Systems

Closed-loop surface water ground-source heat pump systems consist of heat exchangers submerged and anchored at the bottom of the body of water, and piping networks connecting these heat exchangers to reversible brine-to-air and/or brine-to-water geothermal heat pump(s) that transfer(s) heat to (in the heating mode) or from (in the cooling mode) the building from/to the lake by circulating a thermal carrier fluid (usually, a water/antifreeze mixture) (see Figures 20.3 and 20.4).

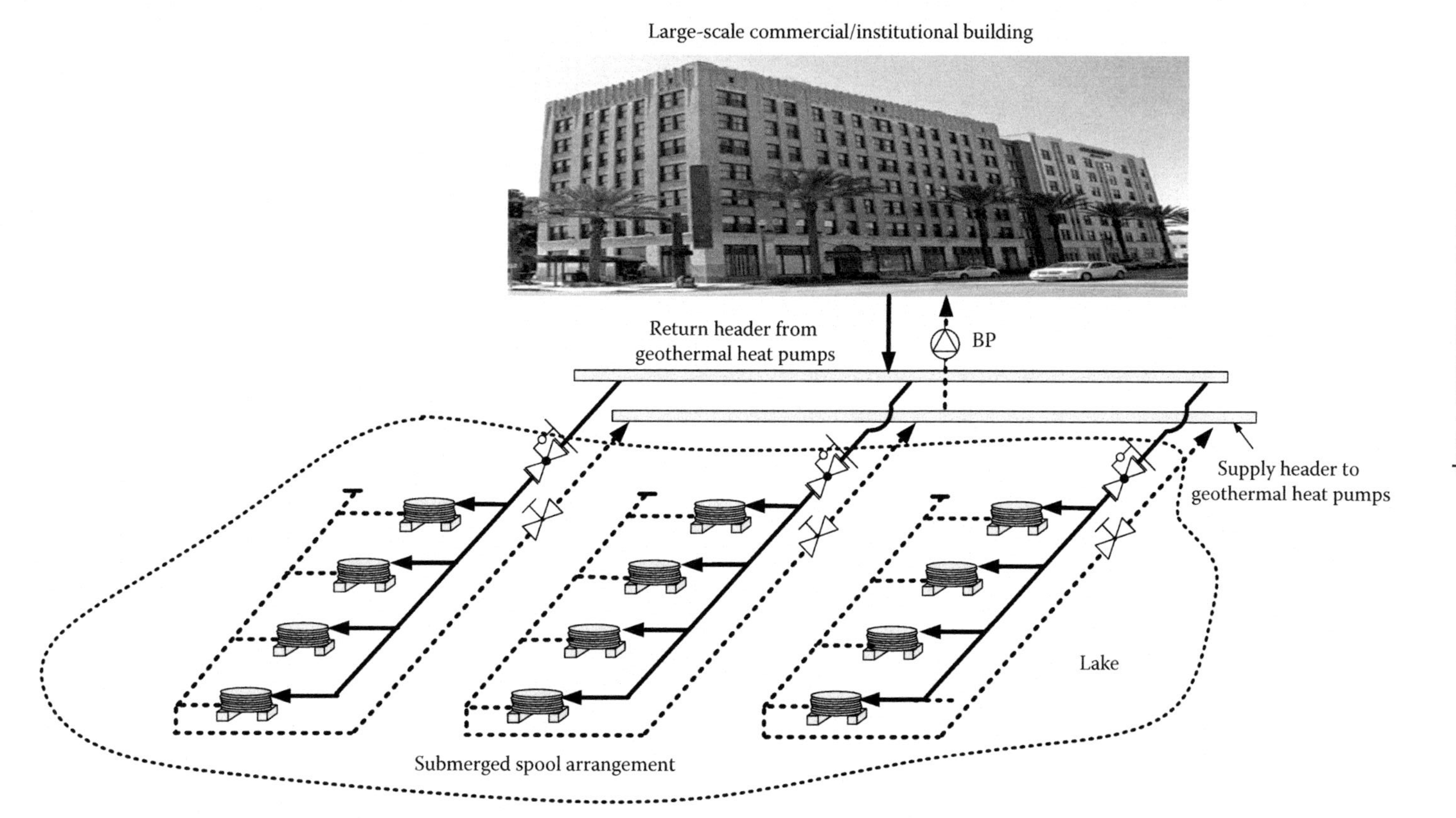

FIGURE 20.4 Schematic representation of an indirect (closed-loop) surface water ground-source heat pump system for large-scale commercial/institutional buildings (Building view retrieved from: https://www.nps.gov/tps/how-to-preserve/briefs/14-exterior-additions.htm, accessed March 9, 2020) (Notes: schematic not to scale; not all components are shown); BP, brine circulation pump.

For residential and light commercial/institutional as well as for large-scale commercial/institutional buildings, the most commonly used surface water heat exchangers are of spool (slinky) type made of high-density polyethylene pipes protected against ultraviolet radiation and positioned at adequate depths in surface water bodies (usually, at 3 m or more below the surface). Among several factors that must be considered when selecting an antifreeze fluid (e.g., propylene glycol) are: (i) corrosivity, (ii) health and fire risks, and (iii) environmental risks from spills or disposal.

The advantages of closed-loop surface water geothermal heat pump systems can be mentioned: (i) may use as heat/sink surface water bodies with poor water quality; (ii) could be operated in locations where the elevation difference, and, thus, static pressure head between the source water and the geothermal heat pumps is significant; in other words, greater elevation differences between the geothermal heat pumps and surface water level are possible; (iii) may operate with surface water temperatures at or below the freezing point of pure water because antifreeze solutions are used as heat transfer fluids.

20.3 MOVING AND STATIONARY SURFACE WATERS

Moving lowland rivers are mainly fed by rain and by melting of snow and ice especially in regions with high pluvial regimes. The flow rate responds quickly to precipitations and is, therefore, high during rainy periods and low during dry periods (typically in the fall). The feeding of rivers by groundwater acts as a buffer, as groundwater can store rain for much longer that rivers. Highland rivers are often primarily fed by the melting of snow and ice, the flow rates being low in winter and increase during the spring and summer.

Typical stationary surface waters (referred here as lakes) are the following: (i) mictic (covered with ice all year where the mixing pattern is typical of lakes in the Antarctic); (ii) cold monomictic (water never gets warmer than 4°C and turnover occurs once in summer, such as the lakes in the Arctic); (iii) dimictic (lakes stratified in summer and winter, and the water turns over once during the spring and once the fall); (iv) warm monomictic (water cools to near 4°C in winter, when turnover may occur); these lakes are stratified during other periods of the year and are not ice covered; (v) oligomictic (water is generally warm and stratified); and (vii) polymictic (frequent mixing throughout the year depending on lake characteristics as depth, size, and shape).

The thermal characteristics of shallow and deep surface water bodies, which could potentially be heat and sink sources for ground-source heat pump systems, are quite different from those of ground/soils and rocks, and can vary along the depth and over the various weather seasons.

20.3.1 THERMODYNAMIC PROPERTIES

At an average atmospheric pressure (101,325 kPa) and up to 100°C (the boiling point), water exists as a liquid. The water mass density, a non-linear function of temperature and an intensive thermodynamic bulk property, is defined as follows:

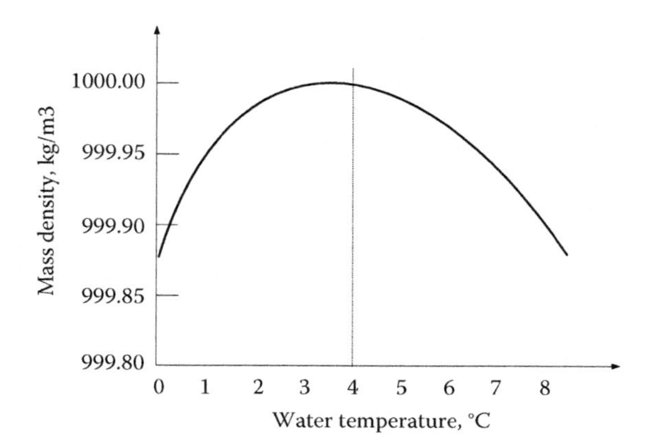

FIGURE 20.5 Water mass density at temperatures around 4°C.

$$\rho_{water} = m/V \tag{20.1}$$

where

ρ_{water} is the water mass density (kg/m^3).
m is the water mass (kg).
V is the water volume (m^3).

The water mass density increases until approximately 4°C and decreases at temperatures above 4°C (Figure 20.5). This means that the maximum density of water occurs at 4°C, not at the freezing point of 0°C. This unique temperature/density relationship of water is the main reason for surface water temperature stratification (see Section 20.2.2).

The enthalpy (latent heat) of fusion (melting) is the change in water enthalpy resulting from providing heat to 1 kg of water to change its state from solid to liquid at constant pressure. For example, when melting 1 kg of ice at 0°C and atmospheric pressure assumed by convention to be 101,325 kPa, 333.55 kJ of heat is absorbed with no temperature change.

The increase in entropy when water is melting (22.0 kJ/kmol·K) is a positive value since the degree of disorder increases in the transition from the organized crystalline ice to the disorganized structure of liquid, is defined as the increase of enthalpy of fusion divided by the (constant) temperature of fusion (melting) point:

$$\Delta s_{fusion} = \Delta h_{fusion}/T_{fusion} \tag{20.2}$$

where

Δs_{fusion} is the entropy increase during fusion (kJ/kg·K).
Δh_{fusion} is the increase of enthalpy of fusion (kJ/kg).
T_{fusion} is the temperature of fusion (K).

The enthalpy (latent heat) of freezing (solidification) is the heat liberated by a unit mass of water at its freezing point (0°C at 101,325 kPa). The temperature at which the water phase transition occurs is named the melting point (for fusion) or the freezing point (for solidification).

Water specific heat (J/kg·K), an extensive thermodynamic property dependent upon the amount of water mass, is defined as the amount of energy required to change the temperature of 1 kg of water by 1 K. Water heat capacity (or thermal capacity) (J/K) is also an extensive physical property defined as the amount of heat to be supplied to a given water mass to produce a temperature change of 1 K.

Water dynamic (i.e., the frictional force that arises between adjacent layers of water that are in relative motion) and kinematic (i.e., absolute viscosity divided by its mass density) viscosity are two physical properties of interest for surface water.

20.3.2 Thermal Stratification, Mixing, and Turnover

Because of the seasonal variation of incoming solar radiation at the lake contact with the atmosphere, in most climates, the average temperature of lakes (as can be seen in Figure 20.2 at 3 m below a lake surface situated in a typical moderate climate), varies significantly during the year, dropping significantly in the winter (Figure 20.6).

Because of incoming solar energy, convection, evaporation, and radiation thermal processes, and the movement of the wind (that takes place at the lake surface), vertical thermal stratification of lakes occurs both in summer and winter inducing large differences in mass density between warm and cold water layers at different depths. When the surface water is cooled until reaches its highest density at 4°C, it sinks down to the depth and reaches the bottom of the lake at temperatures higher than 4°C. On the other hand, if the bottom water is heated, it has a tendency to rise due to lower density.

When deep lakes stratify because of the seasonal temperature variation across the depth, three different layers typically form (Lewis 1983; Wilhelm and Adrian 2007; https://en.wikipedia.org/wiki/Lake_stratification, accessed August 11, 2019)

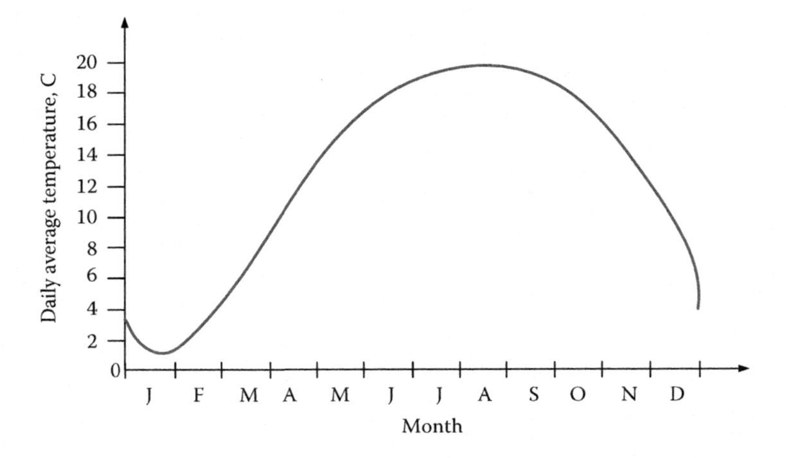

FIGURE 20.6 Schematic of daily average temperatures at 3 m below surface of a lake located in a typical moderate climate (Note: graph not to scale).

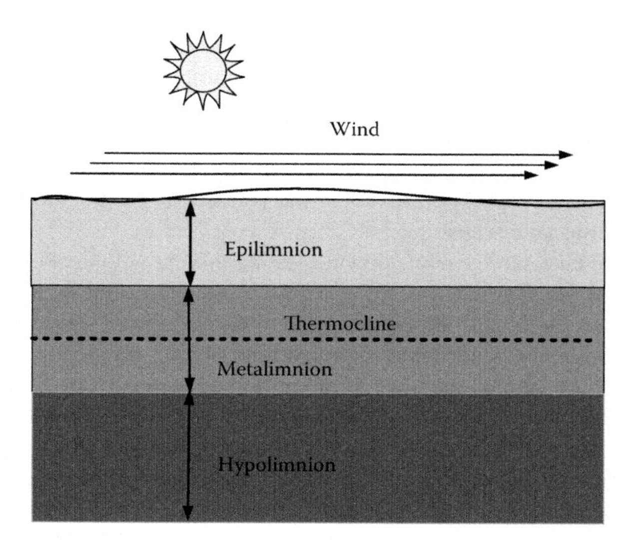

FIGURE 20.7 Typical stratification layers in stationary surface water bodies (lakes).

(Figure 20.7): (i) epilimnion (on the top, a less dense surface water layer that interacts with the wind and sunlight); (ii) metalimnion (in the middle); and (iii) hypolimnion (on the bottom).

In shallow lakes, stratification into epilimnion, metalimnion, and hypolimnion often does not occur (Kirillin and Shatwell 2016).

The epilimnion is the upper well-mixed surface layer with constant temperature profiles and non-constant thickness because of the wind observed from the surface to a 2-m depth. The middle metalimnion transitional layer starts from the point where the water temperature and density change rapidly, the temperature gradients are typically greater than 1°C/m. The thermocline is a horizontal plane within the metalimnion layer where there is a greatest temperature difference and, therefore, a rapid density change. Beneath the metalimnion, and extending to the lake bottom, is the colder (high-density water) and relatively undisturbed hypolimnion layer that often remains around 4°C throughout the year.

The thermal stability of a lake's stratification depends on factors such as (Boehrer and Schultze 2008): (i) climate; and (ii) lake orientation to the dominant wind, depth, shape, salinity, chemical composition of dissolved substances, and size.

In the spring, the lake surface starts to warm up. As water is less dense, it floats on top of the deeper, cold layers. These deeper layers are then isolated and warm up only slowly; for example, when strong wind provokes mixing. In the summer, there is usually a warm water layer (>15°C) down to a 10–25-m depth. Below it, the temperature deceases rapidly and, below 30 m, water remains cold (<8°C). In the summer, because of solar radiation, the warm and less dense water than the water underneath floats on the surface of the lake while the cold and denser water remains below the surface. A density difference is thus established, and a thermal stratification into three layers of water (also called summer stagnation) occurs (see Figure 20.7). The lower zone (hypolimnion) of the lake remains cold because most radiation is absorbed in the upper zone. Circulation loops do not penetrate to the lower zone, and conduction to the ground/soil is quite small. The result is that, in deeper lakes with small or medium inflows, the upper zone exists at 21–32°C, and the lower zone at 4–13°C. In sufficiently deep lakes, summer stratification holds until the fall when the surface water gets cooler (thus, denser than the water below) due to radiation and evaporation losses at the surface, and sinks through the water column mixing the lake water.

In the fall, the lake surface cools down, creating mixing with the deeper layers. As a result, a large layer (20–40 m deep) of ≃10°C water can often be observed at the end of the fall. In the winter, the upper portion begins to cool towards the freezing point and the lower-levels approach the maximum density temperature of 4°C.

In the winter, the lake temperature is rather homogeneous from the bottom to the surface of the lake, usually 4–6°C. When the water is cooled below 4°C, instead of getting denser, it gets less dense. The bottom of deep lakes may stay 3–5°C warmer than the water at the lake surface. This condition is referred to as winter thermal stratification (stagnation). Wind influences much the mixing: strong storms can push this water down to large depths and contributes to homogenizing the lake temperature. If, often in winter, the temperature decreases enough so that it is about the same in the whole lake, vertical mixing can occur, whereby water is exchanged through the whole water column, renewing oxygen and nutrients in the deeper layers. As spring approaches, the surface water warms until the temperature approaches the maximum density point of 4.0°C (see Figure 20.5). Thus, the winter thermal stratification becomes unstable and circulation loops begin to develop from the top to the bottom. As the sun warms the lake surface through late spring and early summer, the temperature differences increase between the surface and bottom of deep lakes. When the surface ice is melted in the spring, solar radiation warms the water at the surface of the lake much faster. As the water warms, it becomes less dense and remains at the surface, floating in a layer above the cooler, denser water below. Later in the spring, as water temperatures rise above 7°C, circulation loops appear in the upper portion of the lake.

The process of lake water mixing is of two types: (i) wind mixing due to the kinetic energy that mixes some portion of the water column; in other words, the kinetic energy imparted by the wind near the surface causes the water to move in the downward direction and, in turn, the bottom water moves towards the surface; and (ii) natural convective mixing due to buoyancy; during the period of lake cooling, the heat is lost from the surface water, becoming denser than the water at the bottom. Therefore, there will be complete inversion: the incoming cool water from the surface replaces the warmer water at the bottom, and vertical mixing starts.

When the whole lake reaches a similar temperature and density (usually once in the spring after the ice melts and once in the fall before ice forms when the water temperature throughout the lake is uniform with a maximum density near 4°C), wind forces are able to mix again the lake water from top to bottom, a process called lake water (spring or fall) turnover. Spring turnover occurs when the ice formed on the surface melts, causing, along with wind action, surface water to sink, freely mix, and match the temperature of deep water. Thus, the water in the lake becomes thermally uniform from top to bottom.

During the fall, solar and thermal radiation from the atmosphere decline, radiative heat losses increase, and the surface waters cool which leads to lake water vertical mixing. In the winter, evaporative and radiative energy losses from the surface temperature of the water continues to drop below 4°C, and water gets less dense, which creates a new thermal stratification that can be upset by wind. In shallow lakes (less than about 3–3.6 m deep), wind forces are usually strong enough to mix the water from top to bottom completely for much of the year if sunlight penetrates to the lake bottom or there is enough wind energy to move water through the entire water column. In lakes deeper than about 3–3.6 m, the temperature differences eventually create a physical force strong enough to resist the wind's mixing forces.

20.3.3 Heat Transfer in Lakes

In lakes, the main heat transfer occurs at the water surface, which receives and exchanges heat with the sun, the ambient air, and precipitations. The heat reaches the deeper layers progressively of which temperatures of these layers become more and more stable with depth, while the temperature of the lake's surface follows approximately the air temperature through the year. The surface energy balance of a freshwater body must incorporate short-wave (solar) and long-wave (terrestrial) radiation, together with evaporation and sensible heat energy fluxes.

Most of the one-dimensional models of heat transfer in surface water bodies that are exposed to multiple heat gains and heat losses are based on assumptions such as (Dake and Harleman 1969; Hamilton and Schladow 1997): (i) volume and specific heat of the surface water body are constant; (ii) changes in time of water volumes and levels due to precipitations and evaporation/condensation, are neglected; (iii) water temperatures are predicted using daily and/or monthly meteorological data; (iv) surface water temperature and density vary only in the vertical direction; (v) surface water body contains no inflows (from streams or rivers) and/or outflows; and (vi) the effect of lake shading due to presence of tall structures or vegetation around the water body is neglected.

The heat transfer in lakes usually consists of (i) solar radiation, and thermal convection, radiation, and evaporation/condensation with the atmosphere at the water (or ice) body surface; (ii) heat transfer inside the vertical water column by penetration of short- and long-wave radiation, advective and turbulent/mixing diffusion processes caused by wind and buoyancy forces; and (iii) conduction from the ground/soil/rocks through sediment layers at the lake bottom.

Heat gains at the lake surface consists of (Figure 20.8) (i) solar incident (direct) short-wave radiation; (ii) convective heat exchange from the surrounding air and the water (or ice plus snow) when the ambient air is warmer; and (iii) heat gains from the condensation of water vapor.

Heat losses occur by (Figure 20.8) (i) short- and long-wave heat reflected; (ii) cooling by thermal convection and surface water evaporation at the lake surface depending on the surface water temperature, wind speed, and ambient air wet-bulb temperature; (iii) conduction to sediments; and (iv) heat transferred to immersed heat exchangers of geothermal heat pump(s).

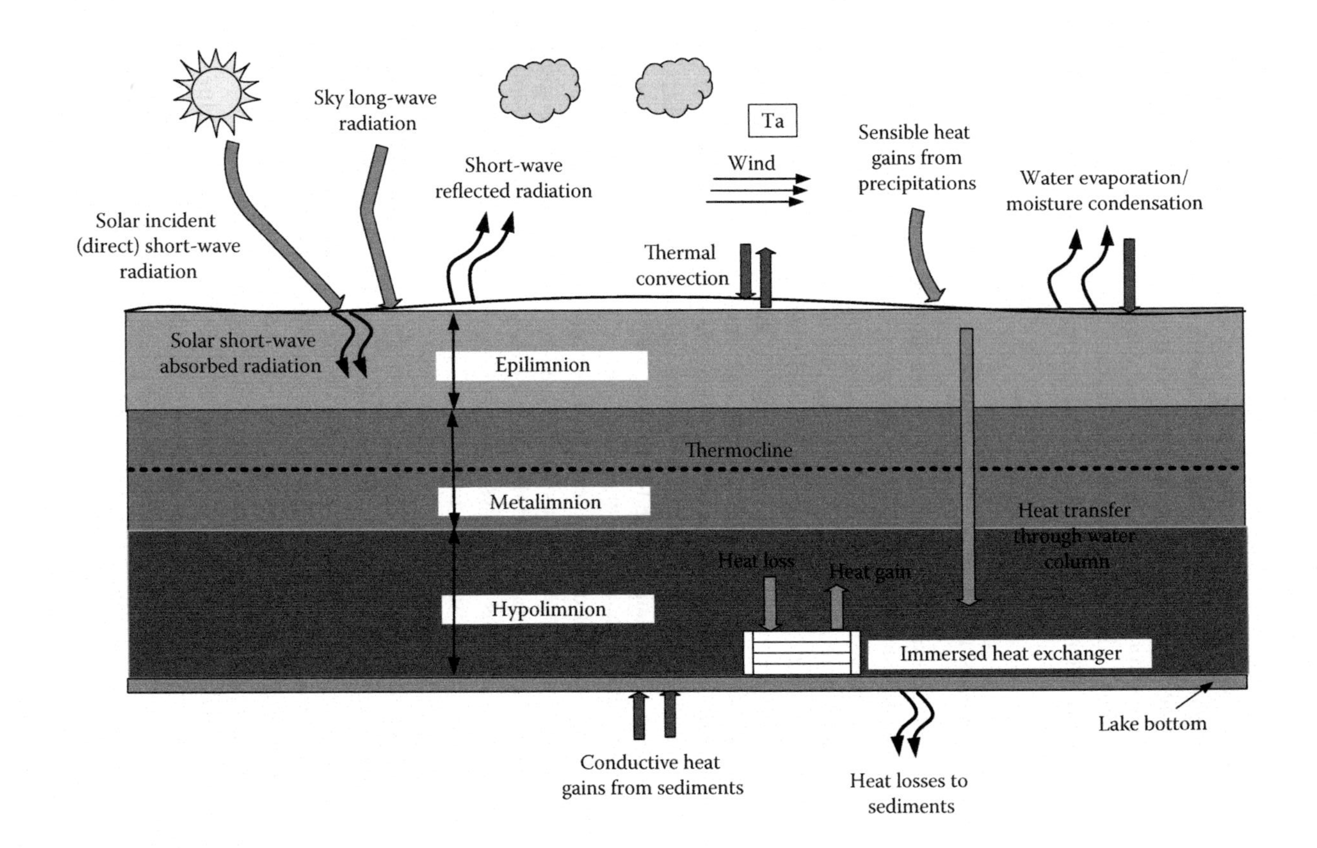

FIGURE 20.8 Main heat transfer mechanisms in lakes (Notes: not shown are other possible heat transfer mechanisms; different mechanisms may occur during winter and summer climate conditions).

The total net thermal energy exchange between the atmosphere and the lake surface energy exchange between the atmosphere and the lake surface can be expressed as:

$$|Q_{tot}| = Q_{s-w,\ solar}^{incident} + Q_{s-w,\ solar}^{absorbed} - Q_{s-w,\ solar}^{reflected} + |Q_{l-w,\ sky}| + |Q_{conv,\ rad}| + |Q_{latent}| + Q_{precip} \quad (20.3)$$

where

$|Q_{tot}|$ is the total heat resulting from the algebraic summation of the right side listed heat amounts transferred into the surface water from the surface (or in the opposite direction, depending on its minus/plus sign) (kJ).

$Q_{s-w,\ solar}^{incident}$ is the incident (direct) short-wave (energy) radiation that reaches the horizontal surface of the lake is absorbed by the lake, and the other part is reflected; it is the dominant heating mechanism that it occurs in the upper portion of the surface water body and varies with cloud amounts, latitude, time of year, and conditions of the surface, i.e., factors not dependent on the difference between surface and air temperature. It can be noted that short-wave incoming radiation from the sun is absorbed and scattered as it traverses the atmospheric gases and suspended particles such as water droplets, ice crystals, and dust; the processes of absorption and scattering depend on the wavelength of the electromagnetic radiation and the size of the obstacle. The average top-of-atmosphere solar irradiance is the solar constant, which is approximately equal to 1,366 W/m^2; the solar constant is not really constant; daily and decadal variations of up to a few watts per square meter are common, as a result of variable solar convective activity. The instantaneous top-of-atmosphere solar irradiance also differs from the solar constant as a function of the Earth-Sun distance; incoming short-wave radiation at the water surface is made up of two main components: direct and diffuse solar radiation (kJ).

$Q_{s-w,\ solar}^{absorbed}$ is the incident short-wave radiation absorbed by the lake surface that can be calculated as:

$$Q_{s-w,\ solar}^{absorbed} = Q_{s-w,\ solar}^{incident}(1 - r_w)\alpha_w \quad (20.4)$$

where

$Q_{s-w,solar}^{absorbed}$ is the solar short-wave radiation absorbed by the surface water (kJ).

$Q_{s-w,\ solar}^{incident}$ is the short-wave incident of solar irradiation on the lake surface (kJ); most portion of solar radiation (which can exceed 950 W per square meter of surface lake) is absorbed by the lake surface (about 97–98%) and the remaining penetrates towards the water column at depths visible to the human eye, mainly influenced by the amount of suspended solids in a lake.

r_w is the water surface reflectivity (a coefficient that determines the fraction of radiation reflected from the surface that usually changes with time of the year and depends on solar altitude angle, cloud cover, latitude, and surface conditions) (–).

α_w is the water surface absorption coefficient that takes into account the degree of solar radiation absorption by the surface water body (–).

$Q^{reflected}_{s-w,\ solar} = Q^{incident}_{s-w,\ solar} \cdot r_w$ is the short-wave radiation reflected (scattered) from the water surface; in general, 2–3% of incident short-wave solar radiation is reflected and scattered on a summer day and the rest will be absorbed (kJ).

$|Q_{l-w,\ sky}|$ is the net sky long-wave radiation from the surface water towards (+) or upward (–) to the sky; it is electromagnetic energy in the spectral band from roughly 3–100 μm; all surface water bodies receive short-wave radiation during daylight and exchange long-wave radiation continuously with the atmosphere. The net rate at which a surface water receives radiation of long and short wavelengths is called net radiation; back radiation typically occurs at night when the sky is clear, and can account for significant amount of cooling. The relatively warm water surface will radiate heat to the cooler sky; for example, on a clear night, a cooling rate of up to 160 W/m^2 is possible from a lake 14 K warmer than the sky. The incoming long-wave radiation from the atmosphere heats the lake surface, while the long-wave radiation is reflected back from the surface (a significant mechanism of heat removal from the lake) cools the lake surface. In accordance with the Stefan-Boltzmann law, the amount of radiation from the water surface depends on the emissivity of that surface and the difference between the fourth power of water and sky absolute temperatures (kJ).

$|Q_{conv,\ rad}|$ is the sensible convective and radiative heat transfer at the water body surface; heat is added or lost from the lake surface by convection, which is driven by the air-water temperature difference (free convection) and wind force acting on the surface (forced convection). Convective heat transfer to the lake occurs when the lake surface is cooler than the air; wind speed increases the rate at which heat is transferred to the lake, but maximum heat gain by convection is usually only 10–20% of maximum solar heat gain. Convective cooling or heating in warmer months will contribute only a small percentage of the total because of the relatively small temperature difference between the air and the lake surface temperature (kJ).

The heat flux density due to convection at the lake surface can be calculated by:

$$\dot{q}_{conv} = h_c(T_{air} - T_{water\ surf}) \tag{20.5}$$

$\dot{q}_{conv}$ is the heat flux density due to convection at the lake surface (W/m^2).

h_c is the surface convective heat transfer coefficient (W/m^{2}K).

T_{air} is the air temperature (°C).

$T_{water\ surf}$ is the surface water temperature (°C).

$|Q_{latent}|$ is the latent heat transfer by both water evaporation and vapor condensation; by free and forced evaporation, heat is removed from the lake surface. The phase change of water from liquid to vapor results in the latent heat loss from the lake surface. The rate of evaporation (that cools the lakes) depends on vapor pressure gradient, wind velocity, and surface water temperature (kJ).

The density of the evaporation heat flux can be calculated as follows:

$$\dot{q}_{evap} = (h_{fg} \cdot h_c / c_{p,\ air})(\omega_{surface} - \omega_{air}) \tag{20.6}$$

where

$\dot{q}_{evap}$ is the density of evaporation heat flux (W/m^2).

h_{fg} is the water latent heat of evaporation (J/kg).

h_c is the convective heat transfer coefficient (W/m^{2}K).

$c_{p,\ air}$ is the air specific heat at constant pressure (J/kg·K).

$\omega_{surface}$ is the absolute humidity of the saturated air at the lake surface ($kg_{water}/kg_{dry\ air}$).

ω_{air} is the absolute humidity of the ambient air ($kg_{water}/kg_{dry\ air}$).

Q_{precip} is the sensible heat due to precipitations (rain, snow) (kJ).

During the winter, when the ambient air temperatures are below °°C, ice forms at lakes' surfaces. The duration and the maximum ice thickness formed for the entire winter depends on the trend of air temperatures. An ice layer formed on the surface partially absorbs the solar short- and net long-wave radiation similar to open surface water. The heat added through this absorption as well as the conductive heat flux from the lake water below ice are major factors for ice melting. The temperature of water beneath the ice layer is above zero (relatively warmer than the surface), which generates heat flux from the water to ice bottom surface. The heat flux at the ice-water interface reduces the ice growth at the bottom of ice and thus counteracts ice growth.

20.4 DESIGN AND INSTALLATION PRINCIPLES

Prior to the selection of stationary surface water (lake, pond, river, or sea, all called lakes here) as a heat and/or sink source for open- and/or closed-loop surface water-source heat pump systems, some of the following preliminary information and actions are recommended: (i) determine the surface water body area, volume and quality, average depth available for stratification, temperature increase and decrease limits, temperature at the immersed heat exchanger's expected depths, maximum ice thickness, wave size, seasonal level fluctuations, wind parameters, and the other existing usages of the water; (ii) avoid using lakes with depths less than 5 m in order to take advantage of seasonal stratifications and turnovers; (iii) investigate the local and/or national existing regulations concerning zoning, ground/soil excavations, and future planned constructions; (iv) assess the environmental impact of extracting and rejecting heat from/to the body of water, their potential impacts on aquatic life, as well as sensitivity of aquatic life to temperature changes; and (v) in the case of moving surface water (rivers), determine the historical high and low water conditions, debris flow, and the impacts on commercial and recreational traffic.

Some typical steps required for the design of surface water ground-source heat pump systems are: (i) determine the surface water required area (the lake should be a minimum of 5,000 m^2 in size with a volume of surface water equal to twice the size of the building being heated/cooled), depth, and temperature; (ii) divide the building into thermal zones; (iii) calculate the thermal loads for each thermal zone; (iv) design building internal water closed loop(s) (material, insulation) and size the piping system for lower pressure drops; provide water (e.g., on/off) flow control through the geothermal heat pumps with isolation motorized (solenoid) valves, select the water circulation pump(s), and specify their control method (e.g., on/off or variable speed); (v) select the geothermal heat pumps for each zone based on their nominal capacity and efficiency, the extended operating temperatures, type, sound, and serviceability; (iv) select and design the building ventilation air system; (vi) choose the type of immersed heat exchanger and of antifreeze (brine) approved by the authority having jurisdiction, and calculate the required lengths; submerged heat exchangers could be conceived as spaced bundles usually fabricated from ¾'' ID high-density polyethylene or polybutylene tubing with central hubs along with horizontal and vertical tube-tube spacers and connections either thermally socket-fused or butt fused with protection from UV radiation, especially when located near the surface; generally, about 100 m of tubing length is required for each 3.5 kW of geothermal heat pump nominal cooling capacity; assume a plan area of 1.5 by 1.5 m for each spool; in cold and moderate climates, the lengths of immersed heat exchangers made of 19–38 mm ID diameters high-density polyethylene or polybutylene pipes approved for lake installation, are recommended between 9 m and 26 m per kW of geothermal heat pump nominal cooling capacity; locate the heat exchanger at proper depth(s) in the water body; at selected deep enough depths, the water temperature should remain above 4°C during the building heating season (winter) in order to avoid lake water freezing conditions; the location of heat exchangers with sufficient numbers of parallel loops must limit the impacts on the land area and shoreline conditions, induce minimum disturbances to lake body, and to aquatic habitats; protections against waves, ice, and boat damages must be ensured; provide minimum distances of 2 m between any part of the lake submerged heat exchanger and any potable water intake; verify that the hydrostatic pressure at the bottom of the lake heat exchanger is within the pressure limits of the pipe used; calculate the friction pressure losses through the longest parallel runout/circuit path and size the brine circulating pumps; (vii) verify that the hydrostatic pressure at the bottom of the pond or lake heat exchanger is within the pressure limits of the pipe used. If necessary, isolate the lake heat exchanger from the building's water loop using an intermediate plate-frame heat exchanger; (viii) the total spool length required can be determined by selecting a spool length per unit of thermal load (e.g., from Table 20.1) to yield an acceptable design entering water temperature (generally in excess of 0°C to avoid significant ice buildup) and dividing this into the thermal load of the submerged heat exchanger; (ix) once the total spool length is determined, it is necessary to determine how many spools of equal length to use (number of circuits), how to group the spools into runouts, and how to arrange the runouts to fit the available body of water; (x) since the density of plastic piping materials is slightly less than water, the spools must be weighted with ballast to avoid floating, and to hold it on or near the bottom of the lake. Care must also be taken that sediment, silt, or growths such as algae do not cover the pipes, as this would reduce heat transfer to the surrounding body of water; (xi) the service pipes between the lake and building should be buried at minimum of 4 feet deep or below the frost line across the shore and kept separated about 2 feet in the trench. Special attention should be given to reduce the possibility of mechanical damage.

TABLE 20.1
Design Entering Liquid Temperatures for Different Surface Water Heat Exchanger Spool Lengths in Northern Cold Climates

	Winter (4.4°C)	Summer (10°C)
Heating		
61 m/ton	0°C	–
91.5 m/ton	0°C	–
Cooling		
30.5 m/ton	–	20°C
61 m/ton	–	18°C
91.5 m/ton	–	15°C

Note: 1 ton = 3.4 kW of geothermal heat pump nominal cooling capacity.

For the system installation, recommendations such as the following should be applied (DOE 1995): (i) multiple parallel submerged heat exchangers must be spread out under the lake surface to limit thermal interference, hot spots, and cold pockets; they should be on the discharge side of the water circulation pump(s) to ensure efficient air purge; (ii) the parallel circuit headers between supply and return runout trenches must be centered to equalize runout lengths and pressure drops as much as possible; (iii) supply and return runouts must have the same number of circuits, balanced lengths and the same diameter and must be buried in separate trenches; (iv) the antifreeze fluid and piping materials must be acceptable for underwater and underground applications; (v) the thermal fusion techniques to make connections in both the trench and the surface-water heat exchanger piping should follow the manufacturers' recommendations; (vi) for the system location, the knowledge of buried services (e.g., surface and subsurface obstructions), the conditions of entrance to the body of surface water and building mechanical room, must be identified; (vii) if the elevation of the surface water is higher than that of the mechanical room, extra care must be taken in trenching and runout installation to prevent surface water from flooding back through the trench and entering the mechanical room through the building penetration; (viii) each spool of submerged heat exchanger should be air-pressure-tested prior to being connected to the headers or runouts; (ix) the thermal fusion connections must be made while the pipes are still coiled in spools on the shore; each spool should be air pressure tested prior to being connected to the header or runout; (x) after the circuit headers have been attached to the spools, the assembly is pressure tested; (xi) ballast, usually in the form of concrete blocks, is attached to each spool; (xi) the central circulation pumps are generally used to flush the air out of each runout circuit and from the surface water heat exchanger; (xii) once the submerged heat exchangers have been purged, flushed, and filled with water, the system can be charged with appropriate antifreeze agent and inhibitor; (xiii) flow testing and final pressure testing is performed prior to backfilling the runout trench; the submerged heat exchanger should be pressurized to the lower of 150% of pipe design pressure rating or 300% of heat exchanger operating pressure. When the pressure testing is complete, lower the pressure in the heat exchanger loop to 69–103 kPa in summer or 172–207 kPa in winter. Once the indoor loop is connected to the submerged heat exchanger, the pressure will be determined by the vertical distance between the highest and lowest points in the system.

20.5 ADVANTAGES AND LIMITATIONS

The advantages of surface water ground-source heat pump systems can be mentioned: (i) when the buildings are located near surface water sources, surface water bodies can be economically viable heat sources and sinks, and at relatively low-cost systems if properly used; (ii) in cold climates, the temperature level of lakes during the winter is relatively high compared to outdoor air and even to soil; therefore, lake water is favorable as a heat source for geothermal heat pump heating; (iii) despite low surface water temperature during winter, lakes are advantageous compared to some other natural heat sources, as outdoor air; because the surface water temperature near the bottom drops to 2–3°C (or even lower), the crucial question is how much heat can be extracted from a lake and what the lowest possible water temperature is as well as the lowest permissible brine temperature; (iv) compared to ground-coupled ground-source heat pump systems, relatively low cost because of reduced drilling or excavation costs; (v) lower energy operating costs and maintenance requirements; (vi) in moderate climates, the use of lake waters instead of the ambient air can be advantageous during the entire heating and cooling seasons, achieving very high levels of energy efficiency; (vii) the closed-loop surface water ground-source heat pump systems present several advantages over the equivalent open-loop systems, as reduced fouling resulting from the circulation of clean water (or water/antifreeze solution) through the geothermal heat pumps; in addition, the closed-loop systems are recommended if the lake temperature could drop below 4°C in order to avoid water frosting on submerged heat exchanger surfaces when the bulk water temperature is in the 1–3°C range; (viii) reduced pumping electrical power requirements because of relatively low elevations from the lake surface to the geothermal heat pumps.

Some limitations of surface water ground-source heat pump systems in cold and moderate climates are as follows: (i) in practice, these are systems not very common because there are not all that many suitable sites where all conditions are met: a large enough body of water close enough in proximity and of suitable depth; (ii) environmental concerns should be investigated before installation; (iii) the surface water must be clean enough to run through the system without much filtration since contaminated water that could produce an undesired buildup of sludge inside the loop pipe and there's always the risk of impelling foreign materials; (iv) in addition, there is the possibility of damage to the submerged heat exchangers and fouling of external heat exchange surfaces; (v) wide variation in water temperature with outdoor conditions if lakes are small and/or shallow; such variation in water temperature would cause undesirable variations in efficiency and capacity, though not as severe as with air-source heat pumps; (vi) in cold climates, where surface water temperatures frequently drop below 4°C, there is a risk of frosting the geothermal heat pumps' evaporators; to avoid such an undesirable situation, the minimum allowed refrigerant evaporation pressure (and temperature) must be strictly controlled, and/or higher as usual surface water or brine flow rates must be provided in order to limit the temperature drop across the geothermal heat pumps' evaporators; (vii) in the case of closed-loop systems has several disadvantages, the performances of geothermal heat pumps could drop slightly because the circulation fluid temperature drops 2–7 K below the lake temperature; and (viii) possibility of carrier fluid leaking into the water resource from the collector and performance degradation by seasonal temperature changes in shallow lakes.

REFERENCES

Boehrer, B., M. Schultze. 2008. Stratification of lakes. *Reviews of Geophysics* 46, Paper number 2006RG000210, 27 pages (10.1029/2006RG000210) (https://pdfs.semanticscholar.org. Accessed August 11, 2019).

Dake, J.M.K., D.R.F. Harleman. 1969. Thermal stratification in lakes: Analytical and laboratory studies. *Water Resources Research* 5(2):484–495.

Hamilton, D.P., S. Schladow. 1997. Prediction of water quality in lakes and reservoirs. Part I - Model description. *Ecological Modelling* 96:91–110.

Kirillin, G., T. Shatwell. 2016. Generalized scaling of seasonal thermal stratification in lakes. *Earth-Science Reviews* 161:179–190.

Lewis Jr., W. M. 1983. A revised classification of lakes based on mixing. *Canadian Journal of Fisheries and Aquatic Sciences* 40(10):1779–1787.

Wilhelm, S., R. Adrian. 2007. Impact of summer warming on the thermal characteristics of a polymictic lake and consequences for oxygen, nutrients and phytoplankton. *Freshwater Biology* 53(2):226–237.

21 Advantages and Limitations of Ground-Source Heat Pump Systems

21.1 INTRODUCTION

Ground-source heat pump systems that use the ground/soil/rocks, groundwater, or surface water as "free," abundant, and renewable heat/sink sources for heating and/or cooling residential, commercial/institutional, and industrial buildings, are relatively simple and reliable alternatives to conventional heating, ventilating, and air-conditioning (HVAC) technologies.

By using reversible geothermal heat pumps, the ground-source heat pump systems eliminate or reduce the need for utilizing separate conventional heating and cooling systems and the need for other energy sources (as fossil fuels). In addition, such systems allow the heat extracted during the winter heating-dominated season to be rejected and, in part, stored in ground/soil/rocks (or aquifers) during the summer cooling-dominated period.

Some of the technical, energetic, operational, economic, and environmental advantages and limitations of ground-source heat pump systems are summarized in this chapter (Minea and Savignac 1996; Kavanaugh and Rafferty, 1997; Minea 2009; Lazzarin 2001; Bloomquist 2001; Berntsson 2002; Self et al. 2013).

21.2 ADVANTAGES

21.2.1 Energy Source Quality and System Efficiency

1. Vertical closed-loop ground-source heat pump systems, mostly using solar energy stored in the shallow underground, are among the most efficient renewable energy–based technologies providing heating and cooling for residential, institutional, and commercial buildings.
2. In large-scale commercial/institutional buildings, in addition to the solar energy stored in the ground/soil/rocks, groundwater, and surface water, ground-source heat pump systems offer opportunities to save additional energy since the heat recovered from the buildings' zones requiring cooling can be transferred to zones with simultaneous heating demands; such an operation provides a highly efficient space-conditioning technology that helps utilities to increase the electricity usage and reduce the

DOI: 10.1201/9781003032540-21

peak demand; in other words, ground-source heat pumps systems offer the possibility of using free, inexhaustible, and ecological energy sources such as direct solar irradiance, as well as heat recovered from building internal gains and exhaust air, engine cooling fluids, combustion gases, etc.; in addition, a part of the rejected cooling energy can be stored in the ground/soil (or aquifers) offering improved thermal conditions at the beginning of the next heating season; in other words, ground-source heat pump systems are able to store thermal energy rejected during summer cooling cycles in the ground /soil/rocks (and/or aquifers) rather than reject it to the atmosphere as is the case with conventional heating and cooling systems, and can then use the energy extracted from the ground/soil/rocks (and/or aquifers) during the winter heating cycle.

3. In cold and moderate climates, at depths higher than 2–3 m, the temperature of undisturbed ground/soil is almost constant year-around; hence, ground-source heat pump systems could operate at higher energy efficiencies compared to air-source heat pumps by using the stable and practically unlimited very low-temperature energy being attractive alternatives for heating and cooling residential as well as small and large commercial/institutional buildings; the boundary between the "shallow zone" where ground/soil temperature fluctuates seasonally and the "deep zone" in which the ground/soil/rocks temperature is relatively constant lies at depths ranging from 2 to 10 m below the ground/soil surface.
4. The ground/soil temperature at depths higher than 2 m is practically independent from the ambient temperature (but is strongly a function of earth formation type), varies very little in temperature and thermal properties, and remains relatively stable, allowing the geothermal heat pumps to operate close to their optimal design points and achieve higher energy efficiencies because they draw heat from the ground/soil, which is at a relatively constant temperature all year round.
5. Ground-source heat pump systems perform better than conventional air-source heat pump systems, especially in climates characterized by large daily temperature fluctuations and swings.
6. Because of the ground/soil/rocks, groundwater, and surface water thermal inertia and higher heat storage capacity of the earth and aquifers, temperature variations of heat/sink sources are damped and phase delayed several weeks and even months, compared to the ambient air temperatures.
7. In cold and moderate climates, ground/soil, groundwater, and surface water are warmer in winter and cooler in summer than the outdoor (ambient) air, making these very low-temperature renewable and sustainable energy reservoirs of higher quality heat and sink sources for geothermal heat pumps serving to space and domestic hot water heating and cooling of residential, commercial/institutional, and industrial buildings.
8. In cold and moderate climates, while the instantaneous heating coefficients of performance ($COP_{heating}$, defined as the ratio of heat supplied to the energy consumed) of standard air-source heat pumps range from 1.2

to 2.5, those of well-designed ground-source heat pump systems range from 3.0 to more than 4.0.

9. Consequently, the seasonal heating performance factors (SHPF, defined as the ratio of the heat delivered by the geothermal heat pumps to the total energy consumed, measured over a given heating season, including energy consumed to circulate brine through the ground-coupled heat exchanger) of ground-source heat pump systems vary between 3 and 4 (and even to 5–6 for groundwater or surface water ground-source heat pump systems), compared to 1.5–2.9 for conventional variable speed air-source heat pump systems.
10. In cold and moderate climates, while the instantaneous cooling energy efficiency ratio (EER) (defined as the ratio of heat extracted from the conditioned spaces to the energy consumed) of standard air-source heat pumps range from 9 to 11, those of well-designed ground-source heat pump systems range from 15 to more than 18.
11. Similarly, the seasonal energy efficiency ratio (SEER), defined as the ratio of the energy extracted from the building to the total energy consumed, measured over a given cooling season, including energy consumed to circulate the brine through the ground-coupled heat exchangers of ground-source heat pump systems) range from 25 to 30, while those of conventional air-source heat pumps usually vary from 12 to 18.
12. At the beginning of the heating season (with ground/soil higher temperatures resulting from the heat rejection during summertime) and at the beginning of the cooling season (with ground/soil lower temperatures resulting from the heat extraction during the heating season), the ground/soil average temperatures are more favorable to perform better heating ($COP_{heating}$, SHPF) and cooling (EER, SEER) energy performances, respectively.
13. Improved energy performances of ground-source heat pump systems compared to air-source heat pumps could be attributed to the fact that they exchange heat with the ground/soil/rocks, groundwater, or surface water that are much more stable heat source/sink thermal reservoirs than the ambient air available at variable temperatures and relative humidity; effectively, at outdoor temperatures lower than –10°C (in heating mode) and higher than 35°C (in cooling mode), the energy performances of air-source heat pumps (COPs, SHPFs, SEERs) drastically decrease. At ambient temperatures below –10°C, air-source heat pumps still operating in the heating mode need electric or fossil back-up energy sources; in other words, ground-source heat pump systems require much less (or any) supplemental (emergency, back-up) heat at extremely low outside (ambient) temperatures in the heating-dominated seasons, or less (or any) heat rejection at extremely high outside temperatures in the cooling-dominated seasons.
14. Although ground-source heat systems require higher initial capital cost, they can reduce energy consumption for space heating (e.g., up to 50% compared with fossil-burned boilers, up to 44% compared with air-source

heat pumps and up to 70% compared with electric resistance heating) and 20–50% of cooling costs, compared to conventional HVAC systems, as well as the power demand (e.g., up to 75% compared with electric baseboards) on the local utility electrical grids.

15. In cold climate countries (such as those of Canada, Norway, and Sweden), by utilizing vertical closed-loop (indirect, secondary fluid) ground-coupled heat exchangers buried in ground/soil/bedrocks, the ground-source heat pump systems may cover up to 80–90% of the annual building heating energy demand and even a considerable part of the building annual cooling demand by means of the free cooling technology.
16. Ground-source heat pumps provide high levels of indoor comfort and relatively low levels of maintenance requirements and costs because of their relatively simple design and control strategies.

21.2.2 Technology Feasibility and Building Integration

1. The technical and economic competitiveness and feasibility of ground-source heat pump systems are established by critical factors such as: (i) borehole depths and construction costs of ground-coupled heat exchangers; (ii) potential energy savings compared to conventional heating, ventilating, and air-conditioning methods; and (iii) availability of tools allowing assessing and optimizing the systems' design and overall control.
2. In cold and moderate climates, ground-source heat pump systems can be installed at any location where drilling or earth trenching are feasible, and/or abundant and appropriate quality groundwater or surface water are available.
3. Vertical U-shaped borehole heat exchangers, typically installed up to 200 m deep in the ground/soil/rocks, are employed where availability of land area is scarce and provide high and steady thermal performance as less temperature fluctuation occurs along with depth.
4. Because vertical ground-coupled heat exchangers require relatively small plots of land areas (e.g., between 1.5 m^2 to 7.5 m^2 per kW of building nominal cooling load, depending on the boreholes' depths), they are the most applied ground-source system configuration for commercial/institutional applications; usually, the buildings' parking lots may accommodate vertical borehole fields up to between 351 and 700 kW of nominal cooling capacities.
5. Because of relatively constant, less variable temperatures of ground/soil/rocks, groundwater, and surface water compared to those of outdoor air, the compressors of geothermal heat pumps operate with more stable and safe thermodynamic parameters (e.g., discharge pressures and temperatures) and quiet conditions than those of air-source heat pumps.
6. Unlike air-source heat pumps, ground-source heat pump systems do not need outdoor fans and/or defrost cycles on the evaporator side in the heating mode.
7. Ground-source heat pump systems are highly reliable and have longer technical life expectancy estimated at more than 25 years for the inside

components (e.g., compressors, fans, and brine/water circulation pumps), and at 50+ years for the outside components (e.g., ground-coupled heat exchangers that replace the evaporators/condensers of conventional air-source heat pumps).

8. Ground-source heat pump systems require simple, low-pressure air ductwork and brine or water pipe network for heat and cool distribution, less mechanical room space, containing only circulating pumps, main headers, and chemical treatment equipment, meaning more available space and flexibility in occupied (or leased) areas.
9. Ground-source heat pump systems require much less outdoor equipment, which implies excellent building aesthetics, reduced outdoor noise levels, high security due to no visible external components to be damaged or vandalized, and no roof penetrations.
10. In well thermally balanced systems, there are no auxiliary heat sources as fossil-fueled boilers (with fuel storage tanks), and combustion or explosive gases within the buildings; in cold and moderate climates, the ground-source heat pump systems do not need cooling towers.
11. Ground-source heat pump systems are optimal for the use of low-temperature water-based systems as underfloor heating and ceiling cooling.

21.2.3 Capital Costs

1. The capital costs of ground-source heat pump systems normally exceed the cost of most, if not all, of the alternative HVAC systems and depend on factors such as: (i) building size and quality of its thermal insulation; (ii) number of thermal zones and the system design complexity; (iii) type (vertical, horizontal) and size (total length according to building annual heating and cooling loads and their distribution over the year) of ground-coupled heat exchangers, and methods and cost of drilling and/or excavation; and (iv) ground/soil/rock structure and thermophysical properties.
2. Vertical ground-coupled heat exchangers are efficient, but expensive as drilling costs are highly dependent on ground/soil/rocks composition, drilling techniques, and availability and experience of drilling contractors.
3. The capital costs of ground-source heat pump systems per unit capacity are usually higher compared, for example, with those of air-source heat pump systems; oversizing ground-coupled heat exchangers incurs higher first costs, while undersizing can result in inadequate performances of the entire system; the capital costs can be significantly reduced by using simple design approaches, implying a limited number of components requiring frequent maintenance work. In other words, unnecessarily complex systems must be avoided.
4. Properly designed ground-source heat pump systems could be more cost-effective (i.e., lower combined installation and operating costs because of high equipment efficiency and eventual annual storage/reuse of geothermal energy, and availability) than conventional HVAC systems in applications such as: (i) new buildings where the technology is relatively easy to incorporate and

(ii) regions where natural gas, propane, or oil are unavailable or available at costs higher than those of local electricity grinds.

5. Estimated heating and cooling efficiencies and typical installation costs of ground-source heat pump systems compared to the best-rated heating and cooling conventional technologies are shown in Table 21.1 (data actualized from https://webstore.iea.org/energy-efficiency-policy-analysis-at-the-iea-2007, accessed March 10, 2021).
6. In practice, the cost of ground-coupled heat exchangers vary depending on type (vertical or horizontal); in the 2000s, the first (capital) construction costs varied from approximately US$30 per meter of borehole in the United States and Canada (Table 21.2), to about US$100/m in Japan.
7. In the case of residential ground-source heat pump systems, the ground-coupled heat exchangers are the most expensive components, accounting

TABLE 21.1
Rated Efficiency and Installed Cost Estimates for Residential Space-Conditioning Technologies (US$ 2020)

Technology	Best-rated heating efficiency	Best-rated cooling efficiency	Typical installed costs
Gas-fired furnace	96% AFUE	n/a	300 US$/kW
Oil-fired furnace	95% AFUE	n/a	340 US$/kW
Air-source central air conditioning	n/a	20–25 SEER	976 US$/kW
Air-source central heat pump	12–15 HSPF	15–20 SEER	1,310 US$/kW
Ground-source heat pump	16–22 HSPF	20–30 SEER	3,000 US$/kW

Notes: n/a, not applicable; AFUE, average gas- or oil-fired fuel furnace utilization efficiency; HSPF, heating seasonal performance factor; SEER, seasonal energy efficiency ratio; heat pump costs are per nominal kW of cooling capacity.

TABLE 21.2
Installation Cost Estimates for Ground-Coupled Heat Exchangers Such as For North America (United States and Canada) (US$ 2020)

Type	US$/kW	Relative cost
Vertical closed-loop heat exchanger	2,500	1.33
Horizontal closed-loop heat exchanger	2,250	1.20
Open-loop groundwater	1,880	1.00

Note: costs are per nominal kW of geothermal heat pump cooling capacity.

for about 30 to 35% of the system total installed cost, including the ductwork (Figure 21.1).

8. For relatively small commercial/institutional space conditioning technologies, the estimated best-rated heating and cooling efficiencies and typical installation costs are shown in Table 21.3.

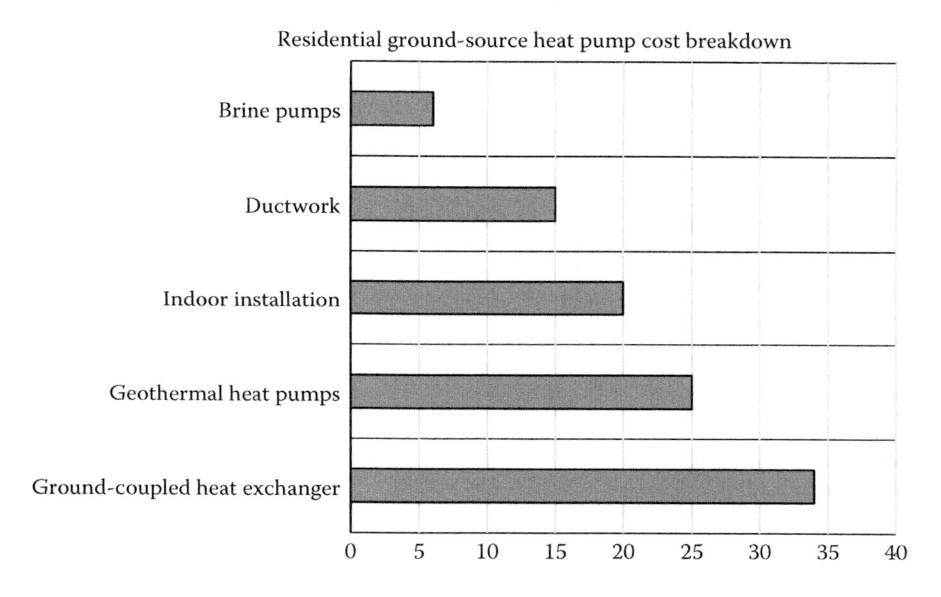

FIGURE 21.1 Approximate cost breakdown by component for residential ground-source heat pump systems in the United States and Canada.

TABLE 21.3
Rated Efficiency and Installed Costs for Commercial Space-Conditioning Technologies (US$ 2020)

Technology	Best-rated heating efficiency	Best-rated cooling efficiency	Typical installed costs
Gas-fired furnace	85% thermal	n/a	60 US$/kW
Oil-fired furnace	85% thermal	n/a	55 US$/kW
Rooftop air conditioner	–		580 US$/kW
Rooftop heat pump	3.3 COP		660 US$/kW

Notes: n/a, not applicable; AFUE, average fuel utilization efficiency; COP, coefficient of performance; HSPF, heating seasonal performance factor; SEER, seasonal energy efficiency ratio; heat pump costs are per nominal kW of cooling capacity (data actualized from https://webstore.iea.org/energy-efficiency-policy-analysis-at-the-iea-2007, accessed March 10, 2021).

9. For larger commercial/institutional constructions, the capital costs vary depending on building and ground-source heat pump systems types (Table 21.4). Table 21.5 shows estimated capital costs of some conventional HVAC systems.

In addition, it can be noted that (DOE 1995; Dooley, 1999; Moore 1999; Allen and Kavanaugh 1999; Bloomquist 2001): (i) there is considerable variability in the capital costs associated with building type, building total annual heating and cooling loads and their distribution over the year, as well as with ground/soil/rock structure and properties (as temperature and moisture content), ground-coupled heat exchanger

TABLE 21.4
Capital Costs by Building and Ground-Coupled Heat Exchanger Type (US $ 2020)

	Capital costs	
Building type	**US$/m^2**	**US$/kW**
School	230	1,890
Office	170	1,850
Apartment/multi-residential	185	1,950
Commercial spaces	67	2,020
Retirement buildings	233	2,060
Correctional facilities	250	2,440
Convenience store/gas stations	430	3,600
Ground-coupled heat exchanger		
Vertical closed-loop	236	2,035
Horizontal closed- loop	110	1,320
Vertical open groundwater	110	1,580
Hybrid (vertical closed-loop and cooling tower)	230	2,700

TABLE 21.5
Capital Costs of Conventional HVAC Systems (Dated from Moore 1999) (US$ 2020)

Type of conventional HVAC system	Capital cost
Rooftop direct expansion with electric heating	95 US$/kW
Rooftop direct expansion with gas heating	115 US$/kW
Air-source heat pump	140 US$/kW
Rooftop variable air volume (VAV)	160 US$/kW
Water-source heat pump with gas boiler and cooling tower	245 US$/kW

(or equivalent system) type (open/closed, vertical/horizontal) and local factors (e.g., drilling/excavation costs); (ii) capital costs for horizontal, closed-loop systems average less than 50% of the cost of the vertical, closed-loop systems; however, for large installations, it may be impossible to find adequate areas for the installation of horizontal closed-loop systems. One exception to this may be schools whose sport or play fields may provide the required open areas for horizontal ground-coupled heat exchangers.

21.2.4 Operating Costs

1. The ground-source heat pump systems, particularly in large-scale commercial/institutional buildings, use simple, low-pressure ductworks and closed-loop fluid distribution networks and, in some countries, tax benefits, low-interest loans, and lower maintenance requirements (mainly due their multiple zoning flexibility, less expensive zone controls, usually, elimination of cooling towers or dry air coolers, and/or heating boilers, and few periodic component replacements), contribute to further reduce the operating costs. In addition, they provide significantly lower operating (running) costs (due to their higher thermodynamic efficiencies); however, largely dependent on the cost of electricity (to operate the geothermal heat pumps, fluid circulation pumps, and controls) compared to the cost of the other conventional fuels, such as natural gas, heating oil, and propane.
2. The new buildings provided with well-designed and constructed ground-source heat pump systems with vertical ground-coupled heat exchangers should not consume more than 108 $kWh/m^2/year$.
3. In many cases, ground-source heat pumps allow cooling the buildings at virtually no cost by directly circulating the geothermal thermal carrier fluid (brine or water) through air-cooling devices (a process known as free or direct cooling).

21.2.5 Maintenance Costs

1. Long-term maintenance costs of buildings equipped with ground-source heat pump systems could be 38 to 56% lower compared to those of buildings provided with conventional HVAC systems (e.g., US$1.45/$m^2$/year versus US$2.58 to US$3.77/$m^2$/year), mainly because: (i) most parts of the ground-source heat pump systems remain protected underground or indoors, while in the case of air-source heat pumps, parts of the system are outdoors, exposed to the elements and potential mechanical damage and the technical lifetime expectance lasts about 15 years; (ii) there is no requirement for annual safety inspections as there is for combustion equipment; (iii) there are few moving parts; (iv) fluid circulation pumps are guaranteed for more than one year and easy to replace; (v) long geothermal heat pumps' compressors technical life of up to 15 years (25 years for scroll compressors, guaranteed for up to 5 years); (vi) hermetically presealed refrigerant circuits of geothermal heat pumps containing relatively

small quantities of refrigerant. For less than 3 kg of refrigerant charge, little maintenance should be required; however, large-capacity geothermal heat pumps containing more than 6 kg of refrigerant, or direct expansion systems containing more than 3 kg of refrigerant, require regular checking for leaks; (vii) the expected technical life of ground-coupled heat exchangers constructed with polyethylene and coated copper tubing is expected to be over 50 years and 30 years, respectively, and be, virtually, maintenance free.

21.2.6 Payback Period

The desirability of initial investments for ground-source heat pump systems is directly related to their payback periods that are simple and convenient parameters that determine the periods necessary for recovering the initial investments with the energy savings obtained. The simple payback periods for ground-source heat pump systems that depend largely on their installed capacity (large-scale commercial/institutional systems achieving shorter payback periods than those of residential and small-scale commercial/institutional systems), can be calculated as follows:

$$\text{Payback period} = \text{First cost/Annual energy savings} \tag{21.1}$$

Annual energy savings can be estimated as the annual energy cost of a conventional (standard) HVAC system minus the annual energy cost of the ground-source heat pump system. Even the payback period is a very simple calculation to assess the risks of projects or investments compared to those of other systems. It does have some drawbacks, mainly because there are many factors it does not take into consideration, such as the life span of the system, additional capital expenses, and variation in time of money value. In cold and moderate climates, the usual simple payback periods for well-sized, vertical closed-loop (indirect, secondary fluid) ground-source heat pump systems may range from 5 to 8 years for large-scale commercial/institutional systems, and from 10 to 15 years for residential and small commercial/institutional-scale.

21.2.7 Life Cycle Costs

Life cycle cost (LCC) analysis that should be performed at an early stage of any project provides a framework of assessing the total cost of ground-source heat pump systems incurred over the course of their lifetime, and the optimal economic points (Figure 21.2). Such analysis is useful because it provides an economic perspective in evaluating the effectiveness and the benefits of energetic systems by appropriately weighing the money spent today compared to money spent in the future.

According to Equation 21.2, the calculation of life cycle cost includes the following: (i) amount of the initial investment, a non-recurring expense; (ii) lifetime operational costs that refer to the charges incurred with running the system itself, including items

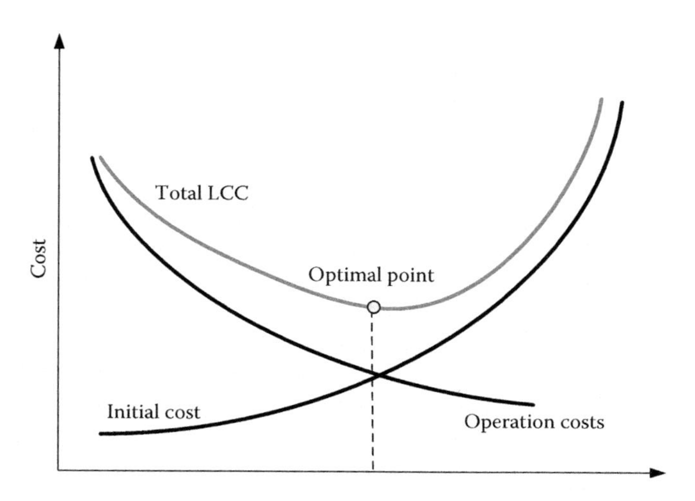

FIGURE 21.2 Principe of life cycle cost (LCC) analysis.

like energy usage, information technology services, taxes, etc.; (iii) lifetime maintenance and repair costs (recurring expenses) that may include annual check-ups, upgrades, specialist professional services, etc.; (iv) financing costs such as interest fees paid throughout the course of the system's life; (v) depreciation costs over the system usable life; (vi) end-of-life costs (also known as disposal or demolition costs) that may include the building charges associated with system removal or scrapping; and (vii) residual value.

$$\begin{aligned} \text{LCC} = {} & \text{Initial Investment} + \text{Lifetime Operating Costs} \\ & + \text{Lifetime Maintenance and Repair Costs} + \text{Financing Costs} \\ & + \text{Depreciation Costs} \\ & + \text{End of Life Costs} - \text{Residual Value} \end{aligned} \tag{21.2}$$

Some of the benefits of life cycle cost analysis are the following (Fuller and Peterson 1996; Yoon and Lee 2015): (i) excellent analytical framework for choosing between two or more systems; instead of focusing on the initial cost, it can be considered the overall costs of different systems; (ii) provides accurate and realistic assessment of costs within a specified life cycle; and (iii) gives an opportunity for total incremental costs over the whole span of time.

Some of limitations of life cycle cost are the following (Fuller and Peterson 1996; Yoon and Lee 2015): (i) it's time consuming because of a large amount of data needed; (ii) needs to estimate particular costs and adjust them over time, which introduces an additional level of uncertainty into the analysis; (iii) costly operation compared to other, more simple and rapid, methods; and (iv) less reliable because some data are assumed for calculations.

21.2.8 Environmental Impacts

Ground-source heat pump systems are alternate energy sources for space heating and cooling since they are environmental friendly and sustainable. With increasing worldwide awareness of the serious environmental problems due to fossil fuel consumption, efforts are being made to develop energy-efficient and environmentally friendly systems by utilization of non-polluting renewable energy sources, such as geothermal and solar energy, and industrial waste heat.

In modern buildings that intend to use more and more renewable energy and material resources, the ground-source hear pump systems contribute to reducing building impacts on human health and the environment during the building's life cycle through better siting, design, construction, operation, and removal.

Among the most relevant environmental issues related to ground-source heat pump systems are the following (Ondreka et al. 2007; Omar 2008; Blum et al., 2010): (i) provide clean, efficient, and energy-saving heating and cooling year round, helping to conserve the natural energy resources; they are among the most environment-friendly technologies playing a significant role in reducing the global fossil energy consumption and greenhouse gas emissions, thus reducing global warming and other environmental impacts; (ii) for designing and installing closed-loop ground-source heat pumps systems, it is necessary to be aware of applicable environmental regulations; for example, within the United States, nearly half of the states have regulations specifying or restricting the use of particular antifreezes or other fluids within the ground-coupled heat exchangers. All of these regulations are based on the need to preserve groundwater and/or aquifer quality; the main concerns are the risk of the underground pipes/boreholes creating an undesirable pathway for water to flow between different water-bearing strata, undesirable temperature changes in the aquifer that may result from the operation of a ground-source heat pumps, and pollution of groundwater that might occur from leakage of additive chemicals used in the system. Where there is a risk of, or actual, releases of polluting matter to groundwater, the agency can serve statutory notices to protect groundwater; (iii) reduce environmental impacts compared with those of fossil-fueled HVAC systems; (iv) emit less greenhouse gases than conventional HVAC systems, given the power generation mix found in most national jurisdictions; (v) since 1990, heat pumping technologies are undergoing several changes in working fluids (refrigerants) and their environmental effects were drastically reduced; (vi) compact brine-to-air and brine-to-water geothermal heat pumps, where the refrigerant charge can be minimized and which can be manufactured and tested in the factory to fulfill the requirements of leak tightness, are cleaner and efficient devices. Leak tightness means few or no greenhouse gases versus conventional heating systems in terms of carbon dioxide emissions; thus, TEWI (total equivalent warming impact) can be significantly reduced; (vii) well-sized ground-source heat pump systems could achieve lower greenhouse gas emission (e.g., up to 40% less CO_2 emissions when compared to conventional gas boilers); (viii) there is a relationship between seasonal efficiency and CO_2 emissions for different domestic fuels; for example, assuming an average CO_2 emission factor for electricity of 0.422 kg/kWh, using a ground-source heat pump with a seasonal efficiency of

3.5 would result in the emission of 0.12 kg of CO_2 for every kWh of useful heat provided. By comparison, a condensing gas boiler (assuming a CO_2 emission factor for gas of 0.194 kg/kWh), operating at a seasonal efficiency of 85%, would result in 0.23 kg CO_2 for every kWh of useful heat supplied (i.e., almost double the CO_2 emissions from the ground-source heat pump). If the geothermal heat pumps are supplied with electrical energy from renewable sources, CO_2 emissions could be substantially reduced or eliminated; (ix) there are benefits to the communities as the consumption of fossil fuels is reduced and the emission of greenhouse gases and other pollutants is decreased and centralized; (x) in the United States, for example, residential fossil fuel heating systems produce anywhere from 1.2 to 36 times the equivalent CO_2 emissions of ground-source heat pump systems; hence, additional CO_2 emissions from 15% to 77% could be avoided through the application of ground-source heat pump systems; (xi) in Germany, the total CO_2 emission for a ground-source heat pump unit for heating is 149 g CO_2/kWh, depending on the considered electrical energy mix compared to 229 g CO_2/kWh for a conventional heating mix, indicating that at least 35% of additional CO_2 emissions could be avoided with the application of ground-source heat pump systems.

21.3 LIMITATIONS

The application potential of ground-source heat pump systems in residential, commercial, and institutional buildings; district heating and cooling systems; or even industrial facilities is in many ways related not only to local climate (e.g., cold or moderate, continental or marine) and Earth hydro-geology circumstances, but also to site-specific conditions, as land availability and building height, social, cultural, and/or political environments and, often, economical or legal requirements.

Because such potential conditions are very variable and may disqualify some types of ground-source heat pump systems, but allow others, it is of great importance that the majority of technical, economical, legal, and environmental constraints be considered via feasibility studies at the beginning of any project in order to preserve the reputation of and the confidence for the geothermal heat pump heating and cooling systems.

Among the most relevant limitations of ground-source heat pump systems are the following (DOE 1995; Sanner et al. 2003; Stene 2004; IEA, 2008; Hughes 2008):

1. Compared with air-source heat pumps (that have lower purchase and installation costs), ground-source heat pump systems involve higher installation costs and longer payback periods.
2. A significant portion of these costs is attributed to (vertical or horizontal) ground-coupled heat exchangers, which may represent up to 50% of the system total costs. In residential applications, the first cost of ground-source heat pump systems is almost double that of standard central equipment; in the residential retrofit market, there are difficulties to install vertical and horizontal ground-cooled heat exchangers. Retrofitted residences without ductwork or adequate ductwork can add US$3,000 to US $10,000 to the cost of any ground-source heat pump system. Compared to rooftop unitary systems in commercial applications, the first cost of a

ground-source heat pump system is 20–40% higher. Added premiums compared to conventional commercial/institutional building HVAC systems are smaller as a result of the fact that HVAC systems in larger buildings are generally much more complicated and more expensive; on the other hand, undersized systems may lead to excessive on-off cycling of the geothermal heat pumps and, thus, reduce their operational efficiency and reliability, and increase the electrical energy consumption. One way of addressing over- or undersizing the geothermal heat pump systems (including the length of ground-coupled heat exchangers and the size of brine or water circulation pumps) is to use enhanced simulation and design tools by skilled and experienced designers, as well as improved installation guidelines, qualified installers, and system commissioning for both residential and non-residential buildings.

3. In the residential sector, the ground-coupled heat pump systems can be about 30–50% more expensive than the conventional heating, ventilating, and air-conditioning systems using air-source heat pumps, mainly due to the cost of vertically buried heat exchangers; therefore, installation costs are the most important barriers to implementation of such systems, especially in the residential sector. However, once installed, the annual energy and maintenance cost are less over their lifetime.
4. On the other hand, optimized designs in large commercial and institutional buildings may lead to relatively small cost premiums compared to those of conventional heating, ventilating, and air-conditioning.
5. Drilling and/or extensive trenching/excavation problems, as well as ground/soil/rocks properties and groundwater movement, may be limiting factors; even closed-loop ground-source heat pump systems (regardless of whether vertical or horizontal ground-coupled heat exchangers are installed) are in general applicable in all types of Earth geology.
6. For open-loop ground-source heat pump systems (based on pumping groundwater) containing one or several aquifers of different configurations and hydraulic properties (e.g., levels, temperatures), well capacities, unfavorable groundwater chemical compositions, or regional supply of drinking water can be some limiting factors.
7. The higher initial capital costs may make the short-term economics of ground-source heat pump systems unattractive, although long-term economics are generally favorable because of lower operating and maintenance costs caused by higher system energy performances in both heating and cooling modes of operation.
8. Historically, the higher initial capital costs, as well as lack of consumer knowledge and confidence in the technology, non-standardized concepts, and uncertainty about actual final energy performances (sometimes, lower than expected compared to other solutions as air-source heat pumps, condensing boilers, or solar technologies), were barriers to growth of the ground-source heat pump market worldwide, especially in countries where conventional energy cost was relatively low. In addition, consumers were sometimes not aware of ground-source heat pump systems.

9. The height of buildings equipped with vertical ground-coupled heat exchangers withhout intermediate heat exchangers isolating the ground piping from the building piping, are limited to about nine stories or less; above this height, the static pressure head may exceed the maximum pressure head ratings of the series 160 or SDR 11 ground heat exchanger pipe. This rule of thumb does not credit the offsetting hydrostatic effect of groundwater, which could allow for a higher building; calculations should be performed to verify that the maximum pressure head has not been exceeded.
10. Vertical ground-coupled heat exchangers, where there is no intermediate heat exchanger isolating the ground piping from the building piping, are limited to buildings of approximately nine stories or less. The static pressure may exceed the maximum pressure head ratings of the series 160 or SDR-11 ground exchanger pipe above this height.
11. Lesser return on investment (i.e., longer payback periods) of ground-source heat pump systems compared to most conventional HVAC alternatives creates economic obstacles to their further penetration into the market. Simple payback periods vary between different building sectors and different countries, but commonly 10–15 years are considered reasonable upper limits.
12. Even damages to the ground-coupled heat exchangers, followed by brine or refrigerant leakages, are extremely rare, if they occur; repairs are difficult and might cost as much as initial installation costs.
13. Ground-source heat pump systems generally contribute to less global emissions of carbon dioxide and other harmful environmental substances; however, country or regional limiting legislations can exist, such as those concerning the contamination of the ground/soil and/or the groundwater by boreholes connected to the surface, boreholes shortcutting different aquifers, the usage of antifreeze fluids or refrigerants changes of the underground temperature that may affect the chemistry and bacterial composition and growth in the underground, damages and/or local disturbances (noise, etc.) caused by drilling and trenching, and damages to fauna and flora.
14. The main costs for the installation are due to borehole drilling (because of expensive equipment needed), digging of trenches for horizontal ground-coupled heat exchangers, and land area requirements.
15. Because of frequent oversized lengths of ground-coupled heat exchangers (sometimes, by 10% and up to 30%), the application of these systems are sometimes comparatively unattractive despite their high energy efficiency.
16. Damages on ground-coupled heat pump systems (leakage and break down due to a freezing incident around the U-tube) can happen during the installation process.
17. Undersizing and/or errors during in-field installation of vertical ground-coupled heat exchangers could lead to their collapse.
18. The heat transfer in both heating and cooling modes of operation may be

affected by bad or missing backfill of the vertical ground-coupled heat exchangers.

19. Further acceptance of the technology will also clearly depend on the availability of accurate, reliable and fast system design and simulation tools.
20. In order to allow ground-source heat pump technology to capture a larger portion of the heating and cooling market, innovations (as hybrid concepts where the downsized ground-coupled heat exchangers is used in conjunction with the cooling towers to meet the heat rejection loads; use of the tower reduces the capital cost of the ground-coupled heat exchangers but somewhat increases maintenance costs; for heavily heating-dominant applications, a downsized loop also can be augmented with an auxiliary heat source such as solar collectors or a boiler) are required to improve the economics of this technology, particularly for heating- or/ and cooling-dominated buildings.
21. Limited availability of skilled contractors to design and perform installation works.
22. High level of system design and management is needed to avoid operational fails.
23. Horizontal systems are not feasible in most urban and suburban zones because of larger ground area requirement.
24. Greater variations in performance of horizontal systems because ground temperatures and thermal properties fluctuate with season, rainfall, and burial depth.
25. Restrictive regulations such as mandating a minimum borehole size, grouting materials, wage rates, and heat exchange method generally increase the cost of such a system.
26. Depths of vertical boreholes are limited, depending the buildings' heights.
27. If properly designed and managed, may achieve limited variation of ground/soil temperatures and thermal properties during the system operation.
28. Even more expensive, because of high drilling costs, they are more efficient in both heating and cooling operating modes, achieving better seasonal energy performances, and often helped by groundwater movement.
29. Have a large portion of length below the groundwater table, giving lower ground thermal resistance.
30. Require less piping, since the ground temperature in vertical wells is more constant than that of horizontal shallow trenches in both winter and summer seasons.
31. Require less land areas and less electrical energy for pumping the secondary fluids (brine or water).
32. After installation, are less likely to suffer mechanical damages and leakages.

REFERENCES

Allen, M., S. Kavanaugh. 1999. Thermal conductivity of cementitious grouts and impacts on heat exchange length design for ground source heat pumps. *HVAC & R Research* 5(2):1–12.

Berntsson, T. 2002. Heat sources-technology, economy and environment. *International Journal of Refrigeration* 25:428–438.

Bloomquist, R. G. 2001. The economics of geothermal heat pump systems for commercial and institutional buildings. In *Proceedings of Conference on Geothermal Energy in Underground Mines*, Ustron, Silesia, Poland, November 21-23.

Blum, P., G. Campillo, W. Munch. 2010. CO_2 savings of ground source heat pump systems – A regional analysis. *Renewable Energy* 35:22–127.

DOE. 1995. Commercial/Institutional Ground-Source Heat Pump Engineering Manual. Prepared by Caneta Research Inc. for U.S. Department of Defense U.S. Department of Energy, Oak Ridge National Laboratory.

Dooley, R. 1999. *Geo-School Cost Estimating Software*. R.J. Dooley & Associates, Geothermal Heat Pump Consortium, Poughkeepsie, New York.

Fuller S., S. Peterson. 1996. *NIST Handbook 135: Life Cycle Costing Manual for the Federal Energy Management Program*, NIST, Gaithersburg, Maryland.

Hughes, P. 2008. *Geothermal (Ground-Source) Heat Pumps: Market Status, Barriers to Adoption and Actions to Overcome Barriers*. Oak Ridge National Laboratory, Oak Ridge, TN.

IEA. 2008. International Energy Agency, Energy Technology Perspectives (https://www.iea.org/reports/energy-technology-perspectives-2008. Accessed March 16, 2021).

Kavanaugh, S.P., K. Rafferty. 1997. *Ground Source Heat Pumps – Design of Geothermal Systems for Commercial and Institutional Buildings*. ASHRAE (American Society of Heating, Refrigerating and Air-Conditioning Engineers) Applications Handbook, Atlanta, GA.

Lazzarin, R. 2001. Ground as a possible heat pump source. *Geothermische Energie*:32–33 (http://www.geothermie.de/gte/gte3233/ground_as_a_possible_heat_pump_s.htm).

Minea, V. 2009. Challenging future of heat pumps. *IEA (International Energy Agency) Heat Pump Centre Newsletters* 27(4):8–12.

Minea, V., P. Savignac. 1996. Tendencies in commercial and institutional building heating and cooling modes. In *Proceedings of Electricity '96 – End-Use Technologies CEA (Canadian Electricity Association) Session*, Montreal, Canada, April 28–May 3.

Moore, A. 1999. *Capital, Operating, and Maintenance Costs of Geo-Exchange and Conventional HVAC Systems*. Princeton Economic Research, Inc., for Lockheed Martin Idaho Technologies Co., Idaho Falls, ID.

Omar A.M. 2008. Ground-source heat pumps systems and applications. *Renewable & Sustainable Energy Review* 12:344–371.

Ondreka, I., M.I. Rusgen, I. Stober, K. Czurda. 2007. GIS supported mapping of shallow geothermal potential of representative areas in south-western Germany – Possibilities and limitations. *Renewable* Energy 32:186–200.

Sanner, B., C. Kaytsas, D. Medrinos, L. Rybach. 2003. Current status of ground source heat pumps and underground thermal energy storage in Europe. *Geothermics* 32:579–588.

Self, S.J., B.V. Reddy, M.A. Rosen. 2013. Geothermal heat pump systems: Status review and comparison with other heating options. *Applied Energy* 101:341–348.

Stene, J. 2004. *IEA HPP Annex 29 – Ground-Source Heat Pumps Overcoming Technical and Market Barriers. Norway Status Report*. SINTEF Energy Research, December.

Yoon, S., S.-R. Lee. 2015. Life cycle cost analysis and smart operation mode of ground source heat pump system. *Smart Structures and Systems* 16(4):743–758.

22 Future R&D Requirements

22.1 INTRODUCTION

Finding new ways to produce and use energy to reduce the cost and minimize environmental impact is one of the key challenges the world faces in the 21st century.

Air-, ground-, and water-source heat pumps recover pollution-free energy from surrounding air, earth, and low-grade industrial waste heat. They diminish primary energy consumption and reduce power demand for building space heating and cooling, and for domestic hot water heating, with high utilization factors and thermal performances. Because about 30% of worldwide carbon emissions are from heating and cooling systems in buildings, new future regulations will require their drastic reduction while converting residential and commercial/institutional buildings into low-energy constructions. During the next decades, building owners will certainly use more and more heat pumps among other energy-efficient devices. In this context of the global warming issues, R&D and technological activities aiming to develop the next generation of heat pumps are numerous and very challenging, which will force new government subsidies and influence consumer behavior (Minea 2009).

Future factors influencing the use of heat pumps will be, among many others: (i) higher demand for air cooling and dehumidification as a consequence of climate changes, and changes in building and living standards (e.g., lower specific energy requirements for heating, higher demands for domestic hot water, and growing demands for a more comfortable indoor climate in both hot and cold seasons); (ii) possible higher production costs for conventional primary energy sources; (iii) global requirements to reduce carbon dioxide emissions by using more carbon efficient heating and cooling technologies will also motivate investments in heat pump technologies to make them more competitive with, for example, wind, hydro, solar, and nuclear energies; (iv) reduce the global energy supply securities risks of net energy importer countries because the heat pumps could contribute to reduce these risks through the use of electricity generated by a wide diversification of fossil and renewable (as hydraulic, wind, biomass, and solar) sources.

22.2 GENERAL HEAT PUMPING CONTEXT

The buildings' HVAC systems' peak electric demands in cold and moderate climates are increasing in both heating (in cold weathers) and cooling/dehumidification (in hot and humid weathers) modes of operation. Heat pumps are used for upgrading low-temperature free heat from renewable sources such as air, water, ground, and industrial/municipal waste heat to useful temperatures.

DOI: 10.1201/9781003032540-22

Past R&D activities have targeted design and control improvements of heat pumps used for building space and hot water preheating, and air cooling/dehumidification in order to provide safe and efficient operations. In cold-climate countries (such as Canada), heat pumps have been successfully implemented as heat recovery devices in advanced supermarkets (Minea 2003), as well as in combined spaces and domestic hot water heating systems (Minea 2007a).

Other studies aimed to overcome the technical and market barriers to ground-source heat pumps (Minea 2007b) and provide economical heating and cooling systems integrated into low and net zero energy houses (Candanedo et al. 2008).

Such studies have contributed to the development and promotion of heat pump technologies in cold and moderate climates. Several works on integration of heat pumps with commercial building HVAC systems, such as retail stores and industrial processes (e.g., drying), have also been successfully performed. Heat pump technologies certainly have a promising, brilliant future in the context of the world's present and future energy contexts.

Some of the global targets for heat pumping technologies are the following: (i) by 2035, institutional and commercial buildings should use 25% less energy; (ii) almost all new buildings and 75% of retrofit buildings should be equipped with heat pumps; (iii) 25% of industrial waste heat should be recovered; (iv) the heat pumps' annual performance factors, including heat pump water heaters, will increase by 50% by 2035 (Europe), and 200% by 2050 (Japan); (v) about 20% of overall energy consumption will be from renewable energies; (vi) the initial cost of heat pumps will be reduced by 25% by 2035 compared to 2000 by at least 50% by 2050; (vii) global greenhouse gas emissions will be drastically reduced, according to future national and international building regulations and subsidy programs; (viii) future heat pumps will use more and more natural (as ammonia, CO_2, and hydrocarbons) and low-GWP (as HFOs) working fluids to reduce the environmental impacts in case of massive leakages; (ix) the energy performances of existing components will be improved and the application field will be extended by developing new types of compressors, heat exchangers, and controls (as inverter-controlled or double-stage scroll and rotary compressors with or without refrigerant injection at intermediate pressures, intelligent on-board fault detection and diagnostics, smart user interfaces, bi-directional connectivity and demand response readiness); (x) new policy instruments (e.g., regulatory requirements and budgetary incentives) will be continuously adopted; (xi) advanced design methods and efficient integration in low/net-zero energy houses (with photovoltaic and solar thermal heat recovery devices, passive thermal storage and cogeneration units) and in industrial processes (e.g., food, pulp and paper, and drying), as well as in district heating and cooling systems will be theoretically developed (e.g., by using pinch analysis of industrial processes, a powerful tool for the efficient integration of heat pumps) and experimentally validated; (xii) combined systems using air-to-air, water-to-air, and water-to-water heat pumps with solar thermal energy as the main or secondary (back-up) heating system will be further developed; (xiii) large-capacity heat pumps will be more and more integrated in future cold and moderate-climate district heating systems to recover heat, for example from disaffected mines, city sewers, rivers, lakes, and seas; (xiv) advanced heating and cooling systems will further include low-temperature distribution systems (e.g., radiant heating floors and cooling ceilings) with little or no fan energy consumption, multi-zones with heat recovery and variable refrigerant flow, and ductless indoor distribution systems; (xv) being efficient to recover exhaust air, heat pump applications will be extended in cold and moderate climates to supermarkets, ice rinks, and cold storage plants; (xvi) the potential for reducing CO_2 emissions by assuming a 30% share of heat pumps in the building sector using technology presently available is of about 6% of the total worldwide CO_2 emissions. With advanced future technologies in power generation, in heat pumps and in integrated control strategies, up to 16% seem to be possible; therefore, heat pumps are one of the key technologies for energy conservation and reducing CO_2 emissions; (xvii) the heat pump retrofit market penetration of 30% by 2030 may reduce total global CO_2 emissions by up to 8% compared to 2000; and (xviii) heating energy efficiency will be improved and electricity generation will become less carbon-intensive; the cost of saving a ton of carbon dioxide emissions through heat pumping technology varies between countries. For heat pumps, the main causes of differences are climate, the nature of the electricity supply system, and the local cost of alternative heating systems and their fuels.

22.3 GROUND-SOURCE HEAT PUMP SYSTEMS

Further improvements of large ground-source heat pump systems by design are still possible and necessary in order to reduce their initial (capital) costs as well as seasonal and annual energy consumptions.

The future growth of geothermal industry in residential and commercial/institutional sectors worldwide has to be supported by more research efforts, especially involving the development of more efficient design methods, advanced ground-coupled heat exchangers, lower cost techniques for boreholes in drilling and trenching, efficient integration of geothermal heat pumps in low/net-zero energy houses, multi-apartment community buildings and cold storage plants, and increased use of natural refrigerants (ammonia, CO_2) and hydrocarbons.

Many other future R&D requirements for modeling, simulation, design, and in-field demonstration of ground-coupled heat exchangers, as well as site characterization, integration, installation, control, and innovative applications of ground-source heat pump systems are the following: (i) develop improved geothermal heat pumps and system control strategies; (ii) develop more cost-effective ground-coupled heat exchangers; (iii) ground/soil/rocks thermal properties are key parameters for the design of large-scale ground-source heat pump systems, but they cannot be reliably estimated without in-situ measurements involving drilling of test boreholes; therefore, development of lower-cost, more accurate, and more convenient/automated methods for faster testing and estimating ground/soil/rock thermal properties will be highly useful in the future; (iv) improve the design algorithms, correlations, and software in order to avoid under- or oversizing of ground-coupled heat exchangers that can lead to excessive first costs and electricity consumptions, and/or poor energy performances of ground-source heat pump systems; (v) develop new, advanced analytical models to design and simulate overall system performances of more common systems as those using horizontal indirect (secondary fluid) ground-coupled heat exchangers, horizontal and vertical direct expansion and thermosiphon systems, building foundation heat exchangers, and surface water geothermal heat pump systems; (vi) in order to use deeper (e.g., deeper than 200 m), vertical (indirect, secondary fluid) ground-coupled heat exchangers, develop materials and technics for grouting procedures, to avoid pipe collapse due to external pressures, maintaining satisfactorily low pressure drops, and limiting short-circuiting effects; (vii) develop 3-D modeling methods of full-scale borehole U-tube heat exchanger arrays, allowing flexible control over the operation of the borehole field with variable brine (or water) flow rates during part load operation; in other words, brine (or water) pumping energy and heat exchanger performance must be further optimized in any type of ground-source heat pump system; (viii) develop statistical and stochastic approaches (i.e., having a random probability distribution or pattern that may be analyzed statistically but may not be predicted precisely), as well as predictive non-physical models (e.g., dynamic neural networks seem particularly attractive for investigating optimal control strategies due to the large dimension, the non-linearity, and the non-convex nature of solutions); (ix) perform in-field continuous measurements of temperatures and heat extraction/rejection of ground-coupled heat exchangers over periods as long as possible in order to determine if they reasonably balance the annual heat extractions and rejections, and if there is or no long-term heat buildup or draw-down problems; (x) improve the market penetration in new applications based on variable economic, regulatory, and political parameters in every national environment; (xi) develop simplified and straightforward guidelines for professional designers providing maps based on a large number of sensitivity analysis relating to some significant parameters (e.g., ground-source thermal characteristics, ground-coupled heat exchangers number and size, heat load profile, nominal efficiency of geothermal heats pump and back-up generators) to the overall performances; (xii) further develop foundation (thermal, energy) piles for which the design problems are quite different from borehole heat exchangers where placement and depth are controlled by the thermal design; (xiii) develop new applications of ground-source heat pump systems in, for example, near-zero energy buildings; in such buildings, photovoltaic-generated electricity (with solar collectors that simultaneously produce electricity and thermal energy) can be used directly as heat sources for the geothermal heat pumps or to (summer) recharge the boreholes in order to prevent any year-to-year decrease in the ground temperature; storage of heat generated using excess photovoltaic electricity could be a better and more cost-effective option than battery storage of electricity; (xiv) extend the applications of hybrid ground-source heat pump systems by design methodology, allowing the simultaneous interactions between the building systems, supplemental heat injecters/heat rejecters, and ground-coupled heat exchangers.

REFERENCES

Candanedo, J., V. Minea, A. Athienitis. 2008. Low-energy houses in Canada – National initiatives and achievements. In *Proceedings of IEA HPP Annex 32 International Workshop*, May, Zürich, Switzerland.

Minea, V. 2003. *Advanced Supermarket Refrigeration/Heat Recovery Systems*. CEA Technologies, CEATI Report No. T011700-7005, August.

Minea, V. 2007a. Simultaneous performance factors for combined multifunctional ground-source heat pumps. *IEA Heat Pump Centre Newsletter* 25(4):16–22.

Minea, V. 2007b. Design improvements of large ground-source heat pump systems in Canada. In Proceedings of the 3rd IEA HPP Annex 29 International Workshop, January 15, Sapporo Japan (http://www.annex29.net).

Minea, V. 2009. Challenging future of heat pumps. *IEA Heat Pump Newsletter* 30(1):36–38.

Index

Note: Page numbers in **bold** represent tables; page numbers in *italics* represent figures.